T0235703

*This mathematical physics contribution
to the Computer Algebra Recipes series
is dedicated to my wife Karen,
who lights my path through life.*

Richard H. Enns

Computer Algebra Recipes for Mathematical Physics

Birkhäuser
Boston • Basel • Berlin

Richard H. Enns
Simon Fraser University
Department of Physics
Burnaby, B.C. V5A 1S6
Canada

AMS Subject Classifications (2000): 15A90, 30-XX, 33-XX, 34-XX, 35-XX, 35Qxx, 40-XX, 42-XX, 44-XX, 49-XX, 65-XX, 68-XX, 70-XX, 97U50

ISBN 0-8176-3223-9   Printed on acid-free paper.

Printed in the United States of America.   (HP)

9 8 7 6 5 4 3 2 1   SPIN 10923559

www.birkhauser.com

Additional material to this book can be downloaded from http://extras.springer.com

Preface

This book is a self-contained guide to problem-solving and exploration in mathematical physics using the powerful Maple 9.5 computer algebra system (CAS). With a CAS one cannot only crunch numbers and plot results, but also carry out the symbolic manipulations which form the backbone of mathematical physics.

The heart of this text consists of over 230 useful and stimulating "classic" computer algebra worksheets or *recipes*, which are systematically organized to cover the major topics presented in the standard Mathematical Physics course offered to third or fourth year undergraduate physics and engineering students. The emphasis here is on applications, with only a brief summary of the underlying theoretical ideas being presented. The aim is to show how computer algebra can not only implement the methods of mathematical physics quickly, accurately, and efficiently, but can be used to explore more complex examples which are tedious or difficult or even impossible to implement by hand.

The recipes are grouped into three sections, the introductory **Appetizers** dealing with linear ordinary differential equations (ODEs), series, vectors, and matrices. The more advanced **Entrees** cover linear partial differential equations (PDEs), scalar and vector fields, complex variables, integral transforms, and calculus of variations. Finally, in the **Desserts** the emphasis is on presenting some analytic, graphical, and numerical techniques for solving nonlinear ODEs and PDEs. The numerical methods are also applied to linear ODEs and PDEs.

No prior knowledge of Maple is assumed in this text, the relevant command structures being introduced on a need-to-know basis. The recipes are thoroughly annotated and, on numerous occasions, presented in a "story" format or in a historical context. Each recipe takes the reader from the analytic formulation or statement of a representative type of mathematical physics problem to its analytic or numerical solution and to a graphical visualization of the answer, where relevant. The graphical representations vary from static 2-dimensional pictures, to contour and vector field plots, to 3-dimensional graphs that can be rotated, to animations in time. For your convenience, all 230 recipes are included on the accompanying CD.

The range of mathematical physics problems that can be solved with the enclosed recipes is only limited by your imagination. By altering the parameter values, or initial conditions, or equation structure, thousands of other problems can be easily generated and solved. "What if?" questions become answerable. This should prove extremely useful to instructor and student alike.

Contents

II   THE ENTREES   125

4   Linear PDEs of Physics   127

Computer Algebra Recipes
for Mathematical Physics

INTRODUCTION

The purpose of computing is insight, not numbers.
R.W. Hamming, *Numerical Methods for Scientists and Engineers* (1973)

*Science means simply the aggregate of all the recipes
that are always successful.*
Paul Valéry, French poet and essayist (1871–1945)

A. Computer Algebra Systems

Computer algebra systems (CASs) are revolutionizing the way we learn and teach those scientific subjects which make extensive use of advanced mathematics. CASs not only allow us to carry out the numerical computations of standard programming languages and to plot the results in a wide variety of ways, but to also perform lengthy and complicated symbolic mathematical manipulations as well. The purpose of this text is to show how a CAS can be used to tackle problem-solving and exploration of concepts and methods in mathematical physics. A CAS can perform a wide variety of mathematical operations, including

- analytic differentiation and analytic/numerical integration,

- analytic/numerical solution of ordinary/partial differential equations,

- Taylor/Laurent series expansions of functions,

- manipulation and simplification of algebraic expressions,

- analytic/numerical solution of algebraic equations,

- production of 2- and 3-dimensional vector field and contour plots,

- animation of analytic and numerical solutions.

The computer algebra worksheets, or *recipes*, in this book are based on the powerful **Maple 9.5** software system. Any reader desiring to use a different release of Maple, or even an alternate CAS, should generally have little difficulty in modifying the recipes to his or her own taste. This is because the Maple input and output is completely annotated for each recipe and the underlying mathematics and physics fully explained.

B. Computer Algebra Recipes

The heart of this text consists of a systematic collection of computer algebra recipes which have been designed to illustrate the concepts and methods of mathematical physics and to stimulate the reader's intellect and imagination. Associated with each recipe is an intrinsically important mathematical physics example and, where feasible, the example is presented in a "story" format wherein real or fictitious characters motivate or explain the recipe.

Every topic or story in the text contains the Maple code or recipe to explore that particular topic. To make life easier for you, all recipes have been placed on the CD-ROM enclosed within the back cover of this text. The recipes are ordered according to the chapter number, the section number, and the subsection (story) number. For example, the recipe **01-1-2**, entitled **The Tale of the Turbulent Tail**, is associated with chapter 1, section 1, subsection 2. Although the recipes can be directly accessed on the CD by clicking on the appropriate worksheet number, it is strongly recommended that you access them through the hyperlinked recipe index file **00recipe**, which provides complete instructions. The computer code exported into the text is accompanied by detailed explanations of the underlying mathematical physics concepts and/or methods and what the recipe is trying to accomplish.

The recommended procedure for using this text is first to read a given topic/story for overall understanding and enjoyment. If you are having any difficulty in understanding a piece of the text code, then you should execute the corresponding Maple worksheet and try variations on the code. Keep in mind that the same objective may often be achieved by a different combination of Maple commands than those that I chose. After reading the topic, you should execute the worksheet (if you have not already done so) to make sure the code works as expected. At this point feel free to explore the topic. Try rotating any three-dimensional graphs or running any animations in the file. See what happens when changes in the model or Maple code are made and then try to interpret any new results. This book is intended to be open-ended and merely serve as a guide to what is possible in mathematical physics using a CAS, the possibilities being limited only by your own background and desires.

At the end of each chapter, **Supplementary Recipes** are presented in the form of problems, their fully annotated solutions (recipes) being included on the CD. These recipes are also hyperlinked to the recipe index file with a simple numbering system. For example, **01-S02** is the second supplementary recipe in Chapter 1. Supplementary recipes can be used in two different ways. They can be regarded as problems to be solved by using the mathematical physics concepts and computer algebra techniques presented in the main text recipes. Your solutions can then be compared with those that I have presented. Even if you are successful, you probably will be interested in the many little computer algebra features that are introduced in my solutions. On the other hand, these additional recipes can be regarded as still more interesting applications of computer algebra to mathematical physics. Enjoy exploring all the recipes!

C. Maple Help

In this text, the Maple commands are introduced on a need-to-know basis. If you wish to learn more about these commands, or about other possible commands which might prove useful in solving a particular mathematical physics problem, Maple's Help should be consulted. The Help system allows you to explore Maple commands and features, listed by name or subject. One can search by topic or carry out a full text search. Both procedures are illustrated by first using the Topic Search to find the correct form of the command for taking a square root, and then using the Full Text Search to find the command for analytically solving an ODE. In either case, begin by using the mouse and clicking on Maple's Help which opens a help window.

(a) **Topic Search**

- Click on Topic Search. Auto-search should then be selected.

- You wish to find the Maple command for taking the square root. Depending on the programming language, the command could be sqr, sqrt, root, ...In this case type sq in the Topic box. Maple will display all the commands starting with sq. Double click on sqrt or, alternately, single click on sqrt and then on OK. A description of the square root command will appear on the computer screen.

(b) **Full Text Search**

- Click on Full Text Search.
- Type ode in the Word(s) box and click on Search.
- Double click on dsolve. A description of the dsolve command for solving ODEs will appear along with several examples as well as hyperlinks to related topics.

To learn more about these search methods as well as other features of Maple's help, open Using Help on the help page.

If on executing a Maple command, the output yields a mathematical function that is unfamiliar to you, e.g., *EllipticF*, you may find out what this function is by clicking on the word to highlight it, then on Help, and finally on Help on EllipticF. You will find that EllipticF refers to the incomplete elliptic integral of the first kind, which is defined in the Help page. The same Help window may also be opened by typing in a question mark followed by the word and a semicolon, e.g., ?EllipticF;

Maple's Help is not perfect and on occasion you might feel frustrated, but generally it is helpful and should be consulted whenever you get stuck with Maple syntax or are seeking just the right command to accomplish a certain mathematical task. Maple learning and programming guides are also available ([Cha03], [MGH+03b], [MGH+03a]). Let us emphasize that in this book we will merely scratch the surface of what can be done with the Maple symbolic computing package, concentrating on those features which are relevant to tackling mathematical physics problems.

D. Introductory Recipes

In the following chapters, recipes will be presented which correlate with the major topics developed in standard undergraduate mathematical physics ([MW71], [Boa83], [AW00]) texts. To give you a preliminary idea of what these recipes will look like and to introduce some basic Maple syntax, consider the following two kinematics examples. These introductory recipes are not on the accompanying CD-ROM, so after reading the following subsections you should open up Maple and type the recipes in and execute them.

D.1 A Dangerous Ride?

A horse is dangerous at both ends and uncomfortable in the middle.
Ian Fleming, British mystery writer, (1908–64)

The vertical displacement (in meters) of a proposed circus ride at t seconds is given by $Y = a t^2 e^{-bt} \cos(ct)/(1 + d\sqrt{t})$. a, b, c, and d are real constants.

(a) Determine the velocity V and acceleration A at arbitrary time t.

(b) Given $a = 2\,\mathrm{m/s^2}$, $b = 3/8\,s^{-1}$, $c = 10\,s^{-1}$, and $d = 1\,s^{-1/2}$, plot V over the time interval $t = 0$ to $T = 20$ seconds.

(c) Find the maximum V in m/s and km/h and the time at which it occurs.

(d) Plot A and V together from $t = 0$ to $T/2 = 10$ seconds and discuss the graph. Do you think that this proposed ride is dangerous? If so adjust the parameter values to make the ride safer.

   To solve this problem, let's first clear Maple's internal memory of any previously assigned values (other worksheets may be open with numerical values given to some of the same symbols being used in the present recipe). This is done by typing in the **restart** command after the opening prompt (>) symbol, ending the command with a colon (:), and pressing Enter (which generates a new prompt symbol) on the computer key board.

```
> restart:
```

All Maple command lines must be ended with either a colon, which suppresses any output, or a semi-colon (;), which allows the output to be viewed.

   The analytic form of the ride's vertical coordinate Y is entered.

```
> Y:=a*t^2*exp(-b*t)*cos(c*t)/(1+d*sqrt(t));
```

$$Y := \frac{a t^2 e^{(-bt)} \cos(ct)}{1 + d\sqrt{t}}$$

Use has been made of the assignment (:=) operator, placing Y on the left-hand side (lhs) of the operator and the time-dependent form of Y on the right-hand side (rhs). Assigned quantities can be mathematically manipulated. The symbols *, /, +, -, and ^ are used for multiplication, division, addition, subtraction, and exponentiation, The Maple forms cos and exp of the cosine and exponential commands are intuitively obvious.

Differentiating Y once with respect to t yields the velocity V,

```
> V:=diff(Y,t);
```

$$V := \frac{2\,a\,t\,e^{(-bt)}\cos(ct)}{1+d\sqrt{t}} - \frac{a\,t^2\,b\,e^{(-bt)}\cos(ct)}{1+d\sqrt{t}} - \frac{a\,t^2\,e^{(-bt)}\sin(ct)\,c}{1+d\sqrt{t}}$$
$$- \frac{1}{2}\,\frac{a\,t^{(3/2)}\,e^{(-bt)}\cos(ct)\,d}{(1+d\sqrt{t})^2}$$

while differentiating twice yields the acceleration A.

```
> A:=diff(Y,t,t);
```

$$A := \frac{2\,a\,e^{(-bt)}\cos(ct)}{1+d\sqrt{t}} - \frac{4\,a\,t\,b\,e^{(-bt)}\cos(ct)}{1+d\sqrt{t}} - \frac{4\,a\,t\,e^{(-bt)}\sin(ct)\,c}{1+d\sqrt{t}}$$
$$- \frac{7}{4}\,\frac{a\,\sqrt{t}\,e^{(-bt)}\cos(ct)\,d}{(1+d\sqrt{t})^2} + \frac{a\,t^2\,b^2\,e^{(-bt)}\cos(ct)}{1+d\sqrt{t}} + \frac{2\,a\,t^2\,b\,e^{(-bt)}\sin(ct)\,c}{1+d\sqrt{t}}$$
$$+ \frac{a\,t^{(3/2)}\,b\,e^{(-bt)}\cos(ct)\,d}{(1+d\sqrt{t})^2} - \frac{a\,t^2\,e^{(-bt)}\cos(ct)\,c^2}{1+d\sqrt{t}}$$
$$+ \frac{a\,t^{(3/2)}\,e^{(-bt)}\sin(ct)\,c\,d}{(1+d\sqrt{t})^2} + \frac{1}{2}\,\frac{a\,t\,e^{(-bt)}\cos(ct)\,d^2}{(1+d\sqrt{t})^3}$$

The form of A would be tedious to derive by hand. With Maple, the calculation is done quickly and without any errors. If the structure of Y is changed, the new forms of V and A are obtained almost immediately by re-executing the above command lines.

The given parameter values are entered. Although not necessary, I like to leave spaces between commands on the same prompt line for easier readability.

```
> a:=2: b:=3/8: c:=10: d:=1: T:=20:
```

The velocity is plotted over the time interval $t=0$ to T and shown in Figure 1.

```
> plot(V,t=0..T);
```

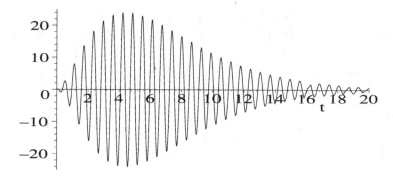

Figure 1: Velocity V (vertical axis) versus time t.

By inspecting the figure, we can see that the maximum velocity occurs around the 5 s mark and is about 24 m/s. A slightly more accurate estimate can be obtained by placing the cursor on the top of the tallest peak in the computer picture and clicking the mouse. The horizontal and vertical coordinates of the cursor location are displayed in a small viewing box at the top left of the computer screen. A much more accurate answer follows on setting the acceleration A equal to zero and applying the floating point solve (`fsolve`) command in a time range which includes the tallest peak, say $t = 4$ to 6 s. This yields an answer $T2$ for the time to 10 digits, Maple's default accuracy.

>   `T2:=fsolve(A=0,t=4..6);`

$$T2 := 4.868771376$$

The maximum velocity occurs at $T2 \simeq 4.87$ seconds. Then, using the `eval` command to evaluate V at $t = T2$,

>   `Vmax:=eval(V,t=T2);`

$$Vmax := 23.81789390$$

yields a maximum velocity $Vmax \simeq 23.8$ m/s. The `convert` command with the `units` option is used to convert $Vmax$ from m/s to km/h.

>   `Vmax:=convert(Vmax,units,m/s,km/h);`

$$Vmax := 85.74441804$$

The maximum velocity is $85\frac{3}{4}$ km/h, which doesn't seem excessively high.

What about the acceleration? Let's plot A and V together in the same figure over the time range $t = 0$ to $T/2 = 10$ seconds. Two plot options (`color` and `linestyle`) are introduced. A red solid line is chosen for V, a blue dashed line for A. Note that `V` and `A` as well as the options have been entered as "Maple lists" (the elements separated by commas and enclosed in square brackets). Maple preserves the order and repetition of elements in a list.

>   `plot([V,A],t=0..T/2,color=[red,blue],linestyle=[SOLID,DASH]);`

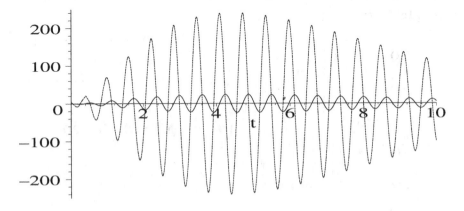

Figure 2: A (dashed curve) and V (solid) versus time t.

If you are printing pictures with multiple plots in black and white, it is particularly important to control the line style so the curves can be distinguished.

In Figure 2, we can clearly see that the acceleration is a maximum when the velocity is zero and zero when the velocity is a maximum. The maximum acceleration is over 200 m/s^2. Since the acceleration due to gravity is about 10 m/s^2, this corresponds to roughly 20 "Gees". Do you think that such an acceleration is possibly dangerous? Justify your answer. Perhaps, do an Internet search on the effects of rapid acceleration on the human body.

Next, we look at a two-dimensional kinematics example which introduces you to the use of a Maple library package. Library packages are very important because they save you the effort of programming specialized plotting and mathematical operations. Approximately 90% of Maple's mathematical knowledge resides in the Maple library. Most of the recipes in this text use one or more library packages.

D.2 The Patrol Route of Bertie Bumblebee

Belief like any other moving body follows the path of least resistance.
Samuel Butler, British author, (1835–1902)

Bertie Bumblebee, intrepid sentry for the central bee hive on the terraformed planet Erehwon[1], flies on a patrol route described t minutes after leaving the central hive by the radial coordinate $r(t) = a\,t^2\,e^{-bt}/(1+t^2)$ sretem (a unit of length on Erehwon) and the angular coordinate $\theta(t) = b + c\,t^{2/3}$ radians. a, b, and c are real constants.

(a) Calculate Bertie's speed V at an arbitrary time t, simplifying the result as much as possible. Attempt to analytically determine the distance Bertie travels in the time interval $t=0$ to an arbitrary time $T > 0$.

(b) Taking $a=3$, $b=\pi/8$ and $c=10$, determine the time it takes for Bertie to make a complete circuit and the total distance flown.

(c) Plot Bertie's path for the complete circuit and superimpose an animation of his motion on this path, representing Bertie as a moving circle.

After clearing Maple's memory with the **restart** command,

```
>  restart:
```

Bertie's radial and angular coordinates are entered.

```
>  r:=a*t^2*exp(-b*t)/(1+t^2); theta:=b+c*t^(2/3);
```

$$r := \frac{a\,t^2\,e^{(-b\,t)}}{1+t^2} \qquad \theta := b + c\,t^{(2/3)}$$

---

[1]In 1872, the British writer Samuel Butler described a fictitious land in the utopian novel *Erewhon*, the title being intended as an anagram for *nowhere*. In this land, the people dealt with disease as a crime and destroyed machinery lest machines destroyed them. This would not be the land for using computer algebra, so in the *Computer Algebra Recipes* series, I have introduced a fictitious planet, *Erehwon*, where names are occasionally spelled backwards, but *Erehwon* is not backward in embracing modern technology.

Note that entering `theta` for the angular coordinate has produced the Greek symbol θ in the output.

Next, Bertie's Cartesian coordinates, $X = r \cos\theta$, $Y = r \sin\theta$, are calculated, the forms of r and θ being automatically substituted in the output.

> `X:=r*cos(theta); Y:=r*sin(theta);`

$$X := \frac{a\,t^2\,e^{(-bt)}\cos(b+c\,t^{(2/3)})}{1+t^2} \qquad Y := \frac{a\,t^2\,e^{(-bt)}\sin(b+c\,t^{(2/3)})}{1+t^2}$$

The speed V at time t is obtained by calculating $V = \sqrt{(dX/dt)^2 + (dY/dt)^2}$.

> `V:=sqrt(diff(X,t)^2+diff(Y,t)^2);`

$$V := \left(\left(\frac{2\,a\,t\,e^{(-bt)}\cos(b+c\,t^{(2/3)})}{1+t^2} - \frac{a\,t^2\,b\,e^{(-bt)}\cos(b+c\,t^{(2/3)})}{1+t^2}\right.\right.$$
$$\left.- \frac{2\,a\,t^3\,e^{(-bt)}\cos(b+c\,t^{(2/3)})}{(1+t^2)^2} - \frac{2}{3}\frac{a\,t^{(5/3)}\,e^{(-bt)}\sin(b+c\,t^{(2/3)})\,c}{1+t^2}\right)^2$$
$$+ \left(\frac{2\,a\,t\,e^{(-bt)}\sin(b+c\,t^{(2/3)})}{1+t^2} - \frac{a\,t^2\,b\,e^{(-bt)}\sin(b+c\,t^{(2/3)})}{1+t^2}\right.$$
$$\left.\left.- \frac{2\,a\,t^3\,e^{(-bt)}\sin(b+c\,t^{(2/3)})}{(1+t^2)^2} + \frac{2}{3}\frac{a\,t^{(5/3)}\,e^{(-bt)}\cos(b+c\,t^{(2/3)})\,c}{1+t^2}\right)^2\right)^{(1/2)}$$

The output looks quite messy, so let's simplify it, making use of the `simplify` command. One of the major difficulties with `simplify` is that the output may not be simplified as much as you would like or not put into a specific form that you are trying to attain. The `simplify` command comes with various optional arguments, e.g., `simplify(V,symbolic)` as in the following command line, which simplifies V assuming that all the parameters are positive.

> `V:=simplify(V,symbolic);`

$$V := \frac{1}{3}a\,t\,e^{(-bt)}(36 - 36\,bt - 36\,t^3\,b + 9\,t^2\,b^2 + 18\,t^4\,b^2 + 9\,t^6\,b^2$$
$$+ 4\,t^{(16/3)}\,c^2 + 8\,t^{(10/3)}\,c^2 + 4\,t^{(4/3)}\,c^2)^{(1/2)}\Big/(1+t^2)^2$$

This last result is certainly simpler than the previous one, all trig terms being eliminated. Whether it's the simplest possible form is a matter of taste. Simplifying with Maple is usually a matter of trial and error and you will see many, many simplification examples as you progress through this book.

To determine the distance d that Bertie flies over a time interval $t = 0$ to some arbitrary time T, an attempt is made to analytically evaluate the integral $d = \int_0^T V\,dt$ using the integration (`int`) command.

> `d:=int(V,t=0..T);`

$$d := \int_0^T \frac{1}{3}a\,t\,e^{(-bt)}(36 - 36\,bt - 36\,t^3\,b + 9\,t^2\,b^2 + 18\,t^4\,b^2 + 9\,t^6\,b^2$$
$$+ 4\,t^{(16/3)}\,c^2 + 8\,t^{(10/3)}\,c^2 + 4\,t^{(4/3)}\,c^2)^{(1/2)}\Big/(1+t^2)^2 dt$$

Maple is unable to find an analytic solution, returning the integral without evaluating it. So, let's enter the given parameter values, $a = 3$, $b = \pi/8$, and $c = 10$. Note that the command Pi for entering π is capitalized. Maple is case sensitive here.

```
>  a:=3: b:=Pi/8: c:=10:
```

The time $T = 12.99$ minutes, which is now entered, is the approximate time for Bertie to complete one circuit. It is determined by trial and error by numerically calculating the total distance to 4 digits, using the floating point evaluation (evalf) command. Increasing T will not change the answer to this accuracy.

```
>  T:=12.99; distance:=evalf(d,4);
```

$$T := 12.99 \qquad distance := 23.98$$

Bertie travels a total distance of about 24 sretem in one complete circuit.

To animate Bertie's flight and superimpose the motion on a plot of the entire route, special plots commands are required. These are contained in the plots library package, which is now "loaded".

```
>  with(plots);
```

Warning, the name changecoords has been redefined

$[animate, animate3d, \ldots display, \ldots polarplot, \ldots textplot3d, tubeplot]$

The with() command is used to load Maple library packages. Normally, I would place a colon on the above command line to suppress the output, but here a partial list of the large number of specialized plot commands that are available in the plots package is shown. The commands *animate*, *polarplot* (to plot the trajectory in polar coordinates), and *display* (to superimpose graphs) in the output list will be used here. There is also a warning message that the name changecoords has been redefined. This warning appears even if a colon is used. If desired, warnings can be removed by using a colon and inserting the command interface(warnlevel=0) prior to loading the library package. From now one, I will generally artificially remove all such warnings in the text.

In the first graph, gr1, an animation of Bertie's motion is created with the animate command. To fit into the width of the page, the lengthy Maple command line is broken over two text lines. Bertie's X and Y coordinates are entered as a Maple list. The time range is taken from $t = 0$ to T. I have chosen to use 500 frames (the default is 25) to make a reasonably smooth animation. A point style is chosen, Bertie being represented by a size 16 blue circle. A line-ending colon is used to prevent the plotting numbers from being displayed.

```
>  gr1:=animate([X,Y],t=0..T,frames=500,style=point,
         symbol=circle,color=blue,symbolsize=16):
```

The polarplot command is used in gr2 to graph the entire route as a thick (the default thickness is 0) orange line. To obtain a smooth curve, a minimum of 500 (the default is 50) plotting points is requested.

```
>  gr2:=polarplot([r,theta,t=0..T],numpoints=500,style=line,
         color=orange,thickness=2):
```

The graphs are now superimposed with the display command, the axis labels x and y being added. The double quotes denote that each enclosed item is a "Maple string". A string is a sequence of characters that has no value other than itself. It cannot be assigned to, and will always evaluate to itself.

```
>  display([gr1,gr2],labels=["x","y"]);
```

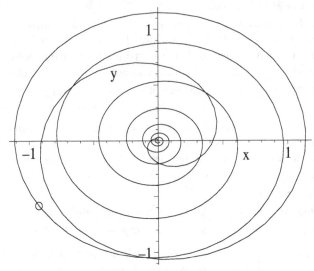

Figure 3: Bertie's patrol route while on sentry duty.

Figure 3 shows the entire path traced out by Bertie and his position (represented by the small circle) two minutes after he starts on his patrol route. The animation can be initiated (the circle starts at the origin and moves along the path, stopping when $t = T = 12.99$ minutes.) by clicking on the computer plot and then on the start arrow in the Maple tool bar at the top of the computer screen. The animation may be made to repeat by clicking on the looped arrow and stopped by clicking on the solid square. Other options are also available.

E. How to Use this Text

Although some of Maple's basic syntax has been provided in these introductory recipes, it is recommended that the computer algebra novice start at the beginning of the **Appetizers**, even if your mathematical physics background is above that of the recipes presented there. It is in these early chapters that more of the basic features of the Maple system are introduced. Further, you might be surprised at how even initially simple problems can be made more interesting and often much more challenging because of the fact that a computer algebra system is being used. Whatever approach you adopt to using this book, I hope that you savor the wide variety of mathematical physics recipes that follow.

Bon Appetit! Your computer algebra chef, Richard.

Part I

THE APPETIZERS

*The last thing one discovers in composing a work
is what to put first.*

Blaise Pascal, French scientist, philosopher (1623–62)

*Each problem that I solved became a rule
which served afterwards to solve other problems.*

René Descartes, French philosopher and mathematician (1596–1650)

*Food probably has a very great influence on the
condition of men....Who knows if a well-prepared soup
was not responsible for the pneumatic pump
or a poor one for a war?*

G. C. Lichtenberg, German physicist, philosopher (1742–99)

Chapter 1

Linear ODEs of Physics

In this chapter, the recipes illustrate how Maple may be used to solve and explore some representative ordinary differential equations (ODEs) from the world of physics. The focus is on linear ODEs (LODEs), i.e., those which are *first order* or *linear* in the dependent variable. Although an example of a nonlinear ODE (NLODE), i.e., one which contains one or more terms which are not linear in the dependent variable, is presented in the second recipe, the study of NLODEs is a much more mathematically challenging topic and will be postponed until the **Desserts**.

This chapter is not intended to teach you all the details of the wide variety of approaches for solving ODEs, but rather to show you how Maple can be used as an auxiliary tool to implement some of the more common methods. However, the series approach to solving an ODE will be postponed until Chapter 2.

As you are probably aware, the subject of solving differential equations (ordinary and partial) is huge. This is exemplified by Daniel Zwillinger's *Handbook of Differential Equations* [Zwi89] which is 700 pages long. This reference book is highly recommended for quickly looking up every known method of solution.

1.1   Linear ODEs with Constant Coefficients

Consider an nth order LODE of the general structure (the spatial coordinate x being replaced by t for time-dependent problems),

$$\frac{d^n y}{dx^n} + a_{n-1}(x)\,\frac{d^{n-1}y}{dx^{n-1}} + \cdots + a_1(x)\,\frac{dy}{dx} + a_0(x)\,y = f(x). \qquad (1.1)$$

If $f(x) = 0$, the ODE is said to be *homogeneous*, otherwise it is *nonhomogeneous*. The first ODEs that physics and engineering students usually encounter are differential equations with constant coefficients (a_0, etc., independent of x) which can be solved in closed form in terms of "elementary" (trigonometric, logarithmic, exponential) functions using a variety of standard methods. In this section, I will illustrate how Maple can be used to implement these methods, bypassing the tedious intermediate steps involved in a hand calculation.

1.1.1   Dazzling Dsolve Debuts

One should never make one's debut with a scandal.
One should reserve that to give an interest to one's old age.
Oscar Wilde, Anglo-Irish playwright, author, *The Picture of Dorian Gray* (1891)

Whether the coefficients are constant or not, the `dsolve` command is dazzling in its ability to solve LODEs. Before we tackle some physical examples, this mathematical recipe briefly looks at what is possible with `dsolve`.

>   `restart:`

Let's begin with the general nonhomogeneous first order LODE given in *ode1*.

>   `ode1:=diff(y(x),x)+a(x)*y(x)=f(x);`

$$ode1 := (\frac{d}{dx}\,y(x)) + a(x)\,y(x) = f(x)$$

A standard method for solving *ode1* is to first find its *integrating factor, IF*. Loading the necessary `DEtools` package, the `intfactor` command yields *IF*.

>   `with(DEtools): IF:=intfactor(ode1);`

$$IF := e^{(\int a(x)\,dx)}$$

The integral in *IF* is performed by specifying $a(x)$, e.g., $a(x) = a$, a constant, and applying the `value` command to `IF`.

>   `a(x):=a: IF:=value(IF);`

$$IF := e^{(a\,x)}$$

Multiplying *ode1* by *IF*, the *first integral I1* is obtained by applying `firint`.

>   `I1:=firint(IF*ode1);`

$$I1 := e^{(a\,x)}\,y(x) - \int e^{(a\,x)}\,f(x)\,dx + \_C1 = 0$$

Since the ODE is first order, one arbitrary coefficient $\_C1$ appears. To evaluate the integral in *I1*, $f(x)$ must be given. Suppose, e.g., that $f(x) = e^{-x}$. Forming `value(I1)`, the first integral is explicitly evaluated.

>   `f(x):=exp(-x): I1:=value(I1);`

$$I1 := e^{(a\,x)}\,y(x) - \frac{e^{(a\,x-x)}}{a-1} + \_C1 = 0$$

The general solution to *ode1*, labeled *y1*, follows on solving *I1* for $y(x)$.

>   `y1:=solve(I1,y(x));`

$$y1 := -\frac{-e^{(a\,x-x)} + \_C1\,a - \_C1}{e^{(a\,x)}\,(a-1)}$$

The above approach has mimicked the major steps of a hand calculation. The same basic result can be obtained more quickly by applying `dsolve` to *ode1*.

>   `dsolve(ode1,y(x));`

$$y(x) = (\frac{e^{(x\,(a-1))}}{a-1} + \_C1)\,e^{(-a\,x)}$$

A common ODE notation is to use primes, each prime (′) standing for d/dx. If desired, this notation can be introduced into the ODE output by first loading the **PDEtools** library package and entering the following **declare** command.

```
>   with(PDEtools): declare(y(x),prime=x);
```

y(x) will now be displayed as y, derivatives with respect to x of functions
of one variable will now be displayed with ′

Then entering, say, the nonhomogeneous second-order linear ODE ode2 with constant coefficients, produces an output with the prime notation.

```
>   ode2:=diff(y(x),x,x)+a1*diff(y(x),x)+a0*y(x)=F(x);
```
$$ode2 := y'' + a1\, y' + a0\, y = F(x)$$

An ODE such as *ode2* can be classified by applying the **odeadvisor** command.

```
>   odeadvisor(ode2);
```
$$[[\_2nd\_order, \_linear, \_nonhomogeneous]]$$

An even more important diagnostic tool is the **infolevel[dsolve]** command, which will give information on what methods are used in attempting to solve the ODE when **dsolve** is applied, even if unsuccessful. An integer between 1 and 5 must be specified, with generally more detailed information being provided as the number is increased. On applying the **dsolve** command to *ode2*, the method of attack is summarized in the following output and, in this case, the general solution $y(x)$ given with two arbitrary coefficients _C1 and _C2.

```
>   infolevel[dsolve]:=5: dsolve(ode2,y(x));
```

Methods for second order ODEs:
— Trying classification methods —
trying a quadrature
trying high order exact linear fully integrable
trying differential order: 2; linear nonhomogeneous with symmetry [0,1]
trying a double symmetry of the form [xi=0, eta=F(x)]
Try solving first the homogeneous part of the ODE
  −> Tackling the linear ODE "as given":
    checking if the LODE has constant coefficients
    <− constant coefficients successful
  <− successful solving of the linear ODE "as given"
  −> Determining now a particular solution to the nonhomogeneous ODE
    building a particular solution using variation of parameters
    particular solution has integrals (!)
    −> trying a d'Alembertian particular solution free of integrals
    <− no simpler d'Alembertian solution was found
  <− solving first the homogeneous part of the ODE successful

$$y = e^{((-\frac{a1}{2} + \frac{\sqrt{a1^2 - 4\,a0}}{2})\,x)}\, \_C2 + e^{((-\frac{a1}{2} - \frac{\sqrt{a1^2 - 4\,a0}}{2})\,x)}\, \_C1$$

$$+ \left(\int e^{(-\frac{(-a1 + \sqrt{a1^2 - 4\,a0})\,x}{2})}\right) F(x)\,dx\, e^{(x\,\sqrt{a1^2 - 4\,a0})}$$

$$- \int F(x)\, e^{(\frac{(a1 + \sqrt{a1^2 - 4\,a0})\,x}{2})}\,dx \right) e^{(-\frac{(a1 + \sqrt{a1^2 - 4\,a0})\,x}{2})} / \sqrt{a1^2 - 4\,a0}$$

The corresponding homogeneous ODE (set $F(x)=0$) was first solved and then a particular solution obtained for $F(x) \neq 0$ using the *variation of parameters* method. The homogeneous solution is obtained by assuming that $y \sim e^{\lambda x}$. This yields $\lambda^2 + a1\,\lambda + a0 = 0$, which has two roots λ_1 and λ_2. So the general solution of the homogeneous ODE is of the form $\_C1\,e^{\lambda_1 x} + \_C2\,e^{\lambda_2 x}$, the specific forms of λ_1 and λ_2 being easily identified from the above Maple output. The variation of parameters method then assumes that $y(x) = \_C1(x)\,e^{\lambda_1 x} + \_C2(x)\,e^{\lambda_2 x}$ for the complete LODE, and solves for the functions $\_C1(x)$ and $\_C2(x)$. See, for example, [Ste87] for the details.

The `infolevel[dsolve]` command can be turned off by setting it to 0. The prime notation can also be turned off by entering the command `OFF`.

```
>   infolevel[dsolve]:=0:  OFF:
```

Boundary value and *initial value* problems are important in engineering and physics. As a simple example of the former, suppose, e.g., that $a1 = 2$, $a0 = 3$, and $F(x) = x + \sin(x)\,e^{-x}$. Then *ode2* reduces to the form given in *ode3*.

```
>   a1:=2:  a0:=3:  F(x):=x+sin(x)*exp(-x):  ode3:=ode2;
```

$$ode3 := (\frac{d^2}{dx^2}\,y(x)) + 2\,(\frac{d}{dx}\,y(x)) + 3\,y(x) = x + \sin(x)\,e^{(-x)}$$

Suppose that the boundary conditions (bcs) are that $y(0) = y(1) = 0$. The boundary value problem is solved for $y(x)$ by entering the equation name and bcs in the `dsolve` command as a "Maple set" (items enclosed in braces, { }). Unlike a list, a set does not preserve order or repetition.

```
>   dsolve({ode3,y(0)=0,y(1)=0},y(x));
```

$$y(x) = -\frac{1}{9}\,\frac{e^{(-x)}\sin(\sqrt{2}\,x)\,(2\cos(\sqrt{2}) + e + 9\sin(1))}{\sin(\sqrt{2})} + \frac{2}{9}\,e^{(-x)}\cos(\sqrt{2}\,x)$$

$$+ \frac{1}{9}\,e^{(-x)}\,(3\,x\,e^x - 2\,e^x + 9\sin(x))$$

Now, an initial value problem is considered. First, let's turn the `declare` command back on and replace x with the time variable t in *ode3*.

```
>   ON:  ode4:=subs(x=t,ode3);
```

$$ode4 := y_{t,t} + 2\,y_t + 3\,y(t) = t + \sin(t)\,e^{(-t)}$$

Because the independent variable is no longer x, the output has used the notation y_t and $y_{t,t}$ to indicate 1st and 2nd time derivatives. Now, suppose that

$y(0) = 100$ and $y_t(0) = 10$. *ode4* is now solved for $y(t)$, subject to the initial conditions. The differential operator D is used to enter the derivative condition.

>   dsolve({ode4,y(0)=100,D(y)(0)=10},y(t));

$$y(t) = \frac{490}{9} e^{(-t)} \sin(\sqrt{2}\,t)\,\sqrt{2} + \frac{902}{9} e^{(-t)} \cos(\sqrt{2}\,t) + \frac{1}{9} e^{(-t)}(3\,t\,e^t - 2\,e^t + 9\sin(t))$$

To simplify $y(t)$, the last output (indicated by %) must first be assigned. If this is not done, entering y(t) would generate the output $y(t)$, not the above answer. Then $y(t)$ is (somewhat) simplified and assigned the name *y4*.

>   assign(%): y4:=simplify(y(t));

$$y4 := \frac{1}{9} e^{(-t)}(490\sin(\sqrt{2}\,t)\,\sqrt{2} + 902\cos(\sqrt{2}\,t) + 3\,t\,e^t - 2\,e^t + 9\sin(t))$$

ODE systems may also be solved with the **dsolve** command. Consider the coupled first-order time-dependent LODEs given in *ode5a* and *ode5b*.

>   ode5a:=diff(x(t),t)=-3*x(t)+4*z(t)+(sin(t))^2*exp(-2*t);

$$ode5a := x_t = -3\,x(t) + 4\,z(t) + \sin(t)^2\,e^{(-2t)}$$

>   ode5b:=diff(z(t),t)=2*z(t)-5*x(t)+(cos(t))^3;

$$ode5b := z_t = 2\,z(t) - 5\,x(t) + \cos(t)^3$$

The set of ODEs, *ode5a* and *ode5b*, is solved for the set of unknown functions, $x(t)$ and $z(t)$, subject to the initial conditions $x(0) = 100$ and $z(0) = 10$.

>   sol:=dsolve({ode5a,ode5b,x(0)=100,z(0)=10},{x(t),z(t)});

The solution *sol*, whose lengthy output has been suppressed here in the text, is assigned, and exponential terms collected in $x(t)$ and $z(t)$, the results being labeled X and Z, respectively.

>   assign(sol): X:=collect(x(t),exp); Z:=collect(z(t),exp);

$$X := (-\frac{57231}{7480}\sin(\frac{\sqrt{55}\,t}{2})\,\sqrt{55} + \frac{203147}{2040}\cos(\frac{\sqrt{55}\,t}{2}))\,e^{(-\frac{t}{2})}$$
$$+ (\frac{1}{6}\cos(2\,t) - \frac{1}{8})\,e^{(-2t)} + \frac{39}{170}\cos(t) + \frac{3}{170}\sin(t) + \frac{3}{34}\sin(3\,t) + \frac{5}{34}\cos(3\,t)$$

$$Z := (\frac{15727}{1632}\cos(\frac{\sqrt{55}\,t}{2}) - \frac{90973}{5280}\sin(\frac{\sqrt{55}\,t}{2}))\,\sqrt{55})\,e^{(-\frac{t}{2})} - \frac{3}{68}\sin(t)$$
$$+ \frac{3}{17}\cos(t) + \frac{3}{17}\cos(3\,t) - \frac{3}{68}\sin(3\,t) + (-\frac{1}{12}\sin(2\,t) + \frac{1}{6}\cos(2\,t) - \frac{5}{32})\,e^{(-2t)}$$

If you are not already dazzled by the power of **dsolve** in obtaining the above result try, e.g., some other nonhomogeneous terms in the ODEs. Unlike with the hand calculation, a new answer can be rapidly and accurately generated.

To this point, we have had no control over the method of solution. As shall be demonstrated in some of the following recipes, various self-evident options can be included in **dsolve**, such as **series**, **laplace**, **numeric**, etc.

1.1.2   The Tale of the Turbulent Tail

Nonsense. Space is blue and birds fly through it.
Felix Bloch, 1952 Nobel laureate in physics, *Heisenberg and the early days of quantum mechanics, Physics Today, December 1976*

As a post-doctoral fellow at the Chadwick Laboratory at the University of Liverpool (England), I enjoyed my leisure time on dark, rainy, often foggy, winter nights playing badminton with my fellow physicists. **The Tale of the Turbulent Tail** is inspired by those games of badminton played so long ago.

In an article contained in a delightful reprint collection entitled *The Physics of Sports* [PLA92], Peastrel, Lynch and Armenti reported on their experimental investigation of the aerodynamics of a badminton shuttlecock or "bird" falling vertically from rest. The relevant data is reproduced in Table 1.1, the distance y in meters that the bird fell in t seconds being given.

Table 1.1: Data for the falling badminton bird.

t	0.347	0.470	0.519	0.582	0.650	0.674	0.717	0.766
y	0.61	1.00	1.22	1.52	1.83	2.00	2.13	2.44
t	0.823	0.870	1.031	1.193	1.354	1.501	1.726	1.873
y	2.74	3.00	4.00	5.00	6.00	7.00	8.50	9.50

From the last pair of entries in the table, Peastrel et al calculated the terminal velocity (occuring when the downward gravitational force balances the upward drag force due to air resistance) to be $(9.50 - 8.50)/(1.873 - 1.726) = 6.8$ m/s. The gravitational acceleration was taken to be $g = 9.8$ m/s^2.

The investigators' goal was to determine which law of air resistance could best account for the experimental data, Stokes's law or Newton's law. For the Stokes model, the drag force on a body of mass m moving with velocity v is given by $F_{\text{Stokes}} = -a\,m\,v$, while for the Newton model, $F_{\text{Newton}} = -b\,m\,|v|\,v$. The positive coefficients a and b can be related to the terminal velocity and g.

After loading the **plots** library package, I will begin this recipe by first

```
>   restart: with(plots):
```

entering the experimental data of Table 1.1 so that the predictions of the two models can be tested for goodness of fit. The data is entered as a "list of lists." The first entry in each list is the time, the second entry the distance.

```
>   data:=[[0.347,0.61],[0.470,1.00],[0.519,1.22],[0.582,1.52],
            [0.650,1.83],[0.674,2.00],[0.717,2.13],[0.766,2.44],
            [0.823,2.74],[0.870,3.00],[1.031,4.00],[1.193,5.00],
            [1.354,6.00],[1.501,7.00],[1.726,8.50],[1.873,9.50]]:
```

Taking the gravitational force on the bird to be mg and assuming that Stokes's law of air resistance prevails (a comment (prefixed by the sharp symbol #) to this effect is added to the end of the command line), the acceleration of the bird is given by *ode1*. This is a first-order linear nonhomogeneous ODE.

```
> ode1:=diff(v(t),t)=g-a*v(t); #Stokes's resistance law
```

$$ode1 := \frac{d}{dt}\, v(t) = g - a\, v(t)$$

After a sufficiently long time, the gravitational and drag forces will balance and the bird will reach its terminal velocity $V1 = g/a$. So the unknown constant a can be eliminated by substituting $a = g/V1$ into $ode1$.

```
> ode1:=subs(a=g/V1,ode1);
```

$$ode1 := \frac{d}{dt}\, v(t) = g - \frac{g\, v(t)}{V1}$$

Since the bird falls from rest, the initial value problem is solved by entering $ode1$ and the initial condition $v(0)=0$ in the dsolve command as a Maple set.

```
> sol1:=dsolve({ode1,v(0)=0},v(t));
```

$$sol1 := v(t) = V1 - e^{\left(-\frac{g\,t}{V1}\right)} V1$$

To compare with the experimental data, we need to calculate the distance that the bird falls in T seconds. For the Stokes model, this distance $y1$ is obtained by integrating the right-hand side (rhs) of $sol1$ from $t = 0$ to T.

```
> y1:=int(rhs(sol1),t=0..T);
```

$$y1 := \frac{V1\left(-V1 + g\,T + V1\, e^{\left(-\frac{g\,T}{V1}\right)}\right)}{g}$$

Let's now look at Newton's model of air resistance. Since the motion is in one direction (down) only, the absolute value sign can be dropped, so that $F_{\text{Newton}} = -b\,m\,v^2$. The acceleration of the bird is now given by $ode2$. This is a first order *nonlinear* nonhomogeneous ODE.

```
> ode2:=diff(v(t),t)=g-b*v(t)^2; #Newton's resistance law
```

$$ode2 := \frac{d}{dt}\, v(t) = g - b\, v(t)^2$$

The terminal velocity is now given by $V2 = \sqrt{g/b}$. The unknown constant b is removed from the equation by substituting $b = g/V2^2$ into $ode2$.

```
> ode2:=subs(b=g/V2^2,ode2);
```

$$ode2 := \frac{d}{dt}\, v(t) = g - \frac{g\, v(t)^2}{V2^2}$$

To see what methods are used in trying to solve $ode2$, the infolevel[dsolve] command is entered and set to 2. Applying the dsolve command to $ode2$ and taking the rhs, an analytic solution $v2$ is generated for the velocity $v(t)$.

```
> infolevel[dsolve]:=2: v2:=rhs(dsolve(ode2,v(t)));
```
Methods for first order ODEs:
— Trying classification methods —
trying a quadrature
trying 1st order linear
trying Bernoulli
trying separable

$<-$ *separable successful*

$$v2 := \frac{2\,V2\,e^{\left(\frac{2\,t\,g}{V2} + \frac{2\,\_C1\,g}{V2}\right)}}{1 + e^{\left(\frac{2\,t\,g}{V2} + \frac{2\,\_C1\,g}{V2}\right)}} - V2$$

Maple has used the fact that *ode2* is *separable*. A separable ([Ste87]) first-order nonlinear ODE can be written in the form $dy/dx = g(x)\,f(y)$. It is solved by separating variables and integrating, i.e., $\int dy/f(y) = \int g(x)\,dx$.

The arbitrary constant $\_C1$ is determined by evaluating *v2* at $t=0$, setting the result to 0, and solving for $\_C1$. On some executions, $\_C1$ involves $I \equiv \sqrt{-1}$, so to ensure a real velocity *v2*, the complex evaluation command (evalc breaks a complex quantity into real and imaginary parts) is applied to *v2*.

>   _C1:=solve(eval(v2,t=0)=0,_C1); v2:=evalc(v2);

$$\_C1 := 0 \qquad v2 := \frac{2\,V2\,e^{\left(\frac{2\,t\,g}{V2}\right)}}{1 + e^{\left(\frac{2\,t\,g}{V2}\right)}} - V2$$

The distance *y2* that the bird falls in T seconds, according to Newton's model, is now calculated by integrating *v2* assuming that $g > 0$, $V2 > 0$, and $T > 0$.

>   y2:=int(v2,t=0..T) assuming g>0,V2>0,T>0;

$$y2 := \frac{V2\,\left(V2\,\ln(1 + e^{\left(\frac{2\,T\,g}{V2}\right)}) - g\,T - V2\,\ln(2)\right)}{g}$$

So, which is the better model equation, *y1* or *y2*? Entering the values of g, $V1$, and $V2$, the forms of *y1* and *y2* are then as follows:

>   g:=9.8: V1:=6.8: V2:=6.8: y1:=y1; y2:=evalf(y2);

$$y1 := -4.718367347 + 6.800000000\,T + 4.718367347\,e^{(-1.441176471\,T)}$$

$$y2 := -3.270523024 + 4.718367347\,\ln(1. + e^{(2.882352940\,T)}) - 6.800000000\,T$$

The experimental data is now plotted in gr1, the data points being represented by size 14 (red by default) circles.

>   gr1:=plot(data,style=point,symbol=circle,symbolsize=14):

The theoretical curves *y1* and *y2* are plotted in gr2 over the time interval $T = 0$ to 2 seconds. *y1* is represented by a thick blue dashed line, *y2* by a thick green solid line. The entry linestyle=[3,1] would produce the same line styles.

>   gr2:=plot([y1,y2],T=0..2,color=[blue,green],
>   linestyle=[DASH,SOLID],thickness=2):

The two graphs are then superimposed with the display command. The minimum number of tickmarks along the horizontal and vertical axes is specified, and axis labels and a title added.

>   display({gr1,gr2},tickmarks=[4,3],labels=["t","y"],title=
>   "Newton's law (solid green), Stokes's law (dashed blue)");

The resulting picture is shown in Figure 1.1. It is clear that Newton's law of resistance is the correct model for the falling badminton bird. Stokes's law is

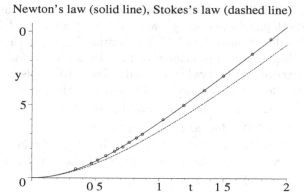

Newton's law (solid line), Stokes's law (dashed line)

Figure 1.1: The experimental points agree with Newton's resistance law.

known to apply to the laminar flow of air about a smooth moving object, while Newton's law applies to the turbulent flow around a non-smooth body. It is the tail "feathers" of the moving badminton bird which create the turbulence.

1.1.3 This Bar Doesn't Serve Drinks

By the time a bartender knows what drink a man will have before he orders, there is little else about him worth knowing.
Don Marquis, American humorist and journalist, (1878–1937)

This recipe involves a bar undergoing rotational motion. No, I don't mean that the local pub appears to be spinning because one has overindulged in liquid refreshments after a tough exam! It has to do with the possible deflection modes ([Mor48], [Hil57], [SR66]) of an originally straight uniform solid bar of length L which rotates in a horizontal plane about one end (say $x = 0$) with uniform angular velocity ω. Below a critical frequency ω_1, the bar remains straight, but on exceeding ω_1 it starts pulsating and its shape changes. On further increasing ω, another critical value ω_2 is reached, above which the shape changes again, and so on. The values of the critical frequencies and the deflection modes depend on how the ends of the bar are supported. Here, I will consider one possibility, leaving it to you to explore other boundary conditions.

For small deflections $y(x)$ from the straight shape, the governing 4th order LODE (each prime denoting an x derivative) is

$$y'''' - k^4 y = 0, \quad \text{where } k^4 \equiv \epsilon \omega^2 / (Y I). \tag{1.2}$$

Here ϵ is the mass per unit length, Y is Young's modulus, and I is the moment of inertia of a cross-section of the shaft about an axis perpendicular to the xy plane. To solve the boundary value problem, four boundary conditions are required for the homogeneous fourth-order equation, two conditions for each end of the bar. I will consider the case when the pivot point at $x = 0$ is a *clamped end* (both y and the slope y' vanish at $x=0$), while $x=L$ is a *free end* (both y'' ($\propto$ bending moment) and y''' ($\propto$ shearing force) vanish at $x=L$).

The recipe begins with the loading of the `plots` library package and the entry of the `infolevel[dsolve]` command. The PDE tools library is also accessed and the `declare` command used to generate the prime notation.

```
> restart: with(plots): infolevel[dsolve]:=2:
```
```
> with(PDEtools): declare(y(x),prime=x);
```

The governing ODE (1.2) is entered. Instead of inputting the fourth derivative as `diff(y(x),x,x,x,x)`, the short-cut entry `diff(y(x),x$4)` is used.

```
> ode:=diff(y(x),x$4)-k^4*y(x)=0;
```

$$ode := y'''' - k^4\, y = 0$$

The general solution of *ode* is obtained using the `dsolve` command and the rhs of the result taken. Maple recognizes that *ode* is a LODE with constant coefficients and makes use of this fact to solve the equation.

```
> y:=rhs(dsolve(ode,y(x)));
```

Methods for high order ODEs:
— Trying classification methods —
trying a quadrature
checking if the LODE has constant coefficients
<— constant coefficients successful

$$y := \_C1\, e^{(-k\,x)} + \_C2\, e^{(k\,x)} + \_C3 \sin(k\,x) + \_C4 \cos(k\,x)$$

Clamped-end boundary conditions are applied at $x = 0$ in *bc1* and *bc2*, and free-end boundary conditions at $x=L$ in *bc3* and *bc4*.

```
> bc1:=eval(y,x=0)=0; bc2:=eval(diff(y,x),x=0)=0;
```

$$bc1 := \_C1 + \_C2 + \_C4 = 0$$
$$bc2 := -\_C1\, k + \_C2\, k + \_C3\, k = 0$$

```
> bc3:=eval(diff(y,x$2),x=L)=0; bc4:=eval(diff(y,x$3),x=L)=0;
```

$$bc3 := \_C1\, k^2\, e^{(-k\,L)} + \_C2\, k^2\, e^{(k\,L)} - \_C3 \sin(k\,L)\, k^2 - \_C4 \cos(k\,L)\, k^2 = 0$$

$$bc4 := -\_C1\, k^3\, e^{(-k\,L)} + \_C2\, k^3\, e^{(k\,L)} - \_C3 \cos(k\,L)\, k^3 + \_C4 \sin(k\,L)\, k^3 = 0$$

Applying the analytic solve command, the set of four boundary conditions are solved for the set of four unknown coefficients. All the coefficients are zero, which corresponds to the *trivial solution* $y(x)=0$, i.e., the straight bar profile.

```
> solve({bc1,bc2,bc3,bc4},{_C1,_C2,_C3,_C4});
```

$$\{\_C1 = 0,\ \_C4 = 0,\ \_C3 = 0,\ \_C2 = 0\}$$

To obtain a non-trivial result, let's first solve the set of three boundary conditions, *bc1*, *bc2*, and *bc4*, for $\_C1$, $\_C2$, and $\_C4$ and assign the solution *sol*.

```
> sol:=solve({bc1,bc2,bc4},{_C1,_C2,_C4}); assign(sol):
```

$\_C1$, $\_C2$, and $\_C4$ are automatically substituted into the remaining boundary condition, *bc3*, which is divided by k^2 and simplified.

```
> bc3:=simplify(bc3/k^2);
```

$$bc3 := -\frac{2\,\_C3\,(e^{(-k\,L)}\cos(k\,L) + 2 + e^{(k\,L)}\cos(k\,L))}{e^{(-k\,L)} - e^{(k\,L)} + 2\sin(k\,L)} = 0$$

For a non-trivial solution, $\_C3 \neq 0$, so the term containing $\cos(kL)$ must be zero. The `select` command is used to extract this term from the lhs of $bc3$.

```
> eq:=select(has,lhs(bc3),cos(k*L))=0;
```

$$eq := e^{(-k\,L)}\cos(k\,L) + 2 + e^{(k\,L)}\cos(k\,L) = 0$$

The expression eq is converted completely to trigonometric form and simplified.

```
> eq2:=simplify(convert(eq,trig)/2);
```

$$eq2 := \cos(k\,L)\cosh(k\,L) + 1 = 0$$

$eq2$ is a *transcendental equation* which must be solved numerically for $k\,L$. The algebraic substitution $kL=K$ is now made in $eq2$.

```
> eq3:=algsubs(k*L=K,eq2);
```

$$eq3 := \cos(K)\cosh(K) + 1 = 0$$

The lhs of $eq3$ is plotted in Figure 1.2, the viewing range being controlled.

```
> plot(lhs(eq3),K=0..12,thickness=2,view=[0..12,-5..5]);
```

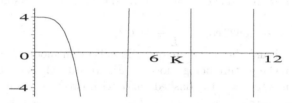

Figure 1.2: Graphically locating the zeros of the transcendental equation.

The approximate values of the zeros of $eq3$ can be determined by visual inspection. A slightly more accurate procedure is to place the cursor on the zero on the computer screen and click the mouse. The horizontal and vertical coordinates of the cursor location are displayed in a small viewing box at the top left of the computer screen. Numerical values of the zeros to 10 digits accuracy can be obtained by applying the numerical solving command,[1] `fsolve`, to $eq3$. If lesser accuracy is desired, e.g., 5 digits, the `Digits` command can be used.

```
> Digits:=5:
```

A "functional", or "arrow" operator[2] f is introduced to systematically search for the K zeros in the range $3\,(n-1)$ to $3\,n$ using the `fsolve` command. Dividing by L then yields the k zeros.

```
> f:=n->fsolve(eq3,K,3*(n-1)..3*n)/L:
```

When a number n is supplied, then subsequently entering `f(n)` will yield the zero (if one exists) in the given range. For $n=1$ the search range is 0 to 3, for

---

[1] This command implements *Newton's method.* [Ste87]
[2] Created on the keyboard with the hyphen (-) followed by the greater than symbol (>).

$n=2$ the range is 3 to 6, and so on. The sequence command, **seq**, is used in *sol2* to generate the first four zeros of *eq2*. Note that the k's have been subscripted, the input syntax being $k[n]$. The zeros are then assigned.

>   `sol2:=seq(k[n]=f(n),n=1..4); assign(sol2):`

$$sol2 := k_1 = \frac{1.8751}{L}, \; k_2 = \frac{4.6941}{L}, \; k_3 = \frac{7.8548}{L}, \; k_4 = \frac{10.996}{L}$$

The transcendental equation *eq3* may be rewritten as $\cos(K) = -1/\cosh(K)$. For large K, $\cosh(K) \to \infty$, so $\cos(K) \to 0$ and $k_m \to (m - 1/2)\,\pi/L$ for large integer m. Setting $a \equiv \sqrt{Y\,I/\epsilon}$, the first four critical frequencies are calculated.

>   `critical_freq:=seq(omega[n]=a*k[n]^2,n=1..4);`

$$critical\_freq := \omega_1 = \frac{3.5160\,a}{L^2}, \; \omega_2 = \frac{22.035\,a}{L^2}, \; \omega_3 = \frac{61.698\,a}{L^2}, \; \omega_4 = \frac{120.91\,a}{L^2}$$

The coefficient *_C3* is collected in y, the result evaluated at $k = k_n$, and the sequence command used to generate the profiles Y_n of the first four deflection modes. For brevity, only Y_1 is shown here in the output of *sol3*.

>   `sol3:=seq(Y[n]=eval(collect(y,_C3),k=k[n]),n=1..4);`

$$sol3 := Y_1 = (0.18111\,e^{\left(\frac{1.8751\,x}{L}\right)} + 1.1811\,e^{\left(-\frac{1.8751\,x}{L}\right)} + \sin(\frac{1.8751\,x}{L})$$
$$- 1.3622\cos(\frac{1.8751\,x}{L}))\_C3,$$

sol3 is assigned and the first four possible deflection modes plotted for $L = 5$ and *_C3* $=1/5$, the picture being shown in Figure 1.3. The solid curve with one zero corresponds to $n=1$, the dashed curve with two zeros to $n=2$, etc.

>   `assign(sol3): L:=5: _C3:=1/5:`
>   `plot([seq(Y[n],n=1..4)],x=0..5,color=[red,green,blue,cyan],`
>   `linestyle=[SOLID,DASH,DOT,SOLID],labels=["x","y"]);`

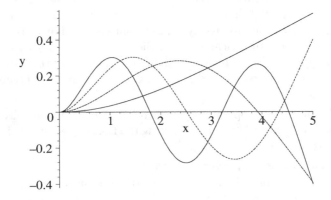

Figure 1.3: First four deflection modes for the rotating bar.

1.1.4   Shake, Rattle, and Roll

Principles aren't of much account anyway, except at election time. After that you hang them up to let them season.
Mark Twain, American author, Municipal Corruption, speech, 4 Jan. 1901

It is election time on the planet Erehwon and the politicians are traveling far and wide to woo potential voters. Presidential candidate Amai Koorc is traveling on his own personal three-coach train which rides along a guide-way on a cushion of air. The middle coach (mass $M > m$) is occupied by Koorc and other political "heavies", while the outer two coaches (each of mass m) contain media "light weights". Each outer coach is connected to the middle one by an identical linear spring (i.e, governed by *Hooke's law*) with spring constant K.

All three coaches were initially sitting stationary on a straight stretch of the guide-way with the coupling springs unstretched, when fringe members of the opposition party imparted an impulse to each coach causing the train to start "rolling" down the track in an erratic fashion which shook up the politicians and rattled the media. This recipe simulates the motion of the train.

To simplify the final results, it is assumed[3] that K, m, and M are positive.

```
>   restart: with(plots): assume(K>0,m>0,M>0):
```
Using Hooke's law for the relative displacements $x(t)$, $y(t)$, and $z(t)$ at time t of the outer left, middle, and outer right coaches from equilibrium, Newton's second law yields the following three equations of motion.

```
>   ode1:=m*diff(x(t),t,t)=K*(y(t)-x(t));
```
$$ode1 := m \left(\frac{d^2}{dt^2} x(t) \right) = K \left(y(t) - x(t) \right)$$

```
>   ode2:=M*diff(y(t),t,t)=K*(z(t)-y(t))-K*(y(t)-x(t));
```
$$ode2 := M \left(\frac{d^2}{dt^2} y(t) \right) = K \left(z(t) - y(t) \right) - K \left(y(t) - x(t) \right)$$

```
>   ode3:=m*diff(z(t),t,t)=-K*(z(t)-y(t));
```
$$ode3 := m \left(\frac{d^2}{dt^2} z(t) \right) = -K \left(z(t) - y(t) \right)$$

This is a system of three coupled second-order LODEs with constant coefficients. Given the general initial condition (dots indicating time derivatives) $x(0) = A$, $y(0) = B$, $z(0) = C$, $\dot{x}(0) = V1$, $\dot{y}(0) = V2$, and $\dot{z}(0) = V3$,

```
>   ic:=(x(0)=A,y(0)=B,z(0)=C,D(x)(0)=V1,D(y)(0)=V2,D(z)(0)=V3):
```
it would be very tedious to solve for $x(t)$, $y(t)$, and $z(t)$ by hand. Even Maple needs some guidance here as to what method is best to use. If no method

---

[3]Note that `assume` applies the assumption throughout the entire worksheet, whereas `assuming` applies it only in the command line that it is used. By default, assumed quantities have attached "trailing tildes" in the output. The tildes can be removed by inserting `interface(showassumed=0)` prior to the assumption. If desired, they can be removed from all worksheets by clicking on File in the tool bar, then on Preferences, I/O Display, No Annotation, and Apply Globally.

option is specified, applying the dsolve command to the three ODEs, subject to the initial condition, would yield a long "messy" result. Including the method=laplace option, which makes use of the Laplace transform method,[4] yields a somewhat simpler, but still lengthy, answer (not shown here).

```
>  sol:=dsolve({ode1,ode2,ode3,ic},{x(t),y(t),z(t)},
          method=laplace);
```

Using the list format [sin,cos], sine terms are now collected first in sol, and then cosine terms. Since the analytic expressions are still lengthy, only $x(t)$ is displayed here in the text.

```
>  sol2:=collect(sol,[sin,cos]);
```

$$sol2 := \{x(t) = \frac{1}{2}\sqrt{\frac{m}{K}}\,(V1 - V3)\sin(\sqrt{\frac{K}{m}}\,t) + (\frac{A}{2} - \frac{C}{2})\cos(\sqrt{\frac{K}{m}}\,t)$$

$$+\frac{1}{2}(\frac{M}{2\,m+M})^{(3/2)}\sqrt{\frac{m}{K}}\sin(\sqrt{\frac{K\,(2\,m+M)}{M\,m}}\,t)\,(V3 - 2\,V2 + V1)$$

$$+\frac{1}{2}\frac{M\,(-2\,B+C+A)\cos(\sqrt{\frac{K\,(2\,m+M)}{M\,m}}\,t)}{2\,m+M}$$

$$+\frac{M\,(2\,B+2\,t\,V2)+2\,m\,(t\,V1+C+A+t\,V3)}{2\,(2\,m+M)}, \ldots\ldots\}$$

Recalling that $x(t)$ represents the displacement of the first outer coach from equilibrium, it can be seen that the motion is built up of two parts. The trig terms represent an oscillatory motion of the coach, while the last term corresponds to a translational motion as time t increases. The oscillatory part is governed by two characteristic frequencies, $\omega = \sqrt{K/m}$ and $\sqrt{K\,(2\,m+M)/(M\,m)}$.

The solution sol2 is now assigned and a functional operator f created to substitute parameter values into each solution and simplify the result.

```
>  assign(sol2): f:=z->simplify(subs(par,z(t))):
```

The three coaches were initially in their equilibrium positions so $A = B = C = 0$. In some suitable set of Erehwonese units, the initial speeds imparted to the coaches by the opposition gang were $V1 = 2/3$, $V2 = -1/3$, and $V3 = 1/6$. Each coach containing light-weight media had a mass $m=1$ while the political heavies' coach had a mass $M=2$. The spring constant $K=4$.

```
>  par:=(A=0,B=0,C=0,V1=2/3,V2=-1/3,V3=1/6,m=1,M=2,K=4):
```

If the first outer coach was initially located at $X=1$, the middle one at $Y=2$, and the other outer coach at $Z=3$, their positions at time t are as follows.

```
>  X:=1+f(x); Y:=2+f(y); Z:=3+f(z);
```

---

[4]The Laplace transform of $x(t)$ is defined as $L(x(t)) \equiv X(s) = \int_0^\infty x(t)e^{-st}dt$. Integrating by parts and assuming $e^{-st}x(t) \to 0$ as $t \to \infty$, then $L(\dot{x}) = s\,X(s) - x(0)$, and $L(\ddot{x}) = s^2 X(s) - sx(0) - \dot{x}(0)$. To solve a LODE with constant coefficients, one can Laplace transform the LODE, solve the resulting algebraic equation for $X(s)$, and then perform the inverse transform to obtain $x(t)$. Laplace transforms are discussed in Chapter 6.

$$X := 1 + \frac{3}{32}\sqrt{2}\sin(2\sqrt{2}\,t) + \frac{1}{8}\sin(2\,t) + \frac{t}{24}$$

$$Y := 2 - \frac{3}{32}\sqrt{2}\sin(2\sqrt{2}\,t) + \frac{t}{24}$$

$$Z := 3 + \frac{3}{32}\sqrt{2}\sin(2\sqrt{2}\,t) - \frac{1}{8}\sin(2\,t) + \frac{t}{24}$$

The positions of the three coaches are now plotted in Figure 1.4 over the interval $t=0$ to 20, the linestyles $1, 2, 3$ generating solid, dotted, and dashed curves.

```
> plot([X,Y,Z],t=0..20,color=[red,blue,green],linestyle=
  [1,2,3],thickness=2,tickmarks=[3,3],labels=["t","X,Y,Z"]);
```

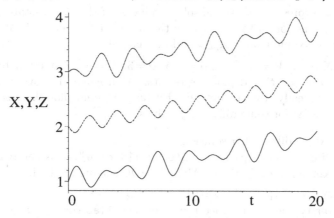

Figure 1.4: Positions of the three train coaches as a function of time.

The motion of the three coaches (represented by size 20 red boxes) is now animated and can be observed by executing the following command line.

```
> animate({[X,0],[Y,0],[Z,0]},t=0..100,frames=200,style=
  point,symbol=box,symbolsize=20,view=[0..8,-0.1..0.1],
  tickmarks=[3,2],labels=["x"," "]);
```

You can explore other possible motions by changing the parameter values and initial conditions. The time range and view may have to be altered.

1.1.5 "Resonances", A Recipe by I. M. Curious

Training is everything. The peach was once a bitter almond; cauliflower is nothing but cabbage with a college education.
Mark Twain, *Pudd'nhead Wilson*, ch. 5, (1894)

Over the years I have been fortunate to have one or more students in each mathematical physics class who stood out due to the quality of their work and their innate curiosity about the subject matter. As a representative of this elite group, let me introduce Ms. I. M. Curious who will discuss in her own words the elegant and thorough computer algebra solution that she generated in solving the following problem involving a familiar nonhomogeneous LODE.

A mechanical system (of unit mass) experiences a Stokes's law drag force $F_{\text{Stokes}} = -2\alpha\dot{x}$, a Hooke's law restoring force $F_{\text{Hooke}} = -\omega^2 x$, and a driving force $F_{\text{Drive}} = f\cos(\Omega t + \delta)$. Here α is a positive constant, ω the natural frequency, f the force amplitude, Ω the driving frequency, and δ a phase angle.

(a) Obtain the solution of the resulting LODE for $\alpha < \omega$ (*underdamping*), subject to the initial condition $x(0) = X$, $\dot{x}(0) = V$. Discuss the method used by Maple to solve the LODE and the structure of the solution.

(b) Derive an expression for the absolute value A of the steady-state amplitude as a function of Ω. Determine the resonant frequency Ω_r at which A has its maximum value A_{max} and evaluate A_{max}.

(c) Taking $X = -1$, $V = 2$, $\omega = 1$, $\alpha = 1/5$, $\Omega = 3$, $f = 5$, and $\delta = 0$, determine $x(t)$ and $v(t) = \dot{x}(t)$ and plot the results. Approximately how long does it take for steady-state to prevail? Evaluate Ω_r and plot the resonance curves A vs. Ω for some different f values.

And here is I. M. to discuss her solution.

"After loading the plots package, `infolevel[dsolve]:=5` is entered so as to provide maximum information on Maple's ODE solving method.

> `restart: with(plots): infolevel[dsolve]:=5:`

To save on later typing, I introduce the following `alias` command. When a, w, W, d, Wr, and Am are subsequently entered, the symbols α, ω, Ω, δ, Ω_r, and A_{max} will be respectively produced in the output.

> `alias(alpha=a,omega=w,Omega=W,delta=d,Omega[r]=Wr,A[max]=Am);`

$$\alpha,\ \omega,\ \Omega,\ \delta,\ \Omega_r,\ A_{max}$$

The Stokes, Hooke's law, and driving forces are entered,

> `F[Stokes]:=-2*a*diff(x(t),t); F[Hooke]:=-w^2*x(t);`
> `F[Drive]:=f*cos(W*t+d);`

$$F_{Stokes} := -2\alpha\left(\frac{d}{dt}x(t)\right) \qquad F_{Hooke} := -\omega^2 x(t) \qquad F_{Drive} := f\cos(\Omega t + \delta)$$

which are automatically substituted into Newton's second law in *ode*.

> `ode:=diff(x(t),t,t)-F[Stokes]-F[Hooke]=F[Drive];`

$$ode := \left(\frac{d^2}{dt^2}x(t)\right) + 2\alpha\left(\frac{d}{dt}x(t)\right) + \omega^2 x(t) = f\cos(\Omega t + \delta)$$

Assuming that $\alpha > 0$, $\omega > 0$, and $\alpha < \omega$, *ode* is solved for $x(t)$ subject to the initial conditions and the rhs of the result taken. The list of mathematical methods tried are given in the output, along with the solution labeled *x1*.

> `x1:=rhs(dsolve({ode,x(0)=X,D(x)(0)=V},x(t)))`
> `        assuming(a>0,w>0,a<w);`
>   *Methods for second order ODEs:*

— Trying classification methods —
trying a quadrature
trying high order exact linear fully integrable
trying differential order: 2; linear nonhomogeneous with symmetry [0,1]
trying a double symmetry of the form [xi=0, eta=F(x)]
Try solving first the homogeneous part of the ODE
−> Tackling the linear ODE "as given":
    checking if the LODE has constant coefficients
    <− constant coefficients successful
<− successful solving of the linear ODE "as given"
−> Determining now a particular solution to the nonhomogeneous ODE
    building a particular solution using variation of parameters
<− solving first the homogeneous part of the ODE successful

$$
\begin{aligned}
x1 := {}& e^{(-\alpha t)}\sin(\sqrt{\omega^2-\alpha^2}\,t)(-2\,V\,\omega^2\,\Omega^2 + \Omega^4\,\alpha\,X + V\,\Omega^4 - \Omega^3\,f\sin(\delta) \\
& + 4\,V\,\Omega^2\,\alpha^2 - \Omega^2\,f\,\alpha\cos(\delta) - \alpha\,\omega^2\,f\cos(\delta) - 2\,\Omega^2\,\alpha\,\omega^2\,X \\
& + \Omega\,f\sin(\delta)\,\omega^2 + 4\,\Omega^2\,\alpha^3\,X - 2\,\Omega\,f\sin(\delta)\,\alpha^2 + \alpha\,\omega^4\,X + V\,\omega^4) \\
& /(\sqrt{\omega^2-\alpha^2}\,(\omega^4 - 2\,\omega^2\,\Omega^2 + 4\,\Omega^2\,\alpha^2 + \Omega^4)) \\
& + e^{(-\alpha t)}\cos(\sqrt{\omega^2-\alpha^2}\,t)(f\cos(\delta)\,\Omega^2 - 2\,f\,\alpha\,\Omega\sin(\delta) - f\cos(\delta)\,\omega^2 + X\,\omega^4 \\
& - 2\,X\,\omega^2\,\Omega^2 + 4\,X\,\Omega^2\,\alpha^2 + X\,\Omega^4)\,/(\omega^4 - 2\,\omega^2\,\Omega^2 + 4\,\Omega^2\,\alpha^2 + \Omega^4) \\
& + \frac{((\omega^2 - \Omega^2)\cos(\Omega t + \delta) + 2\,\alpha\,\Omega\sin(\Omega t + \delta))\,f}{\Omega^4 + (-2\,\omega^2 + 4\,\alpha^2)\,\Omega^2 + \omega^4}
\end{aligned}
$$

Maple has successfully solved the ODE, a task I would not have wished to do by hand. It has recognized that *ode* is a second order nonhomogeneous LODE with the corresponding homogeneous equation having constant coefficients. It has solved *ode* using the method of variation of parameters. The solution *x1* is made up of two parts, the *transient* terms involving the exponentials which vanish (since $\alpha > 0$) as $t \to \infty$, and the remaining *steady-state* contribution which survives in this limit. Note that the steady-state part doesn't depend on the initial conditions. I will now extract the steady-state solution *xss* from *x1* by removing the terms in *x1* which have $e^{-\alpha t}$.

>   xss:=remove(has,x1,exp(-a*t));

$$
xss := \frac{((\omega^2 - \Omega^2)\cos(\Omega t + \delta) + 2\,\alpha\,\Omega\sin(\Omega t + \delta))\,f}{\Omega^4 + (-2\,\omega^2 + 4\,\alpha^2)\,\Omega^2 + \omega^4}
$$

The amplitude A of the steady-state solution can be obtained as follows. The amplitude corresponds to maximum displacement of the oscillator at which time the velocity is zero. I now solve for a time T at which this occurs.

>   T:=solve(diff(xss,t)=0,t);

$$
T := \frac{-\delta + \arctan(\dfrac{2\,\alpha\,\Omega}{\omega^2 - \Omega^2})}{\Omega}
$$

A is then obtained by evaluating *xss* at $t=T$ and simplifying with the `symbolic` option which picks a particular solution branch of the square root, the sign of which can be either plus or minus. The magnitude of A then follows on taking the absolute value of the result. The answer agrees with that found in Fowles and Cassiday[FC99].

> `A:=simplify(eval(xss,t=T),symbolic): Amag:=abs(A);`

$$Amag := \left| \frac{f}{\sqrt{\omega^4 - 2\omega^2 \Omega^2 + 4\Omega^2 \alpha^2 + \Omega^4}} \right|$$

To find the resonance frequency, $Amag$ is differentiated with respect to Ω, set equal to zero, and the result solved for the frequency.

> `sol:=solve(diff(Amag,W)=0,W);`

$$sol := 0, \sqrt{\omega^2 - 2\alpha^2}, -\sqrt{\omega^2 - 2\alpha^2}$$

The resonance frequency Ω_r must correspond to the non-zero positive square root solution. Therefore, the second answer is selected in *sol*. The maximum amplitude A_{max} is then obtained by evaluating $Amag$ at $\Omega = \Omega_r$ and again simplifying with the symbolic option.

> `Wr:=sol[2]; Am:=simplify(eval(Amag,W=Wr),symbolic);`

$$\Omega_r := \sqrt{\omega^2 - 2\alpha^2}$$

$$A_{max} := \frac{1}{2} \left| \frac{f}{\alpha \sqrt{\omega^2 - \alpha^2}} \right|$$

To answer part (**c**) of the problem, the given parameter values are entered.

> `X:=-1: V:=2: w:=1: a:=1/5: W:=3: f:=5: d:=0:`

The displacement $X1$ and velocity $V1$ are calculated, the radical expressions being simplified with the `radsimp` command.

> `X1:=radsimp(x1); V1:=radsimp(diff(x1,t));`

$$X1 := \frac{6737}{9816} e^{\left(-\frac{t}{5}\right)} \sin\left(\frac{2\sqrt{6}t}{5}\right) \sqrt{6} - \frac{159}{409} e^{\left(-\frac{t}{5}\right)} \cos\left(\frac{2\sqrt{6}t}{5}\right)$$
$$- \frac{250}{409} \cos(3t) + \frac{75}{818} \sin(3t)$$

$$V1 := \frac{179}{9816} e^{\left(-\frac{t}{5}\right)} \sin\left(\frac{2\sqrt{6}t}{5}\right) \sqrt{6} + \frac{1411}{818} e^{\left(-\frac{t}{5}\right)} \cos\left(\frac{2\sqrt{6}t}{5}\right)$$
$$+ \frac{750}{409} \sin(3t) + \frac{225}{818} \cos(3t)$$

It was not specified how the results are to be plotted, so I will create a 3-dimensional picture in `gr1` of $X1$ vs. $V1$ vs. t by using the `spacecurve` command. The time range is taken to be $t = 0$ to 40, 1000 plotting points are selected, and the plot is enclosed in a viewing box.

> `gr1:=spacecurve({[t,X1,V1]},t=0..40,numpoints=1000,`
            `thickness=2,axes=box,labels=["t","x","v"]):`

So that the starting point in the 3-dimensional trajectory can be easily identified, the `textplot3d` command is used in `gr2` to place the word `start` (colored red) in the vicinity of the starting point.

```
> gr2:=textplot3d([[0,X-0.15,V,"start"]],color=red):
```

The graphs are superimposed with the `display` command. The resulting picture (Figure 1.5) is enclosed in a viewing box which may be rotated by clicking on the computer plot and dragging with the mouse. In the upper left-hand corner of the computer screen, a small window indicates the angular coordinates of the viewing box. Here the orientation of the box is set to be (45°, 45°).

```
> display({gr1,gr2},orientation=[45,45],tickmarks=[3,3,3]);
```

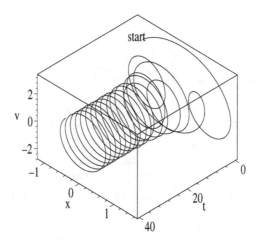

Figure 1.5: Space curve showing *X1* vs. *V1* vs. *t*.

The wild gyrations seen in Figure 1.5 for small t are associated with the transient, while steady-state corresponds to the uniform, cyclic behavior at larger t. To obtain an estimate of the time beyond which steady-state prevails, the orientation of the viewing box can be changed to (-90°, 0°) so as to view *X1* vs. t. Steady-state prevails for t greater than about 25 time units.

The resonance frequency is then evaluated in *W1* and the value *A1* of the absolute amplitude calculated at this frequency.

```
> W1:=evalf(Wr); A1:=evalf(Am);
```
$$W1 := 0.9591663048 \qquad A1 := 12.75775908$$
The resonance frequency of about 0.96 is slightly less than the natural frequency $\omega = 1$. To locate the resonance frequency on the graph of the resonance curves, I will plot a thick, vertical, dashed, blue line of height *A1* at *W1* in L.

```
> L:=plot([[W1,0],[W1,A1]],color=blue,thickness=2,linestyle=3):
```

To free the driving frequency and the force amplitude from their previously assigned values, they are now unassigned.

```
> unassign('W','f'):
```

A functional operator `pl` is created for plotting *Amag* vs. Ω for different f values over the frequency range $\Omega = 0$ to 2ω.

```
>   pl:=i->plot(eval(Amag,f=i),W=0..2*w,thickness=2):
```

Making use of the sequence command, resonance curves are generated for $f = 1, 2, 3, 4, 5$ and displayed along with L in Figure 1.6. As I expected, the resonance curves agree with those drawn in standard texts."

```
>display({L,seq(pl(i),i=1..5)},tickmarks=[2,3],labels=[W,"A"]);
```

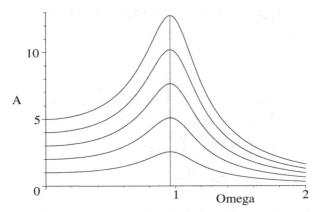

Figure 1.6: Resonance curves for $f = 1$ (bottom curve) to $f = 5$ (top).

1.1.6   Mr. Dirac's Famous Function

The man with a new idea is a crank until the idea succeeds.
Mark Twain, *Following the Equator, ch. 32*, (1897).

Consider a very long uniform horizontal beam glued to an elastic foundation, the foundation exerting a Hooke's law restoring force. Under the influence of a steady point force of magnitude F exerted at $x = 0$, the static displacement $\psi(x)$ of the beam is governed by the nonhomogeneous LODE [Hil57],

$$Y I \frac{d^4\psi}{dx^4} + K\,\psi = F\,\delta(x) \qquad (1.3)$$

where Y is Young's modulus, I the moment of inertia of the beam's cross-section about a horizontal axis, K the spring constant, and $\delta(x)$ the *Dirac delta function*. The delta function (δ function) is defined by

$$\delta(x) = 0,\ x \neq 0; \quad \delta(x) = \infty,\ x = 0; \quad \int_{-\alpha}^{\beta} \delta(x)\,dx = 1,\ (\alpha,\ \beta > 0). \qquad (1.4)$$

The δ function should be regarded as the limit of, e.g, a Gaussian function[5] which is made progressively taller and narrower in such a way that the area

---

[5]For the Gaussian, $\delta(x) = \lim_{A \to \infty}(A/\sqrt{\pi})\,e^{-A^2 x^2}$. The choice of functions is not unique.

under the curve remains equal to 1. Changing the argument of the δ-function from x to $x - \zeta$ shifts the location of the δ-function from $x=0$ to $x=\zeta$.

It should be noted that the δ-function has the dimensions x^{-1}. It has a number of properties, the most important of which is the "selection" property. If $F(x)$ is a smooth function (not another δ-function), then

$$\int_{\zeta-\alpha}^{\zeta+\beta} F(x)\,\delta(x-\zeta)\,dx = F(\zeta), \quad (\alpha,\ \beta > 0). \tag{1.5}$$

The Dirac delta function concept may be generalized to higher dimensions, e.g., $\delta(\vec{r}) \equiv \delta(x)\,\delta(y)\,\delta(z)$ is a 3-dimensional δ-function located at $x=y=z=0$.

A solution of a linear ODE which has a delta function appearing in the nonhomogeneous term is called a *Green function*. The goal of the following recipe is to determine the static displacement of the beam due to the point force, i.e., the Green function, for Equation (1.3) and plot it.

In the ODE (1.3), the moment of inertia I appears. This is a protected symbol in Maple, representing $\sqrt{-1}$. Since I is a commonly used symbol in physics, it is often desirable to unprotect it. The following `interface` command is used to let $j=\sqrt{-1}$, thus freeing up I.

```
>  restart: interface(imaginaryunit=j):
```

The LODE (1.3) is inputted, the command `Dirac(x)` being used to enter $\delta(x)$.

```
>  ode:=Y*I*diff(psi(x),x$4)+K*psi(x)=F*Dirac(x);
```

$$ode := Y\,I\left(\frac{d^4}{dx^4}\,\psi(x)\right) + K\,\psi(x) = F\,\mathrm{Dirac}(x)$$

The ODE *ode* can be simplified by introducing the parameter $\alpha = (Y\,I/K)^{1/4}$ and letting $\psi(x) = (F/K)\,G(x)$, where $G(x)$ is defined to be the Green function. The transformations are substituted into *ode*, producing *ode2*,

```
>  ode2:=subs({Y=alpha^4*K/I,psi(x)=F*G(x)/K},ode);
```

$$ode2 := \alpha^4\,K\left(\frac{\partial^4}{\partial x^4}\left(\frac{F\,\mathrm{G}(x)}{K}\right)\right) + F\,\mathrm{G}(x) = F\,\mathrm{Dirac}(x)$$

which is then divided by F in *ode3*, and the result expanded.

```
>  ode3:=expand(ode2/F);
```

$$ode3 := \alpha^4\left(\frac{d^4}{dx^4}\,\mathrm{G}(x)\right) + \mathrm{G}(x) = \mathrm{Dirac}(x)$$

For $x < 0$ and $x > 0$, the rhs of *ode3* is equal to zero. The general solution of *ode3* is obtained, assuming that $x < 0$.

```
>  sol:=rhs(dsolve(ode3,G(x))) assuming x<0;
```

$$sol := \_C1\,e^{\left(\frac{(1/2-1/2\,j)\,\sqrt{2}\,x}{\alpha}\right)} + \_C2\,e^{\left(\frac{(-1/2-1/2\,j)\,\sqrt{2}\,x}{\alpha}\right)}$$
$$+ \_C3\,e^{\left(\frac{(-1/2+1/2\,j)\,\sqrt{2}\,x}{\alpha}\right)} + \_C4\,e^{\left(\frac{(1/2+1/2\,j)\,\sqrt{2}\,x}{\alpha}\right)}$$

Clearly, the general solution for $x > 0$ will be of a similar structure. Let's look at the region to the left of the origin, assigning this portion of G the name GL. Since the beam is very long, little error is committed by taking the beam to be of infinite length. Physically, the displacement of the beam from equilibrium, and therefore GL, should go to zero as $x \to -\infty$. To satisfy this *asymptotic boundary condition*, only the two exponentials which have $(1/2 - 1/2\,j)$ and $(1/2 + 1/2\,j)$ in the exponents should be kept. The other two exponentials involving $(-1/2 - 1/2\,j)$ and $(-1/2 + 1/2\,j)$ diverge as $x \to -\infty$. The following command selects the two desired exponential terms to keep in GL.

> `GL:=select(has,sol,{1/2-j/2,1/2+j/2});`

$$GL := \_C1\, e^{\left(\frac{(1/2-1/2\,j)\,\sqrt{2}\,x}{\alpha}\right)} + \_C4\, e^{\left(\frac{(1/2+1/2\,j)\,\sqrt{2}\,x}{\alpha}\right)}$$

Similarly, applying the condition $G \to 0$ as $x \to \infty$, the `select` command is used to obtain the form GR of the Green function to the right of the origin.

> `GR:=select(has,sol,{-1/2-j/2,-1/2+j/2});`

$$GR := \_C2\, e^{\left(\frac{(-1/2-1/2\,j)\,\sqrt{2}\,x}{\alpha}\right)} + \_C3\, e^{\left(\frac{(-1/2+1/2\,j)\,\sqrt{2}\,x}{\alpha}\right)}$$

The four constants must now be evaluated. Since the displacement is physically continuous at the origin, the condition $GR - GL = 0$ at $x = 0$ is entered in eq_0.

> `eq[0]:=simplify(eval(GR-GL,x=0))=0;`

$$eq_0 := \_C2 + \_C3 - \_C1 - \_C4 = 0$$

An operator f is formed to simplify the nth derivative of the difference $GR - GL$ evaluated at $x = 0$. Both n and the difference s must be given.

> `f:=(n,s)->simplify(alpha^n*(eval(diff(GR-GL,x$n),x=0)=s));`

$$f := (n,\, s) \to \text{simplify}\left(\alpha^n \left(\left(\frac{\partial^n}{\partial x^n}(GR - GL)\right)\bigg|_{x=0} = s\right)\right)$$

For $n=1$ and 2, $s=0$, i.e., the first and second derivatives of G are continuous at $x=0$. The third derivative $(n=3)$ is discontinuous at $x=0$, however. Mentally integrating *ode3* from the left to the right of the δ function yields $s = 1/\alpha^4$. Making use of these results and f, the remaining three conditions are entered.

> `eq[1]:=sqrt(2)*f(1,0); eq[2]:=f(2,0);`
> `eq[3]:=sqrt(2)*f(3,1/alpha^4);`

$$eq_1 := -\_C2 - \_C2\,j + \_C3\,j - \_C3 + \_C1\,j - \_C1 - \_C4 - \_C4\,j = 0$$

$$eq_2 := (\_C2 - \_C3 + \_C1 - \_C4)\,j = 0$$

$$eq_3 := -\_C2\,j + \_C2 + \_C3 + \_C3\,j + \_C1 + \_C1\,j - \_C4\,j + \_C4 = \frac{\sqrt{2}}{\alpha}$$

The set of four equations are solved for the four constants,

> `sol2:=solve({seq(eq[n],n=0..3)},{_C1,_C2,_C3,_C4});`

$$sol2 := \left\{ \_C1 = \frac{\left(\frac{1}{8} - \frac{1}{8}\,j\right)\sqrt{2}}{\alpha},\; \_C2 = \frac{\left(\frac{1}{8} + \frac{1}{8}\,j\right)\sqrt{2}}{\alpha},\; \_C4 = \frac{\left(\frac{1}{8} + \frac{1}{8}\,j\right)\sqrt{2}}{\alpha},\right.$$

$$\left. \_C3 = \frac{\left(\frac{1}{8} - \frac{1}{8}\,j\right)\sqrt{2}}{\alpha} \right\}$$

and the solution *sol2* assigned. The constants are automatically substituted into *GL* and *GR*, which are put into real forms by applying the `evalc` command.

> `assign(sol2): GL:=evalc(GL); GR:=evalc(GR);`

$$GL := \frac{1}{4}\frac{\sqrt{2}\,e^{(\frac{\sqrt{2}\,x}{2\,\alpha})}\cos(\frac{\sqrt{2}\,x}{2\,\alpha})}{\alpha} - \frac{1}{4}\frac{\sqrt{2}\,e^{(\frac{\sqrt{2}\,x}{2\,\alpha})}\sin(\frac{\sqrt{2}\,x}{2\,\alpha})}{\alpha}$$

$$GR := \frac{1}{4}\frac{\sqrt{2}\,e^{(-\frac{\sqrt{2}\,x}{2\,\alpha})}\cos(\frac{\sqrt{2}\,x}{2\,\alpha})}{\alpha} + \frac{1}{4}\frac{\sqrt{2}\,e^{(-\frac{\sqrt{2}\,x}{2\,\alpha})}\sin(\frac{\sqrt{2}\,x}{2\,\alpha})}{\alpha}$$

Noting that there is a symmetry in the two forms (replacing x with $-x$ in *GL* yields *GR*), the Green function can be put into a more compact form by replacing x by its absolute value, $|x|$, in *GR*, then factoring, and relabeling the result as *G*. The static displacement of the beam is then $\psi(x) = (f/K)\,G(x)$.

> `G:=factor(subs(x=abs(x),GR));`

$$G := \frac{1}{4}\frac{\sqrt{2}\,e^{(-1/2\,\frac{\sqrt{2}\,|x|}{\alpha})}(\cos(\frac{1}{2}\frac{\sqrt{2}\,|x|}{\alpha}) + \sin(\frac{1}{2}\frac{\sqrt{2}\,|x|}{\alpha}))}{\alpha}$$

The Green function *G* is now plotted for the representative value $\alpha = 1$.

> `plot(eval(G,alpha=1),x=-10..10,thickness=2,`
> `labels=["x","G"]);`

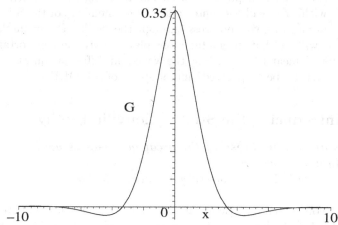

Figure 1.7: Green function *G* for the infinite beam.

The result is shown in Figure 1.7. As expected, *G* and therefore the beam displacement is localized near the origin where the point force is applied.

1.2 Linear ODEs with Variable Coefficients

In this section, linear ODEs with variable coefficients are considered. Our attention will be focused on the second-order *Sturm–Liouville* (S-L) equation,

$$\frac{d}{dx}\left[p(x)\,\frac{dy}{dx}\right] - q(x)\,y + \lambda\,w(x)\,y = 0, \qquad (1.6)$$

where λ is a real parameter while $p(x)$, $q(x)$, $w(x)$ are real functions and $w(x)$ is taken to be non-negative over the range $x = a$ to b of interest. The S-L equation, and its nonhomogeneous counterpart, plays a very important role in the mathematical analysis of many physical problems, particularly for boundary value problems where y or its first derivative vanish at a and b. In this case the y's are referred to as the *eigenfunctions* and the λ's as the *eigenvalues*. For certain choices of p, q, w, and λ, the general S-L equation reduces to specific "well-known" LODEs such as Bessel's equation, Legendre's equation, Hermite's equation, etc. The standard approach to solving these equations is to seek an infinite power series solution. In some cases (e.g., for Bessel's equation) the solution remains as an infinite series, while in other cases the infinite series have to be truncated (e.g., for Legendre's and Hermite's equations) to finite polynomials to avoid divergence problems for large x. In either case, these solutions to special forms of the S-L equation are referred to as *special functions* (e.g., Bessel functions, Legendre functions (polynomials), etc.). The detailed series derivation of these special functions will be postponed until Chapter 2. In this section, we shall be content to see what Maple reveals about some of these special functions and how physical problems involving these functions can be easily dealt with. You will encounter many more members of the S-L "family" of special functions as you progress through the book. For example, in the **Entrees** you will see that special functions play an extremely important role in the solutions of linear PDEs such as the wave and diffusion equations because these equations can be "separated" into systems of S-L ODEs.

1.2.1 Introducing the Sturm–Liouville Family

My books are water; those of the great geniuses is wine.
Everybody drinks water.
Mark Twain, *Mark Twains Notebooks and Journals, Notebook 26*, (1887)

Jennifer, a young mathematician at the Metropolis Institute of Technology (MIT) and a firm believer in using computer algebra to teach applied mathematics to engineers and physicists, has kindly consented to provide us with some of the computer algebra recipes that she has developed. In the following recipe, she introduces two of the more "famous" members of the S-L family, namely the Bessel and Legendre ODEs.

Jennifer creates two functional operators, *SL* for generating a specific ODE from the S-L equation (1.6) on subsequently specifying the forms of p, q, w

and the parameter L (short for λ), and Y for generating the general analytic solution of a specified *ode* and taking the right-hand side of the result.

```
>  restart:
>  SL:=(p,q,w,L)->diff(p*diff(y(x),x),x)-q*y(x)+L*w*y(x)=0;
```

$$SL := (p,\, q,\, w,\, L) \rightarrow (\frac{d}{dx}(p(\frac{d}{dx}y(x)))) - q\,y(x) + L\,w\,y(x) = 0$$

```
>  Y:=ode->rhs(dsolve(ode,y(x))):
```

According to standard texts (e.g., [MW71], [Boa83]), Bessel's' equation should result on choosing $p = x$, $q = -x$, $w = -1/x$, and $L \equiv \lambda = -n^2$. Entering these specific forms as arguments in SL produces *ode1*.

```
>  ode1:=SL(x,-x,1/x,-n^2);
```

$$ode1 := (\frac{d}{dx}y(x)) + x\,(\frac{d^2}{dx^2}y(x)) + x\,y(x) - \frac{n^2\,y(x)}{x} = 0$$

Loading the DEtools library package, Jennifer enters the following **odeadvisor** command line to see if Maple can classify *ode1* and offer any information about this equation. The ODE is identified as Bessel's equation and inclusion of the **help** option causes a help page to be opened with information on this equation. She adds a comment to close the help page when finished reading it.

```
>  with(DEtools): odeadvisor(ode1,help); #close Help page
```

$$[\_Bessel]$$

Entering the **infolevel[dsolve]** command so as to gain information about the methods tried in attempting to solve the ODE, she then uses the arrow operator Y to solve *ode1*.

```
>  infolevel[dsolve]:=2: y1:=Y(ode1);
```

Methods for second order ODEs:
— Trying classification methods —
trying a quadrature
checking if the LODE has constant coefficients
checking if the LODE is of Euler type
trying a symmetry of the form [xi=0, eta=F(x)]
checking if the LODE is missing 'y'
−> Trying a Liouvillian solution using Kovacic's algorithm
<− No Liouvillian solutions exists
−> Trying a solution in terms of special functions:
*   −> Bessel*
*   <− Bessel successful*
<− special function solution successful

$$y1 := \_C1\,\mathrm{BesselJ}(n,\, x) + \_C2\,\mathrm{BesselY}(n,\, x)$$

After identifying the ODE as second order, Maple tries various approaches before seeking and obtaining a general special function solution in terms of Bessel functions. Highlighting, say, BesselJ in the output with the mouse and clicking on "Help on BesselJ", opens a help page on Bessel functions. $\mathrm{BesselJ}(n, x)$ is

identified as a *Bessel function of the first kind of order n*, while BesselY(n, x)
is a *Bessel function of the second kind.* In standard mathematical notation,
they are usually written as $J_n(x)$ and $Y_n(x)$, respectively. Before delving into
some of the properties of Bessel functions, Jennifer notes that on executing the
command ?inifcns, a help page is opened which lists all the functions known
to Maple and has hyperlinks to associated help pages (e.g., for BesselJ).

```
>  ?inifcns;
```

Closing the help page, Jennifer forms an arrow operator f for plotting a sequence of four (the number can be easily changed) Bessel functions with integer
subscript. The name N (BesselJ or BesselY) of the Bessel function must be
supplied along with the horizontal range $x = x1$ to $x2$ over which the function
is to be plotted. The four curves are given different colors and line styles. Line
style 4 produces a DASH-DOT curve. The vertical range is limited with the
view command to -1 to $+1$ because the $Y_n(x)$ become infinitely large at $x=0$
if this point is included in the range of interest.

```
> f:=(N,x1,x2)->plot([seq(N(n,x),n=0..3)],x=x1..x2,thickness=2,
          color=[red,blue,green,cyan],linestyle=[1,2,3,4],
          view=[x1..x2,-1..1],tickmarks=[2,2]):
```

Making use of f, the $J_n(x)$ and $Y_n(x)$ are plotted for $x=x1=0$ to $x2=20$.

```
>  f(BesselJ,0,20);  f(BesselY,0,20);
```

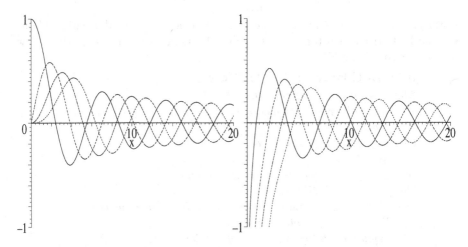

Figure 1.8: Left plot: Left to right, J_0 to J_3. Right: Left to right, Y_0 to Y_3.

The Bessel functions $J_0(x)$, $J_1(x)$, $J_2(x)$, and $J_3(x)$ are ordered from left to
right in the left plot of Figure 1.8. At $x=0$, J_0 has the value 1, while all other
positive integer J_n are equal to zero there. The Bessel functions $Y_0(x)$, $Y_1(x)$,
$Y_2(x)$, and $Y_3(x)$, with the same ordering, are shown on the right of Figure 1.8.
The $Y_n(x)$ diverge to $-\infty$ at $x=0$, so are rejected in physical problems where
the origin is part of the range of interest.

Of particular importance in solving problems involving Bessel functions is knowing the locations of the zeros. The first 5 zeros of, e.g., J_1 are now obtained with the following `BesselJZeros` command. Note that the argument 1 is expressed in floating point form, i.e., 1.0, in order to numerically evaluate the zeros. The command `BesselYZeros` will generate the zeros of the $Y_n(x)$.

> `Zeros_J1:=BesselJZeros(1.0,1..5);`

$Zeros\_J1 := 3.831705970, 7.015586670, 10.17346814, 13.32369194, 16.47063005$

For large x, both the $J_n(x)$ and $Y_n(x)$ have the appearance of slowly decreasing sine or cosine functions. Jennifer confirms this conjecture by Taylor expanding, e.g., $J_1(x)$ about $x=+\infty$ and keeping the first term in the expansion.

> `taylor(BesselJ(1,x),x=infinity,1);`

$$-\frac{\sqrt{2}\cos(x+\frac{\pi}{4})\sqrt{\frac{1}{x}}}{\sqrt{\pi}} + O((\frac{1}{x})^{(3/2)})$$

The "order of" term, $O((1/x)^{3/2})$, is removed with `convert(%,polynom)`, the "ditto operator", %, applying the command to the previously executed result.[6]

> `J1_asymptotic:=convert(%,polynom);`

$$J1\_asymptotic := -\frac{\sqrt{2}\cos(x+\frac{\pi}{4})\sqrt{\frac{1}{x}}}{\sqrt{\pi}}$$

Asymptotically, $J_1(x)$ behaves like a cosine function whose amplitude decreases like $1/\sqrt{x}$. The Bessel functions have many important properties which are too numerous for Jennifer to explore here. For an exhaustive list of the properties of all special functions, she refers the reader to the voluminous Handbook of Mathematical Functions [AS72]. One important property shared by special functions is that they satisfy *recurrence relations*, relating functions of different orders. Here's an example of a recurrence relation for the $J_n(x)$. This recurrence relation relates Bessel functions of orders $n-1$, $n+1$, and n for arbitrary x.

> `recurrence:=BesselJ(n-1,x)+BesselJ(n+1,x)`
> `            =simplify(BesselJ(n-1,x)+BesselJ(n+1,x));`

$$recurrence := \text{BesselJ}(n-1, x) + \text{BesselJ}(n+1, x) = \frac{2n\,\text{BesselJ}(n, x)}{x}$$

Now, Jennifer introduces Legendre's ODE by entering the Sturm–Liouville arrow operator SL with $p=1-x^2$, $q=0$, $w=1$, and $L \equiv \lambda = n(n+1)$. The differential equation *ode2* is then solved and identified as Legendre's equation.

> `ode2:=SL(1-x^2,0,1,n*(n+1));   y2:=Y(ode2);`

$$ode2 := -2x\,(\frac{d}{dx}\,y(x)) + (1-x^2)\,(\frac{d^2}{dx^2}\,y(x)) + n(n+1)\,y(x) = 0$$
Methods for second order ODEs:

....

---

[6]To recall the second previous result, use two percent signs, %%, and so on.

$->$ *Trying a solution in terms of special functions:*
$-> $ *Bessel*
$-> $ *elliptic*
$-> $ *Legendre*
$<- $ *Legendre successful*
$<- $ *special function solution successful*

$$y2 := \_C1 \operatorname{LegendreP}(n,\, x) + \_C2 \operatorname{LegendreQ}(n,\, x)$$

Highlighting LegendreP with the mouse and opening up the associated help page reveals that LegendreP(n, x) is the *Legendre function of the first kind of order n*, while LegendreQ(n, x) is the *Legendre function of the second kind*. In standard mathematical notation, they are written as $P_n(x)$ and $Q_n(x)$, respectively. If the odeadvisor command is applied to *ode2*, the ODE is identified

```
> odeadvisor(ode2);
```

$$[\_Gegenbauer]$$

as a *Gegenbauer* equation, which has the general structure [AS72],

$$(1 - x^2)\, y'' - (2\,\alpha + 1)\, x\, y' + n\,(n + 2\,\alpha)\, y = 0. \tag{1.7}$$

Legendre's ODE is a special case of Equation (1.7) with $\alpha = 1/2$.

The Legendre functions $P_n(x)$ with $n = 0$ or a positive integer play an important role in boundary value problems involving spherical polar coordinates, in which case $x = \cos(\theta)$ where θ is the polar angle measured from the z-axis. The polar angle runs from 0 to π radians, so x spans the range 1 to -1. For $n = 0,\ 1,\ 2, ...,$ the P's reduce to the *Legendre polynomials* which Jennifer now generates for $n = 0$ to 5. The simplify command must be applied to produce the finite polynomial forms.

```
> Ps:=seq(simplify(LegendreP(n,x)),n=0..5);
```

$$Ps := 1,\ x,\ -\frac{1}{2} + \frac{3\,x^2}{2},\ \frac{5}{2}x^3 - \frac{3}{2}x,\ \frac{3}{8} + \frac{35}{8}x^4 - \frac{15}{4}x^2,\ \frac{63}{8}x^5 - \frac{35}{4}x^3 + \frac{15}{8}x$$

The polynomials may also be generated by loading the orthopoly library package and using the syntax P(n,x) to enter the nth order Legendre polynomial.

```
> with(orthopoly): seq(P(n,x),n=0..5);
```

$$1,\ x,\ -\frac{1}{2} + \frac{3\,x^2}{2},\ \frac{5}{2}x^3 - \frac{3}{2}x,\ \frac{3}{8} + \frac{35}{8}x^4 - \frac{15}{4}x^2,\ \frac{63}{8}x^5 - \frac{35}{4}x^3 + \frac{15}{8}x$$

To produce $Q_n(x)$ over the range $x = -1$ to 1, Jennifer enters _EnvLegendreCut: =1..infinity: which selects the desired mathematical branch[7] of the function. The Q's are then produced for $n = 0,\ 1,\ 2.$ They are expressible as combinations of log functions.

```
> _EnvLegendreCut:=1..infinity:
```

```
> Qs:=seq(simplify(LegendreQ(n,x)),n=0..2);
```

---

[7]To plot $Q_n(x)$ for $|x| > 1$, enter _EnvLegendreCut:=-1..1: and adjust the view.

$$Qs := \frac{1}{2}\ln(1+x) - \frac{1}{2}\ln(1-x), \; \frac{1}{2}x\ln(1+x) - \frac{1}{2}x\ln(1-x) - 1,$$

$$-\frac{1}{4}\ln(1+x) + \frac{1}{4}\ln(1-x) + \frac{3}{4}x^2\ln(1+x) - \frac{3}{4}x^2\ln(1-x) - \frac{3x}{2}$$

Using the arrow operator `f`, the Legendre functions of both kinds are now plotted for $n=0$ to 3 over the horizontal range $x=-1$ to 1.

```
>   f(LegendreP,-1,1);  f(LegendreQ,-1,1);
```

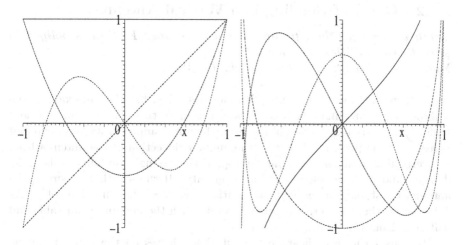

Figure 1.9: Left plot: P_0, P_1, P_2, P_3. Right: Q_0, Q_1, Q_2, Q_3.

$P_0(x)$ (horizontal curve), $P_1(x)$ (diagonal curve), $P_2(x)$ (parabola), and $P_3(x)$ (N-shape) are shown in the left plot of Figure 1.9. $Q_0(x)$ (solid curve), $Q_1(x)$ (parabola), $Q_2(x)$ (N-shape), and $Q_3(x)$ (W-shape) are shown in the right plot of the figure. The P_n are well behaved at $x=\pm1$, but the Q_n diverge at these points and must be rejected in physical problems which include $x=\pm1$ in the range of interest.

An important general property that all solutions $y_n(x)$ of the Sturm–Liouville equation corresponding to a given λ_n possess is *orthogonality*. Provided that $y(x)$, or $y'(x)$, or $p(x)$ vanishes at the end points a and b of the range (referred to as *Sturm–Liouville boundary conditions*), then

$$\int_a^b w(x)\, y_m(x)\, y_n(x)\, dx = 0, \quad \text{for } m \neq n. \tag{1.8}$$

$w(x)$ is referred to as the *weight function*. Noting that $w(x)=1$ for Legendre's equation and that $p = 1 - x^2$ vanishes at $x = a = -1$ and at $b = +1$, the orthogonality conditions $\int_{-1}^{1} P_m(x)\, P_n(x)\, dx = 0$ and $\int_{-1}^{1} Q_m(x)\, Q_n(x)\, dx = 0$ should prevail for $m \neq n$. For specific values of m and n the former condition is easy to prove, while proving the latter condition is more formidable.

Jennifer tackles the second orthogonality condition, taking $m=3$ and $n=2$. She uses the "inert" form `Int` on the left of the following command line to display the form of the integral and the "active" form on the right to evaluate the integral. The answer is 0 as expected.

```
> orthog:=Int(LegendreQ(2,x)*LegendreQ(3,x),x=-1..1)
        =int(simplify(LegendreQ(2,x)*LegendreQ(3,x)),x=-1..1);
```

$$orthog := \int_{-1}^{1} \text{LegendreQ}(2,\,x)\,\text{LegendreQ}(3,\,x)\,dx = 0$$

1.2.2   Onset of Bending in a Vertical Antenna

Truth is stranger than fiction, but it is because Fiction is obliged to stick to possibilities; Truth isn't.
Mark Twain, *Following the Equator, ch. 15*, (1897).

As a follow up to the last recipe, Jennifer has asked her class to examine the stability of an antenna which consists of a thin vertical wire of length L and uniform circular cross-section of radius a which is clamped at its lower end and is free at its upper end. Let θ be the angular deflection of the antenna from the vertical at a distance y from the top, Y the Young's modulus, ρ the mass density, and g the acceleration due to gravity. If the length L is small, the antenna is "stable" in the vertical position, i.e., $\theta = 0$ for all values of y. As L increases, there is a critical value beyond which the antenna is unstable and will bend from the vertical.

By considering the shear and gravitational forces on the wire, it can be shown that the relevant LODE for small angular displacements θ is

$$\frac{d^2\theta}{dy^2} = -c^2\,y\,\theta, \quad \text{where } c = \frac{2}{a}\sqrt{\frac{\rho\,g}{Y}}. \tag{1.9}$$

(a) Noting that $\theta = 0$ at $y = L$ (clamped at bottom end) and $d\theta/dy = 0$ at $y = 0$ (free at top), determine the solution of the ODE, identifying the functions which occur. Express the solution in terms of Bessel functions.

(b) Prove that the critical length for bending is given by $L_{cr} \simeq (2.8/c)^{2/3}$.

(c) Determine L_{cr} for a steel ($Y = 2.1 \times 10^{11}$ N/m^2 and $\rho = 7800$ kg/m^3) wire of radius 1 mm. Take $g = 9.8$ m/s^2.

Jennifer has provided us with the following computer algebra solution taken from her answer key. The governing ODE is entered.

```
>   restart:
>   ode:=diff(theta(y),y,y)=-c^2*y*theta(y);
```

$$ode := \frac{d^2}{dy^2}\,\theta(y) = -c^2\,y\,\theta(y)$$

The ODE is solved for $\theta(y)$ subject to the boundary condition $d\theta/dy = 0$ at $y=0$ and the rhs of the solution taken.

```
>   theta:=rhs(dsolve({ode,D(theta)(0)=0},theta(y)));
```

$$\theta := \_C2\,\sqrt{3}\,\mathrm{AiryAi}(-(c^2)^{(1/3)}\,y) + \_C2\,\mathrm{AiryBi}(-(c^2)^{(1/3)}\,y)$$

Jennifer has not included the other boundary condition in the `dsolve` command because then only the trivial solution $\theta = 0$ would be produced. Highlighting AiryAi in the output with the mouse, clicking on Help, and then on "Help on AiryAi", opens a help page which indicates that AiryAi is an *Airy wave function*. AiryBi is another Airy function. According to Help, the Airy functions are related to the Bessel functions $J_{n/3}$ where n is a positive or negative integer.

Using this information, θ is converted to Bessel functions of the first kind and simplified assuming that $y > 0$ and $c > 0$. θ is expressed in terms of $J_{(-1/3)}$.

```
> theta:=simplify(convert(theta,BesselJ)) assuming y>0,c>0;
```

$$\theta := \frac{2}{3}\,\_C2\,\sqrt{3}\,c^{(1/3)}\,\sqrt{y}\,\mathrm{BesselJ}\!\left(\frac{-1}{3},\,\frac{2\,c\,y^{(3/2)}}{3}\right)$$

The other boundary condition, $\theta(y{=}L) = 0$, is now applied.

```
> bc:=eval(theta,y=L)=0;
```

$$bc := \frac{2}{3}\,\_C2\,\sqrt{3}\,c^{(1/3)}\,\sqrt{L}\,\mathrm{BesselJ}\!\left(\frac{-1}{3},\,\frac{2\,c\,L^{(3/2)}}{3}\right) = 0$$

For θ to have a non-trivial solution, the coefficient $\_C2 \neq 0$, so the boundary condition must reduce to the Bessel function being equal to zero. The `select` command is used to extract the Bessel function from the lhs of the bc and the boundary condition is re-expressed as follows.

```
> bc:=select(has,lhs(bc),BesselJ)=0;
```

$$bc := \mathrm{BesselJ}\!\left(\frac{-1}{3},\,\frac{2\,c\,L^{(3/2)}}{3}\right) = 0$$

Clearly, the critical length for bending must correspond to finding the first zero of the above Bessel function. The following `op` command is used to extract the second operand from the lhs of bc.

```
> X:=op(2,lhs(bc));
```

$$X := \frac{2\,c\,L^{(3/2)}}{3}$$

The first zero of $J_{(-1/3)}$ is obtained.

```
> s:=BesselJZeros(evalf(-1/3),1);
```

$$s := 1.866350859$$

The critical value L_{cr} follows on setting $X = s$ and solving for L. Only one of the answers is real, the other two being complex.

```
> Lcr:=solve(X=s,L);
```

$$Lcr := \frac{1.986352708\,(c^2)^{(2/3)}}{c^2},$$

$$\left(-\frac{0.7046901283\,(c^2)^{(1/3)}}{c} + \frac{1.220559106\,I\,(c^2)^{(1/3)}}{c}\right)^2,$$

$$\left(-\frac{0.7046901283\,(c^2)^{(1/3)}}{c} - \frac{1.220559106\,I\,(c^2)^{(1/3)}}{c}\right)^2$$

The first answer (the real one) is selected and simplified with the `symbolic` option. This is the expression for the critical length.

```
> Lcr:=simplify(Lcr[1],symbolic);
```

$$Lcr := \frac{1.986352708}{c^{(2/3)}}$$

To confirm that the critical length $L_{cr} \simeq (2.8/c)^{2/3}$, the numerator of Lcr is extracted and raised to the $3/2$ power in p. It is then evaluated to 2 digits accuracy in $p2$, yielding 2.8 as desired.

```
> p:=(numer(Lcr))^(3/2); p2:=evalf(p,2);
```

$$p := 2.799526291 \qquad p2 := 2.8$$

To determine L_{cr} for the steel wire, the given parameter values are entered.

```
> g:=9.8: a:=0.001: Y:=2.1*10^11: rho:=7800:
```

The parameter c is evaluated and the critical length for bending of the steel wire antenna determined.

```
> c:=evalf((2/a)*sqrt(rho*g/Y)): Lcr:=Lcr;
```

$$Lcr := 1.752545257$$

The antenna will not bend from the vertical if its length $L < 1.75$ meters.

1.2.3   The Quantum Oscillator

Anybody who is not shocked by this subject has failed to understand it.
Neils Bohr, 1922 Nobel laureate in physics, referring to quantum mechanics

Over the years I have had occasion to teach an introductory quantum mechanics course which concentrates on how to *do* quantum mechanics, leaving it to our departmental "guru" to deal with deeper quasi-philosophical questions of what it *means* in a higher-level course. In the doing category, a standard problem (see Griffiths [Gri95]) is to derive the probability distribution for the quantum mechanical version of the one-dimensional simple harmonic oscillator. This may be done either "algebraically" using "ladder operators" or by the more brute force method of directly solving the time-independent *Schrödinger* equation for the *probability amplitude* ψ. Once ψ is determined, the *probability density* $P(x) = |\psi(x)|^2$ may be calculated. The probability that a particle can be found between, say $x = a$ and $x = b$, is $\int_a^b P(x)\,dx$. Of course, $\int_{-\infty}^{\infty} P(x)\,dx = 1$.

This recipe illustrates how $P(x)$ may be painlessly derived, plotted, and interpreted for the quantum oscillator starting with the Schrödinger equation. The PDEtools package is loaded because it contains the `dchange` command which will be used to change both the dependent and independent variables.

```
> restart: with(PDEtools):
```

The Schrödinger ODE is entered for a particle of mass m. Here E and V are the total and potential energy and $hbar \equiv \hbar = h/(2\pi)$, where h is Planck's constant.

```
> ode:=(hbar^2/(2*m))*diff(psi(x),x,x)+(E-V)*psi(x)=0;
```

$$ode := \frac{1}{2}\frac{hbar^2\left(\dfrac{d^2}{dx^2}\psi(x)\right)}{m} + (E - V)\,\psi(x) = 0$$

For a particle experiencing a Hooke's law restoring force, $V = (1/2)\,m\,\omega^2\,x^2$, where ω is the frequency and x is the displacement of the particle from equilibrium. On entering V, its form is automatically substituted into *ode*.

> `V:=(1/2)*m*omega^2*x^2: ode:=ode;`

$$ode := \frac{1}{2}\frac{hbar^2\left(\dfrac{d^2}{dx^2}\psi(x)\right)}{m} + \left(E - \frac{m\,\omega^2\,x^2}{2}\right)\psi(x) = 0$$

Since $\hbar\,\omega$ has the units of energy, let's express E in these energy units, writing $E = (n+1/2)\,\hbar\,\omega$. The form $n+1/2$ of the "scale factor", where the parameter n remains to be determined, has been chosen for later convenience. At the same time, *ode* is multiplied by $(2/(\hbar\,\omega))$ and the result expanded in *ode2*.

> `E:=(n+1/2)*hbar*omega: ode2:=expand(2*ode/(hbar*omega));`

$$ode2 := \frac{hbar\left(\dfrac{d^2}{dx^2}\psi(x)\right)}{\omega\,m} + 2\,\psi(x)\,n + \psi(x) - \frac{\omega\,\psi(x)\,m\,x^2}{hbar} = 0$$

The constants can be removed from *ode2* by introducing a new independent variable ζ defined by $x = \sqrt{\hbar/(m\omega)}\,\zeta$ and also setting $\psi(x) = f(\zeta)\,e^{-\zeta^2/2}$. This transformation of variables is now entered.

> `tr:={x=sqrt(hbar/(m*omega))*zeta,psi(x)=f(zeta)*exp(-zeta^2/2)}:`

Using the dchange command *ode2* is expressed in terms of the new variables.

> `ode3:=dchange(tr,ode2,[zeta,f(zeta)]);`

$$ode3 := \left(\frac{d^2}{d\zeta^2}f(\zeta)\right)e^{\left(-\frac{\zeta^2}{2}\right)} - 2\left(\frac{d}{d\zeta}f(\zeta)\right)\zeta\,e^{\left(-\frac{\zeta^2}{2}\right)} + 2f(\zeta)\,e^{\left(-\frac{\zeta^2}{2}\right)}n = 0$$

Dividing *ode3* by the common exponential factor, $e^{-\zeta^2/2}$, yields *ode4*,

> `ode4:=expand(ode3/exp(-zeta^2/2));`

$$ode4 := \left(\frac{d^2}{d\zeta^2}f(\zeta)\right) - 2\left(\frac{d}{d\zeta}f(\zeta)\right)\zeta + 2f(\zeta)\,n = 0$$

which is the *Hermite* ODE. Hermite's equation is another S-L ODE, being obtained from (1.6) by setting $p(\zeta) = e^{-\zeta^2}$, $q(\zeta) = 0$, $w(\zeta) = e^{-\zeta^2}$, and $\lambda = 2\,n$. The general solution of *ode4* is now sought.

> `f:=rhs(dsolve(ode4,f(zeta)));`

$$f := \frac{\_C1\,e^{\left(\frac{\zeta^2}{2}\right)}\text{WhittakerM}\left(\dfrac{n}{2}+\dfrac{1}{4},\dfrac{1}{4},\zeta^2\right)}{\sqrt{\zeta}} + \frac{\_C2\,e^{\left(\frac{\zeta^2}{2}\right)}\text{WhittakerW}\left(\dfrac{n}{2}+\dfrac{1}{4},\dfrac{1}{4},\zeta^2\right)}{\sqrt{\zeta}}$$

Surprisingly, the answer is not given in terms of Hermite functions, but instead in terms of another special function, the *Whittaker* functions $M_{\mu,\nu}(z)$ and $W_{\mu,\nu}(z)$, which satisfy *Whittaker's differential equation*,

$$y''(z) + [-1/4 + \mu/z + (1/4 - \nu^2)/z^2]\,y(z) = 0, \qquad (1.10)$$

with $\mu = n/2 + 1/4$, $\nu = 1/4$, and $z = \zeta^2$ here. However, the WhittakerW function can be converted to a Hermite function by using convert(f,Hermite).

>   f:=convert(f,Hermite);

$$f := \frac{\_C1\, e^{(\frac{\zeta^2}{2})} \text{WhittakerM}(\frac{n}{2} + \frac{1}{4}, \frac{1}{4}, \zeta^2)}{\sqrt{\zeta}} + \frac{\_C2\, \text{HermiteH}(n, \sqrt{\zeta^2})\, (\zeta^2)^{(1/4)}}{\sqrt{\zeta}\, 2^n}$$

To avoid divergence problems at $\zeta = \pm\infty$, the WhittakerM function is removed

>   f2:=remove(has,f,WhittakerM);

$$f2 := \frac{\_C2\, \text{HermiteH}(n, \sqrt{\zeta^2})\, (\zeta^2)^{(1/4)}}{\sqrt{\zeta}\, 2^n}$$

from f, and the result simplified with the radsimp command.

>   f2:=radsimp(f2);

$$f2 := \frac{\_C2\, \text{HermiteH}(n, \zeta)}{2^n}$$

In terms of ζ, the probability amplitude is given by $\psi = f2\, e^{-\zeta^2/2}$.

>   psi:=f2*exp(-zeta^2/2);

$$\psi := \frac{\_C2\, \text{HermiteH}(n, \zeta)\, e^{(-\frac{\zeta^2}{2})}}{2^n}$$

To be physically meaningful and to satisfy the normalization condition, ψ must remain finite over the range $\zeta = -\infty$ to $+\infty$. This can only be accomplished if the WhittakerM function is removed (which has already been done) and n takes on the values $n = 0, 1, 2, 3, \ldots$. In this case, the Hermite functions reduce to the *Hermite polynomials* $H_n(\zeta)$. The Hermite polynomials can be readily generated. Here are the first six.

>   seq(H[n]=simplify(HermiteH(n,zeta)),n=0..5);

$$H_0 = 1, \quad H_1 = 2\zeta, \quad H_2 = -2 + 4\zeta^2, \quad H_3 = 8\zeta^3 - 12\zeta,$$
$$H_4 = 12 + 16\zeta^4 - 48\zeta^2, \quad H_5 = 32\zeta^5 - 160\zeta^3 + 120\zeta$$

The Hermite polynomials can also be generated by loading the orthopoly library package and using the syntax H(n,x).

To achieve the normalization condition $\int_{-\infty}^{\infty} P(\zeta)\, d\zeta = 1$, the constant must be chosen to be $\_C2 = \sqrt{2^n/(\sqrt{\pi}\, n!)}$. This form is substituted into ψ and the probability density P calculated and simplified with respect to the exponentials.

>   psi:=subs(_C2=sqrt(2^n/(sqrt(Pi)*n!)),psi);

>   P:=simplify(psi^2,exp);

$$P := \frac{e^{(-\zeta^2)}\, \text{HermiteH}(n, \zeta)^2}{2^n\, \sqrt{\pi}\, n!}$$

To plot the probability density for a given value of n, P is turned into an arrow operator with the unapply command.

>   P:=unapply(P,n):

The probability distribution will now be explored for a specific n value, say $n=25$. Let's check that the right form of $\_C2$ was entered by integrating $P(25)$ over the entire infinite range of ζ. The answer should be 1, which it is.

```
> Prob:=int(P(25),zeta=-infinity..infinity);
```
$$Prob := 1$$

Classically, the energy conservation statement for the harmonic oscillator is

$$\frac{1}{2}m\,v^2 + \frac{1}{2}m\,\omega^2\,x^2 = \frac{1}{2}m\,v^2 + \frac{1}{2}\hbar\,\omega\,\zeta^2 = E = \left(n + \frac{1}{2}\right)\hbar\,\omega,$$

where v is the speed. At the turning points $v = 0$, so that $\zeta = \pm\sqrt{2n+1}$. A functional operator L is now formed for calculating the magnitude of the classical turning point for a specified n value.

```
> L:=n->sqrt(2*n+1):
```

Vertical lines are plotted at the two turning points along with $P(25)$. A list of lists is used for each of the vertical line entries. These lines are dashed and colored blue, while the probability curve is a solid red curve.

```
> plot([[[L(25),0],[L(25),0.24]],[[-L(25),0],[-L(25),0.24]],
  P(25)],zeta=-9..9,numpoints=2000,linestyle=[3,3,1],
  color=[blue,blue,red],thickness=2,labels=["zeta","P"]);
```

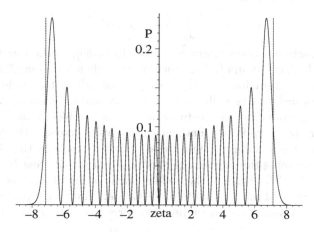

Figure 1.10: Probability distribution for $n=25$.

The resulting picture is shown in Figure 1.10. From the figure, we can see that the quantum oscillator has a non-zero probability of being outside the classical turning points. This probability is now calculated for $n = 25$. The integral is multiplied by 2, because there are two "tails" to the quantum distribution.

```
> Prob2:=evalf(2*int(P(25),zeta=L(25)..infinity));
```
$$Prob2 := 0.04506702177$$

The probability of being outside the classical range is about 0.045. A bizarre feature that occurs for odd values of n is that the probability of finding the particle at $x = 0$ (the center of the potential well) is 0. You are referred to Griffiths for a more thorough discussion of the quantum oscillator. In that reference, the probability distribution is drawn for $n = 100$. It is a trivial task to change n in the above recipe and replot the probability distribution. Try it!

1.2.4   Going Green, the Mathematician's Way

Colorless green ideas sleep furiously.
Noam Chomsky, American linguistic scholar illustrating that grammatical structure is independent of meaning, *Syntactic Structures*, (1957)

Suppose that one desires to solve a *nonhomogeneous Sturm–Liouville equation*,

$$L[y(x)] \equiv \frac{d}{dx}\left[p(x)\,\frac{dy}{dx}\right] - q(x)\,y = f(x). \tag{1.11}$$

subject to S-L boundary conditions at $x = a$ and $b > a$. The Green function method [Zwi89] is to first solve the corresponding Green function ODE,

$$L[G(x|z)] = \delta(x - z), \qquad \text{with } a < (x, z) < b, \tag{1.12}$$

subject to the same boundary conditions, for the Green function $G(x|z)$. Then the solution to (1.11) is given by

$$y(x) = \int_a^b G(x|z)\,f(z)\,dz. \tag{1.13}$$

The Green function approach, which is related ([SR66]) to the variation of parameters method, can be applied to other linear nonhomogeneous ODEs besides the S-L equation. More importantly, it can be generalized to handle nonhomogeneous PDEs such as the nonhomogeneous wave and diffusion equations.

Here's an introductory Green function method problem that I have often assigned over the years in my mathematical physics course. The idea for it, as well as for others, sprang from a table of Green functions that I stumbled across while scanning through an old mathematical physics text by Margenau and Murphy [MM57]. Some of these older books on mathematical techniques for scientists are gold mines of useful information. Now to the problem.

Explicitly determine the Green function corresponding to the ODE,

$$x\,y'' + y' - n^2/x = f(x), \tag{1.14}$$

where n is a positive (non-zero) integer and the boundary conditions are that $y(0)$ must remain finite and $y(1) = 0$. If $n = 1$ and $f(x) = -x^2$ for $0 \le x \le 1/2$ and $f(x) = -(1 - x)^2$ for $1/2 \le x \le 1$, use this Green function to explicitly determine $y(x)$ and plot the result.

The solution is as follows. First let's note that $x\,y'' + y' = (x\,y')'$, so that the ODE is a nonhomogeneous S-L equation with $p(x) = x$ and $q(x) = n^2/x$.

Now the ODE for the Green function is entered.

>   restart:

```
>   ode:=x*diff(G(x),x,x)+diff(G(x),x)-n^2*G(x)/x=Dirac(x-z);
```

$$ode := x\,(\frac{d^2}{dx^2}\,G(x)) + (\frac{d}{dx}\,G(x)) - \frac{n^2\,G(x)}{x} = \text{Dirac}(x - z)$$

The general solution of the ODE is obtained, assuming that $x < z$. The general solution for $x > z$ is of a similar mathematical structure.

```
>   sol:=rhs(dsolve(ode,G(x))) assuming x<z;
```

$$sol := \_C1\,x^{(-n)} + \_C2\,x^n$$

The Green function must satisfy the same boundary conditions as the solution of the original equation. For $x < z$, we must have G_L remain finite at $x = 0$. Since n is a non-zero, positive integer, the term x^{-n} must be removed from *sol* to form *GL*.

```
>   GL:=remove(has,sol,x^(-n));
```

$$GL := \_C2\,x^n$$

In some executions of the recipe, the coefficient $\_C1$ appears in the above output instead of $\_C2$. To avoid difficulty with the **solve** command later, let's introduce our own coefficient C by using the operand command to extract the second operand (x^n) in *GL* and multiply it by C.

```
>   GL:=C*op(2,GL);
```

$$GL := C\,x^n$$

To form *GR* for $x > z$, let's substitute $\_C1 = A$ and $\_C2 = B$ in *sol*.

```
>   GR:=subs({_C1=A,_C2=B}sol);
```

$$GR := A\,x^{(-n)} + B\,x^n$$

To satisfy the bc at $x = 1$, *GR* is evaluated at that point and set equal to zero.

```
>   eq1:=eval(GR,x=1)=0;
```

$$eq1 := A + B = 0$$

At $x = z$, we have the continuity condition, $GL = GR$.

```
>   eq2:=eval(GL=GR,x=z);
```

$$eq2 := C\,z^n = A\,z^{(-n)} + B\,z^n$$

There is a discontinuity in slope at $x = z$, given by $G_R'(z) - G_L'(z) = 1/p(z) = 1/z$.

```
>   eq3:=eval(diff((GR-GL),x),x=z)=1/z;
```

$$eq3 := -\frac{A\,z^{(-n)}\,n}{z} + \frac{B\,z^n\,n}{z} - \frac{C\,z^n\,n}{z} = \frac{1}{z}$$

The set of three equations (*eq1*, *eq2*, *eq3*) are solved for the set of three coefficients (A, B, C) and the solution *sol2* is assigned.

```
>   sol2:=solve({eq1,eq2,eq3},{A,B,C}); assign(sol2):
```

$$sol2 := \left\{ B = \frac{1}{2\,z^{(-n)}\,n},\ C = \frac{-z^{(-n)} + z^n}{2\,z^n\,z^{(-n)}\,n},\ A = -\frac{1}{2\,z^{(-n)}\,n} \right\}$$

On simplifying *GL* and *GR*, the Green function is completely determined.

> `GL:=simplify(GL); GR:=simplify(GR);`

$$GL := \frac{x^n \left(z^n - z^{(-n)}\right)}{2\,n} \qquad GR := \frac{z^n \left(x^n - x^{(-n)}\right)}{2\,n}$$

Using the `piecewise` command, the Green function can be written as a single function G, which displays the general symmetry property $G(x|z) = G(z|x)$.

> `G:=piecewise(x<=z,GL,x>=z,GR);`

$$G := \begin{cases} \dfrac{x^n \left(z^n - z^{(-n)}\right)}{2\,n} & x \le z \\[2ex] \dfrac{z^n \left(x^n - x^{(-n)}\right)}{2\,n} & z \le x \end{cases}$$

To continue with the solution, GL and GR are evaluated at $n=1$,

> `GL1:=eval(GL,n=1); GR1:=eval(GR,n=1);`

$$GL1 := \frac{x \left(z - \dfrac{1}{z}\right)}{2} \qquad GR1 := \frac{z \left(x - \dfrac{1}{x}\right)}{2}$$

and the function $f(z)$ entered as the two pieces $f1$ and $f2$.

> `f1:=-(z^2): f2:=-(1-z)^2:`

Now the integration in (1.13) has to be carried out. The integration is slightly tricky because both G and f are piecewise functions. For $x < 1/2$, we have

$$y1 = \int_0^x GR\,f1\,dz + \int_x^{1/2} GL\,f1\,dz + \int_{1/2}^1 GL\,f2\,dz.$$

This integration is now carried out and the result simplified.

> `y1:=simplify(int(f1*GR1,z=0..x)+int(f1*GL1,z=x..1/2)`
>   `+int(f2*GL1,z=1/2..1));`

$$y1 := \frac{1}{48}\, x \left(-6\,x^2 - 13 + 24\ln(2)\right)$$

Similarly, the integration for $x \ge 1/2$ is performed and simplified.

> `y2:=simplify(int(f1*GR1,z=0..1/2)+int(f2*GR1,z=1/2..x)`
>   `+int(f2*GL1,z=x..1)) assuming x>0;`

$$y2 := -\frac{1}{48}\, \frac{1 + 25\,x^2 + 6\,x^4 - 32\,x^3 + 24\,x^2\ln(x)}{x}$$

The complete solution y to the given nonhomogeneous ODE is pieced together,

> `y:=piecewise(x<=1/2,y1,x>=1/2,y2);`

$$y := \begin{cases} \dfrac{1}{48}\, x \left(-6\,x^2 - 13 + 24\ln(2)\right) & x \le \dfrac{1}{2} \\[2ex] -\dfrac{1}{48}\, \dfrac{25\,x^2 + 1 + 6\,x^4 - 32\,x^3 + 24\,x^2\ln(x)}{x} & \dfrac{1}{2} \le x \end{cases}$$

and plotted over the range $x=0$ to 1.

> `plot(y,x=0..1,thickness=2,labels=["x","y"]);`

The resulting profile is shown in Figure 1.11, thus finishing the problem.

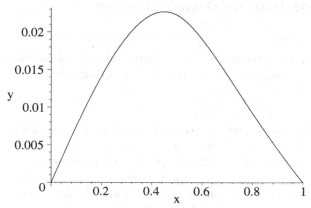

Figure 1.11: Solution of the given nonhomogeneous S-L equation.

1.2.5 In Search of a More Stable Existence

If you pick up a starving dog and make him prosperous, he will not bite you. This is the principal difference between a dog and a man.
Mark Twain, *Pudd'nhead Wilson, ch. 16*, (1894)

Not all LODEs, even if they are of the S-L type, are readily solved in terms of either elementary or special functions. In this case, one can easily generate a numerical solution using the `numeric` option of the `dsolve` command. To illustrate this approach, consider the following mechanics example.

A particle of mass m is fastened to the origin by a spring obeying Hooke's law whose spring constant is $k(t) = k_0 \sin(\omega t)$. The mass is constrained to move on a straight line through the origin. A critical frequency ω_{cr} exists such that for $\omega < \omega_{cr}$ the motion is unstable (displays unbounded growth as time evolves), but is stable (bounded motion) for all frequencies above ω_{cr}. Setting $r = \omega \sqrt{m/k_0}$, determine the critical value r_{cr} corresponding to $\omega = \omega_{cr}$ and plot representative solutions for r just below and just above r_{cr}. Take the initial displacement to be $x(0) = 1$ and the initial velocity to be $\dot{x}(0) = 0.1$.

Applying Newton's second law, the relevant LODE (dots indicating time derivatives with respect to the time variable τ) is

$$m\,\ddot{x}(\tau) + k_0 \sin(\omega\,\tau)\,x(\tau) = 0. \tag{1.15}$$

Introducing the dimensionless time variable $t = (\sqrt{k_0/m})\,\tau$, Equation (1.15) can be rewritten (dots now indicating time derivatives with respect to t) as

$$\ddot{x}(t) + \sin(r\,t)\,x(t) = 0. \tag{1.16}$$

After loading the plots package, a functional operator *ode* is formed for generating Equation (1.16) for different input values of r.

```
>  restart: with(plots):
```

```
>  ode:=r->diff(x(t),t,t)+sin(r*t)*x(t)=0;
```

$$ode := r \to (\frac{d^2}{dt^2} x(t)) + \sin(r\,t)\,x(t) = 0$$

The initial condition is entered and an unsuccessful attempt (no output is generated) is made to analytically solve the ODE for arbitrary r.

```
>  ic:=x(0)=1,D(x)(0)=0.1:
>  dsolve({ode(r),ic},x(t));
```

An arrow operator `sol` is now created to find a numerical solution of the ODE for a specified value of r. To achieve this, the option numeric[8] is included in the `dsolve` command line. The `output=listprocedure` option gives the output as a list of equations of the form variable=procedure, where the left-hand sides are the names of the independent variable, the dependent variable(s) and derivatives, and the right-hand sides are procedures that can be used to compute the corresponding solution components. This output form is most useful when the returned procedure is to be used later as will be the case here.

```
> sol:=r->dsolve({ode(r),ic},x(t),numeric,output=listprocedure):
```

Using `sol`, a "do loop" is introduced for carrying out the repetitive numerical calculation of $x(t)$ at a specific time for systematically increasing values of r. The general syntax for this common programming structure is:

> for *name* from *expression* by *expression* to *expression* while *expression* do
> *statement sequence;*
> end do:

The *statement sequence* is the main body of the do loop giving the steps to be followed in the calculation. In the following opening line of the do loop,

```
>  for i from 1 to 200 do
```

name is the index i, the first *expression* is 1, the **by** *expression* is absent, the third *expression* is 200, and there is no conditional **while** *expression* present. Since the **by** *expression* is missing, the default is to increment i by 1 each time the *statement sequence* is executed.

The statement sequence is now explained. Using `sol`, the following command line evaluates $x(t)$ at arbitrary time t for $r=i/100$. Since i runs from 1 to 200, r will be incremented in steps of 0.01 from 0.01 to 2.00. The concatenation operator || is used in assigning the ith result the name X||i. This operator attaches the numbers 1 to 200 to X, so that the outputs are assigned the names *X1,X2,..., X200*. Concatenated names are useful for employing in the sequence (seq) command.

```
>  X||i:=eval(x(t),sol(i/100));
```

Then each X||i is evaluated at $t=150$, the result being labeled Y||i.

```
>  Y||i:=X||i(150);
```

A large time has been chosen so that it will be easy to distinguish between the unstable solutions which should have grown to large amplitudes and the stable

---

[8]The default numerical method is the Runge–Kutta–Fehlberg 45 ([BF89]) method (discussed in the **Desserts**). Other schemes are listed in the Maple Help under "dsolve,numeric".

solutions which remain bounded. The do loop is now ended, the output being suppressed with a line-ending colon.

```
>  end do:
```

To plot the numerical output, 200 plotting points (pts) are formed using the sequence command. The abscissa of the ith point is given by $i/100$, with i running from 1 to 200. This generates the r values. Since the Y||i for the unstable solutions are extremely large, the log is taken of the absolute value of the Y||i to form the ordinate values. For each plotting point, the abscissa and ordinate values are put into a Maple list format.

```
>  pts:=seq([i/100,log(abs(Y||i))],i=1..200):
```

Choosing a suitable view, the points are now plotted, the default being to join consecutive points with a straight line. The result is shown in Figure 1.12.

```
>  plot([pts],view=[0.01..2,-10..70],tickmarks=[4,3],
     labels=["r",""]);
```

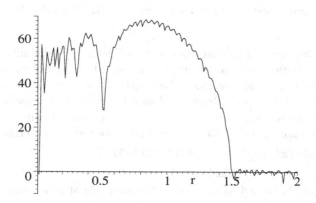

Figure 1.12: Log of the absolute value of $x(t=150)$ versus r.

Remembering that the vertical scale is logarithmic, we can see that the region of instability persists until r reaches slightly less than 1.5, then stability sets in for larger r. That is to say $r_{cr} \simeq 1.5$. More precisely, looking at the numerical output, r_{cr} lies between 1.48 (unstable) and 1.49 (stable).

The odeplot command may be used to plot the numerical solution of an ODE. An arrow operator gr is formed to apply this command to sol(r) for a given r value. $x(t)$ is plotted against the time t, the time range being from 0 to 150. To obtain an accurate curve, 2000 plotting points are chosen.

```
>  gr:=r->odeplot(sol(r),[t,x(t)],0..150,numpoints=2000,
         thickness=2,tickmarks=[3,3]):
```

Then entering gr(1.48) produces the figure shown on the left of Figure 1.13, while gr(1.49) generates the figure on the right.

```
>  gr(1.48); gr(1.49);
```

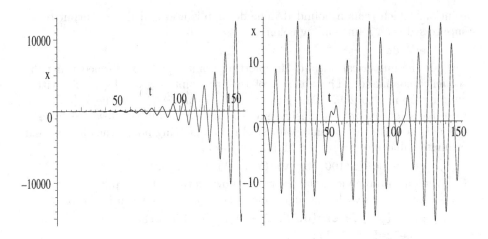

Figure 1.13: Left: Unstable solution for $r=1.48$. Right: Stable for $r=1.49$.

For $r=1.48$ ever-growing oscillations occur, indicative of instability, while for $r=1.49$ the oscillations remain bounded in amplitude, a sign of stability.

The motion of the particle along the x-axis for, e.g., $r=1.49$ can now be animated. The coordinates C (expressed as a list with the vertical coordinate equal to 0) of the particle at arbitrary time t are evaluated using sol(1.49). Then, e.g., the coordinates at time $t=100$, are obtained[9] by entering C(100).

```
>  C:=eval([x(t),0],sol(1.49)):  C(100);
```
$$[1.52446967145678580, 0]$$

An operator gr2 is formed to plot a size 20 blue circle at coordinates C(i).

```
>  gr2:=i->plot([C(i)],style=point,symbol=circle,
   symbolsize=20,color=blue,labels=["x"," "]):
```

The command seq(gr2(i),i=0..250) plots graphs for $t = i = 0, 1, 2, ...250$. Displaying this sequence with the option insequence=true generates a picture which is animated by clicking on the computer plot and then on the start arrow.

```
>  display(seq(gr2(i),i=0..250),insequence=true,
   tickmarks=[3,3]);
```

1.3 Supplementary Recipes

01-S01: Newton's Law of Cooling

Newton's law of cooling is governed by the ODE, $\dot{T}=-k\,(T-Ts)$, where T is the temperature of an object at time t, k is the positive cooling constant, and Ts is the temperature of the object's surroundings. If $T(0) = T0$, determine $T(t > 0)$. What information does including infolevel[dsolve]:=2: provide?

---

[9]Note that the numeric option of the dsolve command yields more digits in the output than the "standard" default of 10 digits.

A horseshoe is originally 100 °C hotter than the surrounding air. After 15 minutes, the temperature difference has fallen to 60 °C. How long will it take for the horseshoe to reach a temperature 10 °C above the surrounding air?

01-S02: Charging a Capacitor

A capacitor C is connected in series with a resistor R and a voltage source $V = V_0(t/\tau)^2 e^{-t/\tau}$, where τ is a characteristic time.

(a) Making use of Ohm's law for the voltage drop across R and Kirchoff's voltage sum rule, derive the ODE for the charge $q(t)$ on C at time t. Then, analytically solve the ODE for $q(t)$, given that $q(0)=0$.

(b) Taking $R = 5$ ohms, $C = 2$ farads, $V_0 = 3$ volts, and $\tau = 1$ second, plot $q(t)$ for $t = 0$ to 20 seconds. At what time is $q(t)$ a maximum and what is the maximum charge? Answer this first qualitatively by clicking the mouse cursor on the maximum, and then quantitatively.

01-S03: Radioactive Chain

An important radioactive chain involves the disintegration of the unstable ^{238}U nucleus. It decays via α-emission into ^{234}Th, which in turn β-decays into ^{234}Pa, and so on until the stable isotope ^{206}Pb is created. The decay rates of the first three species (N is the number of atoms and λ the decay constant) in a radioactive chain at time t can by described by

$$\dot{N}_1 = -\lambda_1 N_1, \quad \dot{N}_2 = \lambda_1 N_1 - \lambda_2 N_2, \quad \dot{N}_3 = \lambda_2 N_2 - \lambda_3 N_3.$$

(a) If $N_1(0)=N$, $N_2(0)=0$, and $N_3(0)=0$, determine $N_1(t)$ $N_2(t)$, and $N_3(t)$.

(b) The λ values for the uranium chain are vastly different, so for plotting purposes consider a hypothetical radioactive chain for which $N = 1000$, $\lambda_1 = 1$, $\lambda_2 = 2$, and $\lambda_3 = 3$. Plot $N_1(t)$ $N_2(t)$, and $N_3(t)$ in the same figure for $t = 0..5$ using different colors and line styles for each curve. At what time is N_3 a maxima? What is N_3 at this time?

01-S04: Stokes and Newton Join Forces

The drag force ([FC99]) on a sphere of diameter d moving with speed v is, in general, given in SI units by $F_{\text{drag}} = -a\,v - b\,v^2$, with $a = 1.55 \times 10^{-4}d$ and $b = 0.22\,d^2$. I.e., it is a combination of Stokes's and Newton's resistance laws

(a) A spherical mass m is dropped from rest. Derive and solve the relevant nonlinear ODE for $v(t)$. What method does Maple use to solve the ODE?

(b) A basketball ($d=0.25, m=0.60$), a raindrop ($d=10^{-4}, m=0.52 \times 10^{-9}$), and a soap bubble ($d=10^{-2}, m=10^{-7}$), are all dropped from rest. Taking the gravitational acceleration $g=9.8$ m/s^2, determine $v(t)$ for each body.

(c) Plot each $v(t)$, showing the approach to the terminal velocity V. Determine V for each body and the time it takes to come within 1% of V?

01-S05: Exploring the RLC Series Circuit

(a) Derive the ODE for a series circuit consisting of a resistor R, an inductor L, and a capacitor C and cast it into the form $\ddot{q} + 2\alpha\dot{q} + \omega^2 q = 0$, where $q(t)$ is the charge on C at time t, $\alpha = R/(2\,L)$ and $\omega = 1/\sqrt{LC}$.

(b) Given $q(0) = A$, $\dot{q}(0) = B$, solve the ODE for overdamping $(\alpha > \omega)$, underdamping $(\alpha < \omega)$, and critical damping $(\alpha = \omega)$. What fact does Maple use in successfully arriving at each solution? Calculate the currents.

(c) Taking $A = -1$, $B = 2$, $\omega = 1$, $\alpha = \alpha 1 = 1/5$, and $\alpha = \alpha 2 = 2$, evaluate $q(t)$ and the currents for the three cases. Plot the $q(t)$ together using different colors and linestyles for each case. Do the same for the currents.

01-S06: The Whirling Bar Revisited

Suppose that the rotating bar of recipe **01-1-3** has *hinged-end* (y and y'' (the bending moment) are both zero) boundary conditions at each end. Obtain the first four critical frequencies and corresponding shapes and plot the latter in a single figure. Compare the lowest critical frequency to that in **01-1-3**.

01-S07: Driven Couple Oscillators

Masses m_1 and m_2, with equilibrium positions at $x = 1$ and $x = 5$, are free to move horizontally on a smooth surface (the x-axis). m_1 is connected to a fixed wall on its left by a linear spring (spring constant k) and on its right to m_2 with an identical spring. A driving force $F = f \sin(\omega t)$ acts to the right on m_2. Air resistance is present, given by Stokes's drag law, $F_{\text{drag}} = -a v$.

(a) Derive the governing ODEs for the displacements $x_1(t)$ and $x_2(t)$ of m_1 and m_2 from equilibrium. Taking $m_1 = 2$, $m_2 = 1$, $k = 1$, $a = 1/100$, $\omega = 2$, $f = 2$, $x_1(0) = 1/10$, $x_2(0) = 0$, $\dot{x}_1(0) = 0$, and $\dot{x}_2(0) = 0$, solve the ODEs using the Laplace transform option. Express $x_1(t)$ and $x_2(t)$ in real forms.

(b) Extract the steady-state parts of x_1 and x_2 and plot them. Discuss the results. Plot the transient expressions over a time interval for which the transients become small. This will reveal the *envelope* of the transients.

(c) Animate the motion of m_1 and m_2 about their equilibrium positions over a time interval for which the transients become small.

01-S08: Some Properties of the Delta Function

(a) Two common representations of the δ function are $G(x) = (\alpha/\sqrt{\pi}) e^{-\alpha^2 x^2}$ and $F(x) = \sin(\alpha x)/(\pi x)$ with $\alpha \to \infty$. Confirm that these representations are reasonable by (i) plotting $G(x)$ and $F(x)$ for increasing α, (ii) showing that $\int_{-\infty}^{\infty} G(x)\, dx = 1$ and $\int_{-\infty}^{\infty} F(x)\, dx = 1$ for $\alpha > 0$.

(b) Using `Dirac(x-a)`, confirm the following: (i) $\int_{-\infty}^{\infty} \delta(x - a)\, dx = 1$, (ii) $\int_{-\infty}^{\infty} f(x)\, \delta(x - a)\, dx = f(a)$, (iii) $\int_{-\infty}^{\infty} f(x)\, \delta'(x - a)\, dx = -f'(a)$, (iv) $\int_{-\infty}^{\infty} \delta(b(x - a))\, dx = 1/|b|$.

(c) Using the command `Heaviside(x)`, create a rectangle $h(x)$ of unit height and with edges at $x = \pm 1$. Plot $h(x)$ for $x = -3$ to 3, using constrained scaling. Differentiate $h(x)$ with respect to x and interpret the result.

01-S09: A Green Function

Derive the Green function G which is the solution to $G'' + k^2 G = -\delta(x - \zeta)$, subject to the boundary conditions $G(-1) = G(1)$ and $G'(-1) = G'(1)$. Simplify

G as much as possible and write it as a piecewise function. Check that G obeys the general symmetry property $G(x|\zeta) = G(\zeta|x)$. At what k values does G have singularities where G diverges?

01-S10: A Potpourri of General Solutions
Classify the following LODEs as to order, homogeneneous or nonhomogeneous, and constant or variable coefficients. Obtain the general solution for each and identify which part of the solution is the particular solution and which part is the solution to the corresponding homogeneous ODE. Identify any non-elementary functions in the solution. What method of attack does Maple use?

(a) $x\,y' + (1+x)\,y = e^x$;

(b) $y''' + x\,y = 0$;

(c) $y'' + 3\,y' + 2\,y = e^x$;

(d) $x^2\,y'' - 6\,y = x^3\,\ln x$;

(e) $y'' - (3/x)\,y' + (15/(4\,x^2) + \sqrt{x})\,y = 0$;

(f) $x^2\,y'' + (1 - 2\,\alpha)\,x\,y' + (k^2\,\beta^2\,x^{2\beta} + (\alpha^2 - p^2\,\beta^2))\,y = 0$.

01-S11: Chebyshev Polynomials
Taking $p = \sqrt{1 - x^2}$, $q = 0$, $w = 1/\sqrt{1 - x^2}$, and $\lambda = n^2$, show that the S-L ODE yields the *Chebyshev equation*, $(1 - x^2)\,y'' - x\,y' + n^2\,y = 0$. For n a non-negative integer, Chebyshev's ODE is known to have polynomial solutions (the *Chebyshev polynomials* $T_n(x)$) defined over the range $x = -1$ to $+1$. Using the command `expand(ChebyshevT(n,x))`, generate the $T_n(x)$ for $n = 0$ to 10. Then use the `dsolve` command to obtain the general solution of the Chebyshev ODE and show that it can be expressed in terms of the $T_n(x)$. Demonstrate that, e.g., $\int_{-1}^{1} T_9(x)\,T_{10}(x)\,w(x)\,dx = 0$, i.e., the Chebyshev polynomials form a set of orthogonal functions. Plot the first five polynomials for $x = -1$ to 1.

01-S12: The Growing Pendulum
Consider a pendulum which consists of a point mass m at the bottom end of a light supporting rod of length L which is allowed to move in a vertical plane about a pivot point at its top end. Suppose that L increases at a steady rate, i.e., $L = L_0 + v\,t$, where $v > 0$ is a constant speed and t the time.

(a) Letting the rod make an angle $\theta(t)$ with the vertical and neglecting drag, use Newton's second law for angular motion $(d(I\,\omega)/dt = T$, where I is the moment of inertia, ω the angular velocity, and T the torque) to derive the LODE for small θ. Solve the ODE for $\theta(t)$, given $\theta(0) = \Theta$, $\dot{\theta}(0) = 0$. Simplify $\theta(t)$ and identify the functions which occur.

(b) Taking $L_0 = 1$ m, $g = 10$ m/s^2, $\Theta = \pi/6$ rads, and $v = 0.5$ m/s, plot $\theta(t)$ for $t = 0$ to 100 seconds. Describe the behavior of the period. Then animate the motion of the pendulum arm (representing it as a thick line) over this time interval taking 100 frames and using constrained scaling.

01-S13: Another Green Function

Assuming that $m \neq 0$, derive the Green function G which is the solution to $(1 - x^2)\,G'' - 2\,x\,G' - m^2\,G/(1 - x^2) = \delta(x - z)$, subject to the boundary conditions that G remains finite at the end points $x = \pm 1$ of the range. Simplify G as much as possible and write it as a piecewise function. Plot G over the range $x = -1$ to 1, taking $m = 3$ and $z = 1/2$.

01-S14: Going Green, Once Again

A light elastic string, under tension T and fixed at $x = \pm L$, is horizontal in the absence of any applied forces. The string is embedded in an elastic membrane which exerts a Hooke's law (spring constant k) restoring force if the string is displaced from equilibrium. If a force $f\,e^{-\beta x^2}$ is applied to the string, derive the ODE for its shape $y(x)$, assuming that $y(x)$ is small. Using the Green function method, determine $y(x)$. The answer will involve the *error function*, $\mathrm{erf}(z) \equiv (2/\sqrt{\pi}) \int_0^x e^{-t^2}\,dt$. Check your result by solving the original ODE with Maple's dsolve command, subject to the given boundary condition. Taking $T = 100$ N, $k = 10$ N/m, $L = 1$ m, $\beta = 1/100$ m^{-2}, and $f = 50$ N, plot $y(x)$.

Chapter 2

Applications of Series

In this chapter, it is assumed that you are already familiar with the basic concepts of infinite series covered in standard calculus texts such as *Calculus* by James Stewart [Ste87]. The emphasis here is on applications of series, the topics being Taylor series, series solutions of LODEs, Fourier series, Legendre and Bessel series, and summing series. Laurent series are covered in Chapter 5.

2.1 Taylor Series

The Taylor expansion of a function $f(x) = f(x_0 + h)$ about x_0 is given by

$$f(x) = f(x_0) + h f'(x_0) + \frac{1}{2!}h^2 f''(x_0) + \frac{1}{3!}h^3 f'''(x_0) + \cdots. \qquad (2.1)$$

Thus, e.g., taking $x_0 = 0$, so $h = x - x_0 = x$,

$$\sin(x) = x - \frac{x^3}{3!} + \frac{x^5}{5!} - \frac{x^7}{7!} + \cdots = \sum_{n=0}^{\infty} (-1)^n \frac{x^{2n+1}}{(2n+1)!} \equiv \sum_{n=0}^{\infty} u_n.$$

Applying the *ratio test*, $r \equiv |u_{n+1}/u_n| = x^2/((2n+3)(2n+2)) \to 0$ as $n \to \infty$. Since $r < 1$, the series converges for all x, i.e., the *radius of convergence* is ∞.

The extension to functions of more than one variable is straightforward, e.g., for a function $f(x, y) = f(x_0 + h, y_0 + k)$, the Taylor series is of the form,

$$f(x, y) = f(x_0, y_0) + \left(h \frac{\partial}{\partial x} + k \frac{\partial}{\partial y} \right) f(x_0, y_0) + \frac{1}{2!} \left(h \frac{\partial}{\partial x} + k \frac{\partial}{\partial y} \right)^2 f(x_0, y_0) + \cdots.$$

2.1.1 Polynomial Approximations

Every man sees in his relatives, and especially in his cousins, a series of grotesque caricatures of himself.
H. L. Mencken, American journalist, (1880–1956)

In this recipe, I will demonstrate how the Taylor series expansion may be used to obtain the smallest polynomial approximation to the integral

$f = \int_0^x u^4 e^{u^2} \, du$, valid within ± 0.000001 over the range $0 \le x \le 1$.

The desired accuracy A is entered, and then the integral f is displayed and evaluated in the following command line.

```
>   restart: A:=0.000001:
>   f:=Int(u^4*exp(u^2),u=0..x)=int(u^4*exp(u^2),u=0..x);
```

$$f := \int_0^x u^4\, e^{(u^2)}\, du = \frac{1}{2}\, x^3\, e^{(x^2)} - \frac{3}{4}\, x\, e^{(x^2)} - \frac{3}{8}\, I\, \sqrt{\pi}\, \mathrm{erf}(x\, I)$$

The answer is given in terms of the *error function*, $\mathrm{erf}(z) = (2/\sqrt{\pi}) \int_0^z e^{-u^2}\, du$, with $z = I\, x = \sqrt{-1}\, x$. The result can be expressed in terms of real quantities by converting the rhs to an *imaginary error function*, $\mathrm{erfi}(x) = (2/\sqrt{\pi}) \int_0^x e^{u^2}\, du$.

```
>   f:=convert(rhs(f),erfi);
```

$$f := \frac{1}{2}\, x^3\, e^{(x^2)} - \frac{3}{4}\, x\, e^{(x^2)} + \frac{3}{8}\, \sqrt{\pi}\, \mathrm{erfi}(x)$$

Then f is simplified by collecting exponential terms.

```
>   f:=collect(f,exp);
```

$$f := (\frac{1}{2}\, x^3 - \frac{3}{4}\, x)\, e^{(x^2)} + \frac{3}{8}\, \sqrt{\pi}\, \mathrm{erfi}(x)$$

Although f can be evaluated numerically, a polynomial approximation to f is now obtained. An operator F is formed which Taylor expands f in powers of x about $x = 0$, dropping terms of order x^{N+1}.

```
>   F:=N->taylor(f,x=0,N+1):
```

A second arrow operator $F2$ is created to remove the order of term in the series.

```
>   F2:=N->convert(F(N),polynom):
```

Using F2(2*n+1), a sequence of polynomial approximations to the integral is generated in T for $n = 0$ to 10 (not all polynomials are shown here). Note that the Taylor series expansion of the integral generates only odd order polynomials, hence the argument $2\,n + 1$. For plotting purposes, T is assigned.

```
>   T:=seq(t||(2*n+1)=F2(2*n+1),n=0..10); assign(T):
```

$$T := t1 = 0,\ t3 = 0,\ t5 = \frac{x^5}{5},\ t7 = \frac{1}{5}\, x^5 + \frac{1}{7}\, x^7,\ t9 = \frac{1}{5}\, x^5 + \frac{1}{7}\, x^7 + \frac{1}{18}\, x^9,$$

$$t11 = \frac{1}{5}\, x^5 + \frac{1}{7}\, x^7 + \frac{1}{18}\, x^9 + \frac{1}{66}\, x^{11}, \dots\dots\dots\dots\dots\dots\dots\dots\dots\dots\dots\dots\dots,$$

$$t21 = \frac{1}{5}\, x^5 + \frac{1}{7}\, x^7 + \frac{1}{18}\, x^9 + \frac{1}{66}\, x^{11} + \frac{1}{312}\, x^{13} + \frac{1}{1800}\, x^{15} + \frac{1}{12240}\, x^{17} + \frac{1}{95760}\, x^{19}$$

$$+ \frac{1}{846720}\, x^{21}$$

The difference between the exact integral f and each polynomial for $n = 2$ (yielding $t5$) and 10 (yielding $t21$) is plotted over the range $x = 0$ to 1.

```
>   plot([seq(f-t||(2*n+1),n=2..10)],x=0..1,view=[0..1,0..A],
        axes=box,thickness=2);
```

Since the differences turn out to be positive, the vertical view has been taken to be between 0 and A, the upper limit setting the desired accuracy. The sequence of differences is shown in Figure 2.1, the difference corresponding to $n=2$ on the far left and that corresponding to $n=10$ on the far right. Since the first difference curve that remains completely within the vertical viewing range is the curve corresponding to $n=10$, the smallest polynomial which is within the desired accuracy over the range $x=0$ to 1 is given by $t21$.

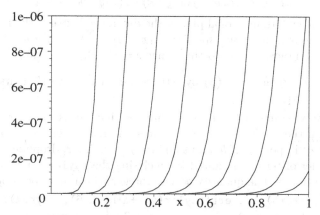

Figure 2.1: Differences between exact integral and polynomial approximations.

2.1.2 Finite Difference Approximations

I'm not a teacher: only a fellow-traveler of whom you asked the way.
I pointed ahead–ahead of myself as well as you.
George Bernard Shaw, Anglo-Irish playwright (1856–1950)

Replacing ordinary and partial derivatives by their *finite difference approximations* (FDAs) is useful in numerically solving ODEs and PDEs as will be illustrated in Chapter 9. These approximations are based on a Taylor series approach. For example, an FDA to $y''(x)$ can be obtained by Taylor expanding $y(x \pm h)$ about x for small h, viz.,

$$y(x \pm h) = y(x) \pm hy'(x) + \frac{1}{2!}h^2y''(x) \pm \frac{1}{3!}h^3y'''(x) + O(h^4), \qquad (2.2)$$

where $O(h^4)$ indicates that terms of order h^4 (and higher) are being neglected. Forming the sum $y(x+h) + y(x-h)$, and solving for y'', yields the FDA,

$$y''(x) = [y(x+h) + y(x-h) - 2y(x)]/h^2 + O(h^2). \qquad (2.3)$$

Eq. (2.3), known as the *central difference approximation* to the second derivative, was easily derived by hand. However, a computer algebra approach can sometimes prove useful in deriving other FDAs as our mathematical physics student, I. M. Curious, will now illustrate in answering the following problem.

(a) Show that to $O(h^4)$ an alternate FDA to the second derivative is given by

$$y''(x) = [-y(x+2h) + 16y(x+h) - 30y(x) + 16y(x-h) - y(x-2h)]/(12h^2).$$

Taking 12 digits accuracy, plot the difference between the FDA and the exact y'' for $y=x^{5.1}$ over the range $x=0.1$ to 10 for $h=0.1$ and $h=0.05$.

(b) Show that an FDA for $f_{xxyy} \equiv \partial^4 f(x,y)/\partial x^2 \partial y^2$ is given by

$$f_{xxyy} = [f(x+h, y+k)+f(x-h, y+k)+f(x+h, y-k)+f(x-h, y-k)$$
$$-2\,(f(x+h, y)+f(x-h, y)+f(x, y+k)+f(x, y-k))+4\,f(x, y)]/(h^2 k^2),$$

and produce a 3-dimensional plot of the difference between the FDA and the exact fourth derivative for $f = \sin(x-y)\,e^{-x^2 y^2}$. Take $h=k=0.05$, 12 digits accuracy, and a plotting range $x=-3...3$, $y=-3...3$.

I. M. begins her recipe for part **(a)** by setting the accuracy to 12 digits.

```
>   restart: Digits:=12:
```

Noting that the wording of the problem implies that terms of order h^6 have been dropped in the Taylor series, she creates an arrow operator t to Taylor expand $y(x+a\,h)$ in powers of h about an arbitrary point x, dropping terms of $O(h^6)$. The order of term is removed by enclosing the **taylor** command with **convert(,polynom)**. The quantity a is an integer which must be supplied.

```
>   t:=a->y(x+a*h)=convert(taylor(y(x+a*h),h,6),polynom):
```

As a test, I. M. chooses $a=2$ and generates the Taylor series for $y(x+2h)$.

```
>   t(2);
```

$$y(x+2\,h) = y(x) + 2\,D(y)(x)\,h + 2\,(D^{(2)})(y)(x)\,h^2 + \frac{4}{3}\,(D^{(3)})(y)(x)\,h^3$$
$$+ \frac{2}{3}\,(D^{(4)})(y)(x)\,h^4 + \frac{4}{15}\,(D^{(5)})(y)(x)\,h^5$$

In the output, $(D^{(4)})(y)(x)$, for example, stands for the 4th derivative, $y''''(x)$. Making use of the operator t, the proposed FDA is entered in *eq1*.

```
>   eq1:=-t(2)+16*t(1)-30*t(0)+16*t(-1)-t(-2);
```

$$eq1 := -y(x+2\,h) + 16\,y(x+h) - 30\,y(x) + 16\,y(x-h) - y(x-2\,h)$$
$$= 12\,(D^{(2)})(y)(x)\,h^2$$

On the rhs of the output, only the second derivative term survives, all other terms up to $O(h^6)$ having canceled. Using the **isolate** command, the second derivative (entered as D[1,1](y)(x)) is isolated to the lhs of *eq2*,

```
>   eq2:=isolate(eq1,D[1,1](y)(x));
```

$$eq2 := (D^{(2)})(y)(x)$$
$$= -\frac{1}{12}\,\frac{y(x+2\,h) - 16\,y(x+h) + 30\,y(x) - 16\,y(x-h) + y(x-2\,h)}{h^2}$$

thus confirming the FDA.

An arrow operator y is formed to evaluate $y=x^{5.1}$ at an arbitrary x value. So that the FDA can be evaluated for different h values, the rhs of *eq2* is turned into a functional operator A, depending on h, using the **unapply** command.

```
>   y:=x->x^5.1: A:=unapply(rhs(eq2),h):
```

The exact $y''(x)$ is calculated and the FDAs generated for $h=0.1$ and $h=0.05$.

> `exact:=diff(y(x),x,x); approx1:=A(0.1): approx2:=A(0.05):`

$$exact := 20.91\, x^{3.1}$$

The differences between the FDAs *approx1* and *approx2* and *exact* are plotted for $x=0.1$ to 10 as solid red and dashed green curves, respectively, the resulting plot being shown in Figure 2.2.

> `plot([approx1-exact,approx2-exact],x=0.1..10,color=`
> `[red,green],thickness=2,linestyle=[1,3],tickmarks=[2,2]);`

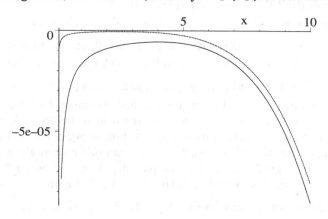

Figure 2.2: FDA minus the exact result for $h=0.1$ (solid) and $h=0.05$ (dashed).

As expected, the accuracy of the FDA is improved with the smaller value of h. Both curves have greatest accuracy over a limited range of x.

Now, I. M. tackles part **(b)** of the problem. The plots library package is loaded and 12 digits accuracy entered.

> `restart: with(plots): Digits:=12:`

The FDA involves a Taylor expansion in two variables. I. M. uses the multivariate Taylor series command, `mtaylor`, to create an operator t for performing the expansion of $f(x + ah, y + bk)$ in powers of h and k, neglecting terms of sixth order. `mtaylor` does not generate the order of term, so `convert( ,polynom)` needn't be applied. Two integers a and b must be entered as arguments in t.

> `t:=(a,b)->f(x+a*h,y+b*k)=mtaylor(f(x+a*h,y+b*k),[h,k],6):`

The given FDA is entered in *eq1*.

> `eq1:=t(1,1)+t(-1,1)+t(1,-1)+t(-1,-1)`
> `-2*(t(1,0)+t(-1,0)+t(0,1)+t(0,-1))+4*t(0,0);`

$$eq1 := f(x + h,\, y + k) + f(x - h,\, y + k) + f(x + h,\, y - k) + f(x - h,\, y - k)$$
$$- 2f(x + h,\, y) - 2f(x - h,\, y) - 2f(x,\, y + k) - 2f(x,\, y - k)$$
$$+ 4f(x,\, y) = h^2\, k^2\, D_{1,1,2,2}(f)(x,\, y)$$

Only a single term involving $D_{1,1,2,2}$ results on the rhs of the output, the subscripts 1 and 2 denoting derivatives with respect to x and y, respectively. Dividing *eq1* by h^2k^2, and simplifying, yields the desired FDA.

```
> eq2:=simplify(eq1/(h^2*k^2));
```

$$eq2 := (\mathrm{f}(x + h, \ y + k) + \mathrm{f}(x - h, \ y + k) + \mathrm{f}(x + h, \ y - k) + \mathrm{f}(x - h, \ y - k)$$
$$- 2\,\mathrm{f}(x + h, \ y) - 2\,\mathrm{f}(x - h, \ y) - 2\,\mathrm{f}(x, \ y + k) - 2\,\mathrm{f}(x, \ y - k)$$
$$+ 4\,\mathrm{f}(x, \ y))/(h^2\,k^2) = D_{1,1,2,2}(f)(x, y)$$

The function $f(x, y) = \sin(x - y)\,e^{-x^2\,y^2}$ is entered, as well as $h = k = 0.05$.

```
> f:=(x,y)->sin(x-y)*exp(-x^2*y^2): h:=0.05: k:=0.05:
```

The exact 4th derivative is calculated and the FDA given by the lhs of *eq2*.

```
> exact:=diff(f(x,y),x,x,y,y); approx:=lhs(eq2):
```

The **plot3d** command is used to create a 3-dimensional plot of *approx* minus *exact* for $x = -3...3$, $y = -3...3$. The grid is taken to be 25 ×25, the shading to be XYZ (i.e., the color varies in the 3 directions), the axes boxed, the plot "illuminated" with a light source at a certain orientation given by entering **lightmodel=light2**, the number of plotting points to be 2000, and the orientation specified. The resulting picture is shown in Figure 2.3.

```
> plot3d(approx-exact,x=-3..3,y=-3..3,grid=[25,25],
  shading=XYZ,axes=box,lightmodel=light2,numpoints=2000,
  tickmarks=[3,3,3],orientation=[20,40]);
```

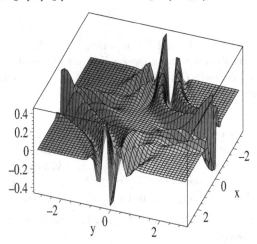

Figure 2.3: Difference between FDA and exact 4th derivative.

The regions where the FDA and exact result differ appreciably can be seen as a series of ridges and valleys. The 3-dimensional plot can be rotated on the computer screen by dragging with the mouse.

2.2 Series Solutions of LODEs

Consider an nth-order homogeneous linear ODE with variable coefficients,

$$\frac{d^n y}{dx^n} + a_{n-1}(x) \frac{d^{n-1} y}{dx^{n-1}} + \cdots + a_1(x) \frac{dy}{dx} + a_0(x)\, y = 0. \qquad (2.4)$$

If $a_0(x), \ldots, a_{n-1}(x)$ are *regular* (single-valued and analytic) at a point $x = x_0$, then x_0 is referred to as an *ordinary point* of the ODE. Near x_0, the general solution of the LODE can be written as a Taylor series, $y = \sum_{m=0}^{\infty} c_m (x - x_0)^m$, whose radius of convergence is the distance to the nearest *singular point* (a non-ordinary point) of the ODE. The coefficients c_m are found by substituting the series into the ODE and setting the coefficients of each power of x to zero.

If x_0 is not an ordinary point, but $(x - x_0)\, a_{n-1}(x)$, $(x - x_0)^2\, a_{n-2}(x), \ldots$, $(x - x_0)^n\, a_0(x)$ are regular at x_0, then x_0 is a *regular singular point*. Near such an x_0, a *Frobenius series* solution of the form $y = (x - x_0)^s \sum_{m=0}^{\infty} c_m (x - x_0)^m$, with $c_0 \neq 0$ and s not necessarily an integer, can always be found. The radius of convergence is again the distance to the nearest singular point outside x_0.

In this section, series solutions to two familiar Sturm–Liouville ODEs are obtained, illustrating the two types of expansions mentioned above.

2.2.1 Jennifer Renews an Old Acquaintance

We need two kinds of acquaintances, one to complain to,
while to the others we boast.
Logan Pearsall Smith, American essayist, *Afterthoughts, "Other People", 1931*

In recipe **01-2-1**, Jennifer introduced the Legendre functions. In this recipe, she renews her acquaintance with these important special functions, showing us how they arise as a series solution to the Legendre ODE,

$$(1 - x^2)\, y'' - 2\, x\, y' + n\, (n + 1)\, y = 0. \qquad (2.5)$$

Casting (2.5) into the "standard" form (2.4), Equation (2.5) has regular singular points at $x = \pm 1$. Jennifer seeks a Taylor series solution about the ordinary point $x_0 = 0$. She decides that it will suffice for calculation purposes to work with a finite number of terms, say $N = 7$, in the series.

```
>  restart: N:=7:
```
The left-hand side of Legendre's equation is now entered.
```
>  ode:=(1-x^2)*diff(y(x),x,x)-2*x*diff(y(x),x)+n*(n+1)*y(x);
```

$$ode := (1 - x^2)\, (\frac{d^2}{dx^2}\, y(x)) - 2\, x\, (\frac{d}{dx}\, y(x)) + n\, (n + 1)\, y(x)$$

The power series $y(x) = \sum_{m=0}^{N} c_m x^m$ is inputted using the **add** command.
```
>  y(x):=add(c||m*x^m,m=0..N);
```

$$y(x) := c0 + c1\, x + c2\, x^2 + c3\, x^3 + c4\, x^4 + c5\, x^5 + c6\, x^6 + c7\, x^7$$

$y(x)$ is automatically substituted into *ode* and powers of x are collected.

```
> ode2:=collect(ode,x);
```

$$ode2 := (-56\,c7 + n\,(n+1)\,c7)\,x^7 + (-42\,c6 + n\,(n+1)\,c6)\,x^6$$
$$+ (42\,c7 - 30\,c5 + n\,(n+1)\,c5)\,x^5 + (n\,(n+1)\,c4 - 20\,c4 + 30\,c6)\,x^4$$
$$+ (-12\,c3 + n\,(n+1)\,c3 + 20\,c5)\,x^3 + (-6\,c2 + 12\,c4 + n\,(n+1)\,c2)\,x^2$$
$$+ (6\,c3 + n\,(n+1)\,c1 - 2\,c1)\,x + 2\,c2 + n\,(n+1)\,c0$$

Since *ode2* represents the lhs of the Legendre equation, it must be set equal to zero to form the complete equation. But x is arbitrary, so the coefficient of each power of x must separately be equal to zero. A functional operator `eq` is created to set the coefficient of x^m in *ode2* equal to zero.

```
> eq:=m->coeff(ode2,x,m)=0:
```

The set of equations `eq(m)` is solved for the `c||(m+2)` for $m=0$ to $N-2$.

```
> sol:=solve({seq(eq(m),m=0..N-2)},{seq(c||(m+2),m=0..N-2)}):
```

The solution *sol* is assigned, the coefficients factored and the series formed,

```
> assign(sol): y:=add(factor(c||m)*x^m,m=0..N):
```

and the coefficients of *c0* and *c1* collected in y.

```
> y:=collect(y,{c||0,c||1});
```

$$y := (1 - \frac{n\,(n+1)\,x^2}{2} + \frac{n\,(n-2)\,(n+3)\,(n+1)\,x^4}{24}$$
$$- \frac{n\,(n-2)\,(n-4)\,(n+5)\,(n+3)\,(n+1)\,x^6}{720})c0$$
$$+ (x - \frac{(n+2)\,(n-1)\,x^3}{6} + \frac{(n-1)\,(n-3)\,(n+4)\,(n+2)\,x^5}{120}$$
$$- \frac{(n-1)\,(n-3)\,(n-5)\,(n+6)\,(n+4)\,(n+2)\,x^7}{5040})c1$$

y gives the first few term in the general series solution of Legendre's ODE. It consists of an even (coefficient of *c0*) and odd (coefficient of *c1*) series in x. Noting that the denominator of the x^m term is just $m!$ (e.g., for $m=6$, $6! = (6)(5)(4)(3)(2)(1) = 720$), the structure of higher order terms in the infinite series is easy to deduce. In deriving y, Jennifer has mimicked a hand calculation. The same result could be more easily obtained by using the **series** option in the **dsolve** command, as Jennifer will now demonstrate. She unassigns y and sets the order of the first term to be neglected in the series solution. Since she has taken $N = 7$, the order here is 8, i.e., terms of $O(x^8)$ are dropped in the series. If the order is not specified, the default is to neglect terms of $O(x^6)$.

```
> unassign('y'): Order:=N+1;
```

$$Order := 8$$

Then *ode* is solved for $y(x)$, subject to the initial condition $y(0) = c0$ and $y'(0) = c1$, using the **series** option in the **dsolve** command.

```
> Y:=dsolve({ode,y(0)=c||0,D(y)(0)=c||1},y(x),series);
```

The output (suppressed here in the text) contains the "order of" term $O(x^8)$ which is then removed from the rhs of Y using `convert( ,polynom)`.

> `Y:=convert(rhs(Y),polynom):`

Collecting the coefficients of *c0* and *c1* yields a series solution Y equivalent to that obtained in y.

> `Y:=collect(Y,{c||0,c||1});`

$$Y := (1 - \frac{n(n+1)x^2}{2} + (\frac{1}{24}n^4 + \frac{1}{12}n^3 - \frac{5}{24}n^2 - \frac{1}{4}n)x^4$$

$$+ (-\frac{1}{720}n^6 - \frac{1}{240}n^5 + \frac{23}{720}n^4 + \frac{17}{240}n^3 - \frac{47}{360}n^2 - \frac{1}{6}n)x^6)c0$$

$$+ (x + (\frac{1}{3} - \frac{1}{6}n^2 - \frac{1}{6}n)x^3 + (-\frac{13}{120}n^2 + \frac{1}{120}n^4 + \frac{1}{60}n^3 - \frac{7}{60}n + \frac{1}{5})x^5$$

$$+ (\frac{41}{5040}n^4 - \frac{1}{5040}n^6 - \frac{1}{1680}n^5 + \frac{29}{1680}n^3 - \frac{5}{63}n^2 - \frac{37}{420}n + \frac{1}{7})x^7)c1$$

Often, in cases of physical interest, $x \equiv \cos(\theta)$ where the angle θ varies from 0 to π radians. So x then ranges from 1 to -1. It turns out that both series diverge at $x = \pm 1$, unless n is a positive integer in which case the even or odd series in Y, or y, terminates with the power x^n when n is even or odd. To explicitly demonstrate this, Jennifer turns Y into an operator depending on n with the `unapply` command,

> `YY:=unapply(Y,n):`

and then uses a do loop to generate the solutions *y0,y1,...,y5* for $n=0, 1, ..., 5$

> `for n from 0 to 5 do y||n:=YY(n); end do;`

$$y0 := c0 + (x + \frac{1}{3}x^3 + \frac{1}{5}x^5 + \frac{1}{7}x^7)c1$$

$$y1 := (1 - x^2 - \frac{1}{3}x^4 - \frac{1}{5}x^6)c0 + c1\,x$$

$$y2 := (1 - 3x^2)c0 + (x - \frac{2}{3}x^3 - \frac{1}{5}x^5 - \frac{4}{35}x^7)c1$$

$$y3 := (1 - 6x^2 + 3x^4 + \frac{4}{5}x^6)c0 + (x - \frac{5}{3}x^3)c1$$

$$y4 := (1 - 10x^2 + \frac{35}{3}x^4)c0 + (x - 3x^3 + \frac{6}{5}x^5 + \frac{2}{7}x^7)c1$$

$$y5 := (1 - 15x^2 + 30x^4 - 10x^6)c0 + (x - \frac{14}{3}x^3 + \frac{21}{5}x^5)c1$$

In *y0*, the coefficient of *c0* is 1, which is just the zeroth order Legendre polynomial $P_0(x)$, while the coefficient of *c1* involves the leading terms of an infinite series. In *y1*, the *c0* coefficient is an infinite series, while the *c1* coefficient is x, which is just the first order Legendre polynomial $P_1(x)$. In *y2*, the *c0* coefficient is proportional to $P_2(x)$, the coefficient of *c1* an infinite series, and so on for increasing n values. Jennifer creates a functional operator **F** to extract the coefficient of either *c0* (set $p=0$) or *c1* (set $p=1$) for each solution *yn*.

> `F:=(n,p)->coeff(y||n,c||p):`

The Legendre polynomials $P_n(x)$ are traditionally normalized so that each poly-nomial has the value 1 at $x=1$. Making use of `F(2*n,0)` and including this normalization, the first few even subscript Legendre polynomials are generated.

> `Peven:=seq(P||(2*n)=F(2*n,0)/eval(F(2*n,0),x=1),n=0..2);`

$$Peven := P0 = 1, \; P2 = -\frac{1}{2} + \frac{3\,x^2}{2}, \; P4 = \frac{3}{8} - \frac{15}{4}x^2 + \frac{35}{8}x^4$$

The odd subscript polynomials P_1, P_3, and P_5 can be generated by replacing `F(2*n,0)` with `F(2*n+1,1)` in the last command line.

The odd exponent infinite series are similarly extracted by using `F(2*n,1)`. These are labeled $Q0$, $Q2$, $Q4$ for reasons which will now be explained.

> `Qeven:=seq(Q||(2*n)=F(2*n,1),n=0..2);`

$$Qeven := Q0 = x + \frac{1}{3}x^3 + \frac{1}{5}x^5 + \frac{1}{7}x^7, \; Q2 = x - \frac{2}{3}x^3 - \frac{1}{5}x^5 - \frac{4}{35}x^7,$$

$$Q4 = x - 3\,x^3 + \frac{6}{5}x^5 + \frac{2}{7}x^7$$

Each of the above infinite series can be summed and are found to be proportional to the Legendre functions $Q_0(x)$, $Q_2(x)$, $Q_4(x)$ of the second kind. Jennifer concludes her recipe by demonstrating, for example, that the Taylor expansion of $Q_0(x)$ is the same as $Q0$ given above. The solution branch between $x=-1$ and 1 is selected by entering the following command.

> `_EnvLegendreCut:=1..infinity:`

Then Taylor expanding `LegendreQ(0,x)` about `x=0`, dropping terms of $O(x^8)$, and simplifying yields a series expansion for Q_0 which agrees with that for $Q0$.

> `Q[0]:=simplify(LegendreQ(0,x)=taylor(LegendreQ(0,x),x=0,8));`

$$Q_0 := \frac{1}{2}\ln(x+1) - \frac{1}{2}\ln(1-x) = x + \frac{1}{3}x^3 + \frac{1}{5}x^5 + \frac{1}{7}x^7 + O(x^8)$$

In physical problems where x varies from -1 to 1, the $Q_n(x)$ must be rejected because they diverge at the end points of the range.

2.2.2   Another Old Acquaintance

Acquaintance. A person whom we know well enough to borrow from, but not well enough to lend to.
Ambrose Bierce, American author, *The Devils Dictionary*, (1842–1914)

After revisiting the Legendre functions, it's not too surprising that Jennifer renews another old acquaintance, the Bessel functions, again showing how they arise as a series solution to Bessel's equation (with p non-negative),

$$x^2\,y'' + x\,y' + (x^2 - p^2)\,y = 0. \tag{2.6}$$

Putting (2.6) into the standard form (2.4), the second-order LODE has a regular singular point at $x=x_0=0$, so Jennifer will now seek a Frobenius power series solution expanded about $x=0$, viz., $y=\sum_{m=0}^{\infty} c_m\, x^{m+s}$. The total number N

of terms which will be kept in the series for plotting purposes is taken to be 100. The left-hand side of Bessel's equation is entered in *ode*.

```
> restart: N:=100:
```
```
> ode:=x^2*diff(y(x),x,x)+x*diff(y(x),x)+x^2*y(x)-p^2*y(x);
```

$$ode := x^2 \, (\frac{d^2}{dx^2} \, y(x)) + x \, (\frac{d}{dx} \, y(x)) + x^2 \, y(x) - p^2 \, y(x)$$

For $y(x)$, Jennifer enters the summand of the Frobenius series.

```
> y(x):=c[m]*x^(m+s);
```

$$y(x) := c_m \, x^{(m+s)}$$

Noting that $y(x)$ is automatically substituted into *ode*, *ode* is divided by x^s and the result simplified. The resulting summand is given by the output of *eq*.

```
> eq:=simplify(ode/x^s);
```

$$eq := c_m \, x^m \, m^2 + 2 \, c_m \, x^m \, m \, s + c_m \, x^m \, s^2 + x^{(2+m)} \, c_m - p^2 \, c_m \, x^m$$

The sum $\sum_{m=0}^{\infty}$ over the five terms of *eq* is to be set equal to zero which implies that the coefficients of equal powers of x must also be equal to zero. Now all the terms in *eq* involve x^m, except for the fourth one which contains x^{m+2}. But, on summing, the fourth operand in *eq* can be transformed as follows,

$$\sum_{m=0}^{\infty} c_m \, x^{m+2} = \sum_{m=2}^{\infty} c_{m-2} \, x^m = \sum_{m=0}^{\infty} c_{m-2} \, x^m. \qquad (2.7)$$

if we agree to define $c_{-1} = 0$, $c_{-2} = 0$. Jennifer enters these coefficient values.

```
> c[-1]:=0:   c[-2]:=0:
```

The transformation (2.7) can be accomplished in *eq* by using the **subsop** command to replace the 4th operand with $c_{m-2} \, x^m$.

```
> eq2:=subsop(4=c[m-2]*x^m,eq);
```

$$eq2 := c_m \, x^m \, m^2 + 2 \, c_m \, x^m \, m \, s + c_m \, x^m \, s^2 + c_{m-2} \, x^m - p^2 \, c_m \, x^m$$

Now each term in *eq2* involves x^m. Dividing this equation by x^m, equating the result to 0, and simplifying, yields a *recurrence relation* relating c_m to c_{m-2}.

```
> eq3:=simplify(eq2/x^m)=0;
```

$$eq3 := c_m \, m^2 + 2 \, c_m \, m \, s + c_m \, s^2 + c_{m-2} - p^2 \, c_m = 0$$

Terms involving c_m are now collected in the recurrence relation *eq3*.

```
> eq4:=collect(eq3,c[m]);
```

$$eq4 := (m^2 + 2 \, m \, s + s^2 - p^2) \, c_m + c_{m-2} = 0$$

Setting $m=0$ in *eq4* yields the so-called *indicial equation*, *eq5*.

```
> eq5:=eval(eq4,m=0);
```

$$eq5 := (s^2 - p^2) \, c_0 = 0$$

Assuming that the coefficient $c_0 \neq 0$, *eq5* has two solutions for s, which are explicitly extracted using the **solve** command. Jennifer will consider the series solution (the second one in *sol*) to Bessel's equation corresponding to $s = p$. The other series solution is obtained by replacing p with $-p$.

>   `sol:=solve(eq5,s); s:=sol[2];`

$$sol := -p, \ p$$
$$s := p$$

Evaluating the recurrence relation *eq4* for $m = 1$, and noting that for non-negative p, $1+2\,p$ can never be zero, it follows from *eq6* below that the coefficient $c_1 = 0$. This coefficient is extracted from *eq6* with the `solve` command.

>   `eq6:=eval(eq4,m=1); c[1]:=solve(eq6,c[1]);`

$$eq6 := (1 + 2\,p)\,c_1 = 0$$
$$c_1 := 0$$

For $m \geq 2$, one must work with the full recurrence equation *eq4*. The recurrence relation can be rewritten by isolating the coefficient c_m on the left side of the equation and then factoring the result as in *eq7*.

>   `eq7:=factor(isolate(eq4,c[m]));`

$$eq7 := c_m = -\frac{c_{m-2}}{m\,(m + 2\,p)}$$

The coefficients c_2 to $c_N = c_{100}$ are now explicitly evaluated by first using the `unapply` command to turn *eq7* into a functional operator in terms of the argument m and then using a do loop to iterate the recurrence relation.

>   `eq8:=unapply(eq7,m):`
>   `for m from 2 to N do c[m]:=rhs(eq8(m)); end do:`

Jennifer has used a command line ending colon to suppress the very long output. All the even subscript coefficients are proportional to c_0, while odd subscript coefficients are proportional to c_1 and therefore are all equal to zero.

To study the behavior of the series as more and more terms are retained, an arrow operator JJ is introduced to add the terms in the series for $m=0$ to M.

>   `JJ:=M->add(c[m]*x^(m+s),m=0..M):`

Jennifer now calculates the sum Jp for $M=6$ and collects the c_0 coefficients.

>   `Jp:=collect(JJ(6),c[0]);`

$$Jp := \left(x^p - \frac{x^{(2+p)}}{2\,(2+2\,p)} + \frac{x^{(4+p)}}{8\,(2+2\,p)\,(4+2\,p)} - \frac{x^{(6+p)}}{48\,(2+2\,p)\,(4+2\,p)\,(6+2\,p)} \right) c_0$$

With the proper assignment (which will be done shortly) of the arbitrary coefficient c_0, the infinite series corresponding to Jp is the Bessel function J_p of the first kind of order p. Not surprisingly, the same result could have been much more easily obtained by using the series option of the `dsolve` command. To demonstrate this, Jennifer now unassigns $y(x)$. The order of the series is set as well as the information level on the methods used in the `dsolve` command.

>   `unassign('y(x)'): Order:=7: infolevel[dsolve]:=5:`

Jennifer replaces $y(x)$ with $z(x)$ in *ode* and obtains the general series solution of *ode2*. She has deleted some of the unsuccessful methods in the output.

```
>   ode2:=subs(y(x)=z(x),ode): dsolve(ode2,z(x),series);
```
....

dsolve/series/ordinary: trying Newton iteration
dsolve/series/direct: trying direct subs
dsolve/series/froben: trying method of Frobenius
dsolve/series/froben: indicial eqn is -p^2+r^2
dsolve/series/froben: roots of indicial eqn are [[p], [-p]]

$$z(x) = \_C1\ x^p(1 - \frac{1}{4p+4}\ x^2 + \frac{1}{(8p+16)(4p+4)}\ x^4$$

$$- \frac{1}{(12p+36)(8p+16)(4p+4)}\ x^6 + O(x^7))$$

$$+\_C2\ x^{(-p)}(1 + \frac{1}{4p-4}\ x^2 + \frac{1}{(8p-16)(4p-4)}\ x^4$$

$$+ \frac{1}{(12p-36)(8p-16)(4p-4)}\ x^6 + O(x^7))$$

The method of Frobenius has been successfully used, the roots of the indicial equation obtained, and the general series solution constructed for $z(x)$. The first series, involving the arbitrary coefficient $\_C1$, is exactly the same as in Jp. The second series, involving the other arbitrary coefficient $\_C2$, is the second independent solution of the second order LODE obtained by replacing p with $-p$. I.e., the second independent solution is $J_{-p}(x)$.

This conclusion about the mathematical form of the second solution is true providing that p is not zero or an integer. For $p = 0$, there is only one root, $s = 0$, to the indicial equation and the second series is identical with the first. For integer values of p, all denominators in the second series beyond a certain term vanish, so this series becomes meaningless. Thus for $p = 0, 1, 2, ...$ a more general second solution must be sought. Jennifer has decided not to go into this issue here, referring her students to standard ODE texts which cover the topic. At this stage it suffices to note that this second independent solution is the Bessel function Y_p of the second kind of order p.

Returning to her "hand mimicking" calculation, Jennifer completes the identification of the infinite series solution as $J_p(x)$ by making the "standard" choice for c_0, viz. $c_0 = 1/(2^p p!)$.

```
>   c[0]:=1/(2^p*p!):
```

To see how may terms have to be retained in the series over a certain range of x, e.g., $x = 0$ to 50, to obtain a reasonably correct curve, Jennifer will now plot the sequence of results obtained from JJ for $p = 0$ and $N/5 = 100/5 = 20$, 40, 60, 80, and 100 terms along with the "exact" infinite series result for $J_0(x)$. Because the finite series results diverge to ∞ before $x = 50$ is reached, the vertical view is limited to be between -1 and 1.

```
>   plot([seq(eval(JJ(k*N/5),p=0),k=1..5),BesselJ(0,x)],
    x=0..50,thickness=2,numpoints=500,labels=["x","J"],
    tickmarks=[3,3],view=[0..50,-1..1]);
```

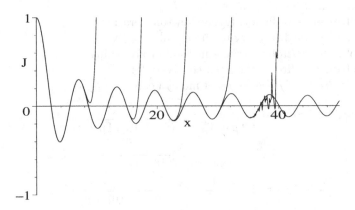

Figure 2.4: Divergent curves from left to right: 20, 40, 60, 80, and 100 terms.

From left to right in Figure 2.4, the divergent curves represent the finite series representations of J_0 for 20, 40, 60, 80, and 100 terms. The non-divergent oscillatory curve is the infinite series representing the exact $J_0(x)$.

2.3   Fourier Series

Consider a single-valued function $f(x)$ defined over the *fundamental interval* $-L \leq x \leq L$ and satisfying the boundary conditions $f(-L) = f(L)$. If $f(x)$ has a finite number of discontinuities and maxima and minima, and $\int_{-L}^{L} |f(x)|\, dx$ is finite,[1] then $f(x)$ can be expanded in the *Fourier series* ([MW71], [Boa83])

$$f(x) = \frac{1}{2}a_0 + \sum_{n=1}^{\infty} \left[a_n \cos\left(\frac{n\pi x}{L}\right) + b_n \sin\left(\frac{n\pi x}{L}\right) \right], \qquad (2.8)$$

with $a_n = \dfrac{1}{L} \displaystyle\int_{-L}^{L} f(x) \cos\left(\frac{n\pi x}{L}\right) dx, \quad b_n = \dfrac{1}{L} \displaystyle\int_{-L}^{L} f(x) \sin\left(\frac{n\pi x}{L}\right) dx.$

The forms of a_n and b_n can be derived from Equation (2.8) by noting that $y_n(x) = \cos(n\pi x/L)$ (or $\sin(n\pi x/L)$) satisfies the orthogonality condition $\int_{-L}^{L} w(x)\, y_m(x)\, y_n(x)\, dx = 0$, for $m \neq n$, with $w(x) = 1$. The identification of $w(x)$ follows on noting that the $y_n(x)$ are solutions of $y''(x) + (n\pi/L)^2 y(x) = 0$. This equation is a Sturm–Liouville ODE, (1.6), with $p = 1$, $q = 0$, $w = 1$, and $\lambda = -(n\pi/L)^2$. The a_n and b_n follow on multiplying (2.8) by $\cos(n\pi x/L)$ (or $\sin(n\pi x/L)$), integrating from $-L$ to L, and using the orthogonality condition.

   If $f(x)$ is an odd function, that is to say $f(-x) = -f(x)$, then $a_n = 0$ and $b_n = (2/L) \int_0^L f(x) \sin(n\pi x/L)\, dx$, so $f(x)$ is expressed as a *Fourier sine series*.

---

[1] These are sufficient, but not necessary, conditions.

On the other hand, if $f(x)$ is an even function, i.e., $f(-x) = f(x)$, then $a_n = (2/L) \int_0^L f(x) \cos(n \pi x/L) \, dx$ and $b_n = 0$, so $f(x)$ is a *Fourier cosine series*.

Since each term in (2.8) is periodic with period $2L$, then $f(x + 2L) = f(x)$. Thus, the Fourier series may either represent an $f(x)$ defined in the fundamental interval $(-L, L)$, or a periodic $f(x)$ with period $2L$ for all of x.

When $f(x)$ is defined only in the range 0 to L, it can be written as a Fourier sine series by including the range to $-L$ to 0 and considering $f(x)$ to be an odd function about $x = 0$. Alternately, it can be written as a Fourier cosine series by considering $f(x)$ to be an even function about the origin. It may turn out that one series fits $f(x)$ better than the other for a finite number of terms.

To this point, the fundamental interval has been taken to be $2L$. This can be easily changed. For example, consider $f(t)$ defined in the range $t = 0$ to T and we want the fundamental interval to be T, not $2T$. To accomplish this, set $x = t$ and $L = T/2$ in (2.8) and a_n and b_n and change the range of the integrals from $-T/2 \dots T/2$ to $0 \dots T$. In this case, the general Fourier expansion becomes

$$f(t) = \frac{1}{2}a_0 + \sum_{n=1}^{\infty} \left[a_n \cos\left(\frac{2n\pi t}{T}\right) + b_n \sin\left(\frac{2n\pi t}{T}\right) \right], \qquad (2.9)$$

$$\text{with} \quad a_n = \frac{2}{T} \int_0^T f(t) \cos\left(\frac{2n\pi t}{T}\right) dt, \quad \text{and} \quad b_n = \frac{2}{T} \int_0^T f(t) \sin\left(\frac{2n\pi t}{T}\right) dt.$$

Again, this series may be used to represent either a function defined in the fundamental interval 0 to T or a periodic function whose period is T.

The concept of expanding a function $f(x)$ in terms of sines and cosines can be extended to other special functions $y_n(x)$ satisfying a S-L type equation. If the $y_n(x)$ satisfy the same boundary conditions at a and b as $f(x)$, then

$$f(x) = \sum_n A_n y_n(x), \quad \text{with } A_n = \frac{\int_a^b w(x) f(x) y_n(x) \, dx}{\int_a^b w(x) y_n(x)^2 \, dx}. \qquad (2.10)$$

The functions $y_n(x)$ are said to form a *complete set*. Often, they are *normalized* so that $\int_a^b w(x) y_n(x)^2 \, dx = 1$. Since they have the orthogonality property, they then satisfy the *orthonormality condition*

$$\int_a^b w(x) y_m(x) y_n(x) \, dx = \delta_{mn}, \qquad (2.11)$$

where δ_{mn}, the *Kronecker delta*, is defined by $\delta_{mn} = 1$ for $m = n$ and 0 for $m \neq n$.

As an example of expanding in terms of special functions, the *Legendre–Fourier series* (or simply the *Legendre series*) is given by

$$f(x) = \sum_{n=0}^{\infty} A_n P_n(x), \quad \text{with } A_n = \frac{(2n + 1)}{2} \int_{-1}^1 f(x) P_n(x) \, dx. \qquad (2.12)$$

A mathematical example of this series is given in Recipe **02-3-3**.

2.3.1   Madeiran Levadas and the Gibb's Phenomenon

The idealist walks on tiptoe, the materialist on his heels.
Malcolm de Chazal, French writer, (1902–81)

On the island of Madeira, water is transported from the mountains by a net-work of levadas (irrigation canals) which often cling to the mountain side with vertigo-inducing drop offs and pass through pitch-black tunnels. This recipe is inspired by some interesting hikes that I have taken on Madeiran levada re-taining walls. A levada retaining wall is described by the piecewise function $f = (L + x)/2$ for $-L \leq x \leq -L/2$, $f = L/4$ for $-L/2 \leq x < 0$, and $f = 0$ for $0 < x \leq L$, with $L = \pi$. Determine the Fourier series representation of $f(x)$ and plot it and f together. Calculate the cross-sectional area of the retaining wall using $f(x)$ and then the Fourier series. Discuss the various results.

To simplify the command entries, let's set $X = \pi x/L$.

```
>  restart: X:=Pi*x/L:
```

Using the inert **Sum** command, an operator F is formed to calculate the Fourier series, keeping N terms.

```
>  F:=N->a[0]/2+Sum(a[n]*cos(n*X)+b[n]*sin(n*X),n=1..N);
```

$$F := N \to \frac{1}{2} a_0 + (\sum_{n=1}^{N} (a_n \cos(n\,X) + b_n \sin(n\,X)))$$

The formal expressions for the coefficients are entered using the inert form of the integral command. In the outputs, X is replaced with $\pi x/L$.

```
>  a[0]:=(1/L)*Int(f,x=-L..L);
```

$$a_0 := \frac{1}{L} \int_{-L}^{L} f\, dx$$

```
>  a[n]:=(1/L)*Int(f*cos(n*X),x=-L..L);
```

$$a_n := \frac{1}{L} \int_{-L}^{L} f \cos(\frac{n\pi x}{L})\, dx$$

```
>  b[n]:=(1/L)*Int(f*sin(n*X),x=-L..L);
```

$$b_n := \frac{1}{L} \int_{-L}^{L} f \sin(\frac{n\pi x}{L})\, dx$$

The value $L = \pi$ is specified and the piecewise function f entered.

```
>  L:=Pi: f:=piecewise(x<-L/2,(L+x)/2,x<0,L/4,x<L,0);
```

$$f := \begin{cases} \dfrac{\pi}{2} + \dfrac{x}{2} & x < -\dfrac{\pi}{2} \\ \dfrac{\pi}{4} & x < 0 \\ 0 & x < \pi \end{cases}$$

The value of the coefficient a_0 is obtained.

```
>  a[0]:=value(a[0]);
```

$$a_0 := \frac{3\pi}{16}$$

The coefficients a_n and b_n can be simplified by assuming that n is an integer. The "type match command" (::) is used in the assumption.

```
>  a[n]:=simplify(value(a[n])) assuming n::integer;
```

$$a_n := -\frac{1}{2} \frac{(-1)^n - \cos(\frac{\pi n}{2})}{\pi n^2}$$

```
>  b[n]:=simplify(value(b[n])) assuming n::integer;
```

$$b_n := -\frac{1}{4} \frac{\pi n + 2\sin(\frac{\pi n}{2})}{\pi n^2}$$

The Fourier series is generated in FF, with $N=15$. (Only the leading terms in the output are shown here in the text.) You can increase the value of N, but you might then wish to suppress the very lengthy output by putting a colon on the end of the command line.

```
>  FF:=value(F(15));
```

$$FF := \frac{3\pi}{32} + \frac{1}{2}\frac{\cos(x)}{\pi} - \frac{1}{4}\frac{(\pi+2)\sin(x)}{\pi} - \frac{1}{4}\frac{\cos(2x)}{\pi} - \frac{1}{8}\sin(2x)$$
$$+\frac{1}{18}\frac{\cos(3x)}{\pi} - \frac{1}{36}\frac{(3\pi-2)\sin(3x)}{\pi} - \frac{1}{16}\sin(4x) + \frac{1}{50}\frac{\cos(5x)}{\pi} + \cdots$$

Finally, the function f and the Fourier series FF are plotted over the range $x=-L=-\pi$ to $x=L=\pi$, being represented by thick blue and red lines. The resulting picture is shown on the left of Figure 2.5.

```
>  plot([f,FF],x=-L..L,color=[blue,red],thickness=2,axes=box);
```

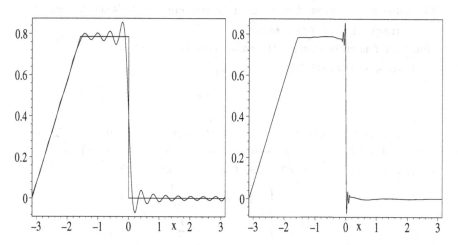

Figure 2.5: Left: Fourier series for $N=15$ and f. Right: Series for $N=100$.

The Fourier series oscillates around the exact f. The fit can be improved (the size of the oscillations reduced) by increasing N. The plot on the right of Figure 2.5 shows the Fourier series result for $N = 100$. The overshoot in the vicinity of the step function at $x = 0$ persists, however, no matter how large an N is chosen. This is called the *Gibbs' phenomenon*. Notice also in the left plot that the Fourier series curve passes approximately through the midpoint of the step. As $N \to \infty$, the Fourier curve will pass exactly though the midpoint, a general property of Fourier series at step discontinuities.

The exact cross-sectional area of the retaining wall is calculated in *Area1* by integrating f from $x = -L = -\pi$ to π. The cross-sectional area is also calculated in *Area2* by integrating the Fourier series FF over the same range.

```
>  Area1:=int(f,x=-Pi..Pi);  Area2:=int(FF,x=-Pi..Pi);
```

$$Area1 := \frac{3\pi^2}{16} \qquad Area2 := \frac{3\pi^2}{16}$$

The two areas are identical. The "wiggles" in the Fourier series about the f curve exactly cancel. You might like to confirm that this is still true for $N = 100$.

2.3.2   Sine or Cosine Series?

I see it all perfectly; there are two possible situations – one can either do this or that. My honest opinion and my friendly advice is this: do it or do not do it – you will regret both.
Søren Kierkegaard, Danish philosopher, (1813–55)

Consider the function $f(x) = x$ for $0 \le x \le L/2$ and $f(x) = L - x$ for $L/2 \le x \le L$, with $L = 1$. Extending the range to $-L \le x \le L$, derive a Fourier sine series and a cosine series representation of $f(x)$. Plot $f(x)$ and the two series together and discuss the results.

The value of L is entered and, again for convenience, let's set $X = \pi x/L$.

```
>   restart: L:=1: X:=Pi*x/L:
```

The function f is entered using the piecewise command.

```
>   f:=piecewise(x<L/2,x,x<L,L-x);
```

$$f := \begin{cases} x & x < \dfrac{1}{2} \\ 1-x & x < 1 \end{cases}$$

An odd function $f1$ is introduced which is the same as f for $0 \le x \le L$, but is equal to $-f$ for $-L \le x \le 0$. $f1$ will be used to generate the sine series.

```
>   f1:=piecewise(x<-L/2,-(L+x),x<L/2,x,x<L,L-x);
```

$$f1 := \begin{cases} -1-x & x < \dfrac{-1}{2} \\ x & x < \dfrac{1}{2} \\ 1-x & x < 1 \end{cases}$$

An even function $f2$ is introduced which is the same as f for $0 \leq x \leq L$ and symmetrical about the origin. $f2$ is used to generate the cosine series.

```
>   f2:=piecewise(x<-L/2,L+x,x<0,-x,x<L/2,x,x<L,L-x);
```

To confirm that $f1$ and $f2$ are odd and even extensions of f to the region $x < 0$, they are plotted as solid and dashed curves in Figure 2.6. For $x > 0$, the two curves are identical, and therefore indistinguishable.

```
>   plot([f1,f2],x=-L..L,linestyle=[SOLID,DASH],thickness=2);
```

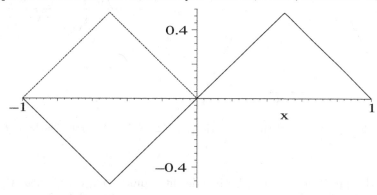

Figure 2.6: Solid curve, $f1$. dashed curve, $f2$.

For a given f, functional operators A, B and F are introduced to calculate the series coefficients a_n and b_n, and to produce the series out to N terms.

```
>   A:=(f,n)->(1/L)*int(f*cos(n*X),x=-L..L);
```

$$A := (f,\, n) \rightarrow \frac{1}{L} \int_{-L}^{L} f \cos(n\,X)\, dx$$

```
>   B:=(f,n)->(1/L)*int(f*sin(n*X),x=-L..L);
```

$$B := (f,\, n) \rightarrow \frac{1}{L} \int_{-L}^{L} f \sin(n\,X)\, dx$$

```
> F:=(f,N)->A(f,0)/2+sum(A(f,n)*cos(n*X)+B(f,n)*sin(n*X),n=1..N);
```

$$F := (f,\, N) \rightarrow \frac{1}{2} A(f,\, 0) + \left(\sum_{n=1}^{N} (A(f,\, n) \cos(n\,X) + B(f,\, n) \sin(n\,X)) \right)$$

Taking $N=10$, the Fourier sine and cosine series are produced in $F1$ and $F2$.

```
>   F1:=F(f1,10); F2:=F(f2,10);
```

$$F1 := \frac{4 \sin(\pi x)}{\pi^2} - \frac{4}{9}\frac{\sin(3\pi x)}{\pi^2} + \frac{4}{25}\frac{\sin(5\pi x)}{\pi^2} - \frac{4}{49}\frac{\sin(7\pi x)}{\pi^2} + \frac{4}{81}\frac{\sin(9\pi x)}{\pi^2}$$

$$F2 := \frac{1}{4} - \frac{2\cos(2\pi x)}{\pi^2} - \frac{2}{9}\frac{\cos(6\pi x)}{\pi^2} - \frac{2}{25}\frac{\cos(10\pi x)}{\pi^2}$$

The two series are plotted along with f and shown on the left of Figure 2.7.

```
>   plot([f,F1,F2],x=0..L,color=[blue,red,green],thickness=2);
```

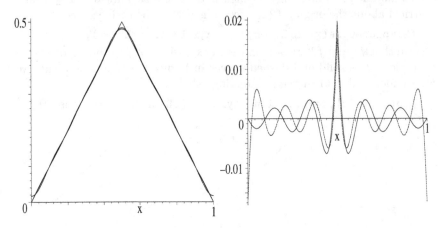

Figure 2.7: Left: f and the two series. Right: Solid, $f - F1$. Dashed, $f - F2$.

Even though N is not large, both series fit f quite well, except near the apex of the triangle and, for the cosine series, near $x = 0$ and 1. To magnify the difference between the series results and f, the differences $f - F1$ and $f - F2$ are plotted as solid and dashed curves in the right graph of Figure 2.5.

```
>  plot([f-F1,f-F2],x=0..L,linestyle=[1,3],thickness=2);
```

The cosine series clearly fits less well at the end points of the x range. Can you suggest why this is the case?

2.3.3   How Sweet This Is!

Few things are harder to put up with than the annoyance
of a good example.
Mark Twain, American author, *Pudd'nhead Wilson, 1894*

If you have calculated series expansions by hand, you know what a tedious task it can be to explicitly calculate a large number of terms in the series and then have to plot the results. By now, you should have gotten a clear idea that using computer algebra is the way to go in handling such problems. The following recipe for the Legendre–Fourier series is a particularly "sweet" example that derives a "beautiful" result very quickly.

Consider a step function, $f(x) = 0$ for $-1 < x < 0$ and $f(x) = 1$ for $0 < x < 1$. Derive the Legendre–Fourier series for f and plot the series and f together over the range $-1 < x < 1$. Calculate the area between the x-axis and the series curve and compare with the exact result for f. Discuss the plot and area results.

```
>  restart:
```

Functional operators A and F are formed to calculate the coefficients A_n, and the Legendre series out to N terms, for a given function f.

```
>   A:=(f,n)->((2*n+1)/2)*int(f*LegendreP(n,x),x=-1..1);
```

$$A := (f, n) \to \frac{1}{2} (2n + 1) \int_{-1}^{1} f \, \mathrm{LegendreP}(n, x) \, dx$$

```
>   F:=(f,N)->sum(A(f,n)*LegendreP(n,x),n=0..N);
```

$$F := (f, N) \to \sum_{n=0}^{N} A(f, n) \, \mathrm{LegendreP}(n, x)$$

The given f is entered with the `Heaviside(x)` command and the Legendre-Fourier series calculated in *F1* for $N = 15$.

```
>   f:=Heaviside(x); F1:=F(f,15);
```

$$f := \mathrm{Heaviside}(x)$$

$$
\begin{aligned}
F1 := {} & \frac{1}{2} + \frac{3x}{4} - \frac{7}{16} \, \mathrm{LegendreP}(3, x) + \frac{11}{32} \, \mathrm{LegendreP}(5, x) \\
& - \frac{75}{256} \, \mathrm{LegendreP}(7, x) + \frac{133}{512} \, \mathrm{LegendreP}(9, x) - \frac{483}{2048} \, \mathrm{LegendreP}(11, x) \\
& + \frac{891}{4096} \, \mathrm{LegendreP}(13, x) - \frac{13299}{65536} \, \mathrm{LegendreP}(15, x)
\end{aligned}
$$

This is a formidable looking series, which has been generated in three command lines. If I now wanted to see what the series looks like for, say $N = 100$, changing 15 to 100 in *F1* and executing the command would generate the new result almost instantaneously.

How formidable the result really is, even for 15 terms, can be appreciated by expanding *F1*. The **sort** command is used to order the polynomial expansion from the highest exponent to the lowest. Wow, what a result!

```
>   F1:=sort(expand(F1));
```

$$
\begin{aligned}
F1 := {} & -\frac{128931743655}{134217728} x^{15} + \frac{503889568875}{134217728} x^{13} - \frac{800852700375}{134217728} x^{11} \\
& + \frac{664630841875}{134217728} x^9 - \frac{307629132525}{134217728} x^7 + \frac{78646056489}{134217728} x^5 \\
& - \frac{10402917525}{134217728} x^3 + \frac{703956825}{134217728} x + \frac{1}{2}
\end{aligned}
$$

Now f and *F1* are plotted in Figure 2.8 over the range $x = -1$ to 1, and are represented by dashed and solid curves, respectively.

```
>   plot([f,F1],x=-1..1,thickness=2,linestyle=[3,1]);
```

Two features of the Legendre series are quite clear from the picture, and can be confirmed by taking N larger. (You will have to adjust the view and number of digits and plotting points.) The series curve displays a Gibb's phenomenon and also passes through the midpoint of the step.

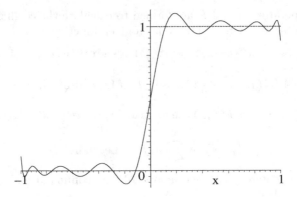

Figure 2.8: Dashed curve, Step function f. Solid curve, Legendre series $F1$.

The area between the Legendre series curve and the x-axis is now calculated and found to be exactly the same as the area for the step function.

```
>  Area:=int(F1,x=-1..1);
```

$$Area := 1$$

2.4 Summing Series

In this section, two different approaches to summing infinite series are presented. The first recipe makes use of ideas already introduced in the earlier Fourier series examples. In the second recipe, the series to be summed is replaced by a complex series which Maple is able to sum.

2.4.1 I. M. Curious Sums a Series

Once, I thought I made a mistake, but I was mistaken.
From the diary of I. M. Curious

In this recipe, Ms. Curious answers the following question:

By expanding $f(x) = x(L - x)$, defined in the interval $(0, L)$ with $L=\pi$, in a Fourier sine series of period $2L$ and setting $x = L/2$, prove that

$$\sum_{n=0}^{\infty} \frac{(-1)^n}{(2n + 1)^3} = \frac{\pi^3}{32}.$$

Confirm this result by directly summing the series with Maple, first showing that the sum can be expressed as either a generalized *hypergeometric* function or as a *polylogarithm* function.

I. M. begins her solution by assuming that the Fourier series summation indices m and n are integers. She then sets $L=\pi$ and enters $f = x(L - x)$.

```
>   restart: assume(m::integer,n::integer):
>   L:=Pi: f:=x*(L-x):
```

To expand f in a Fourier sine series of period $2L$, she forms the following odd piecewise function, pw, defined in the interval $-L < x < L$.

```
>   pw:=piecewise(x<0,x*(L+x),x>0,f);
```

$$pw := \begin{cases} x\,(\pi + x) & x < 0 \\ x\,(\pi - x) & 0 < x \end{cases}$$

I. M. then calculates the Fourier coefficients a_0, a_m for $m \neq 0$, and b_m.

```
>   a[0]:=(1/L)*int(pw,x=-L..L);
```

$$a_0 := 0$$

```
>   a[m]:=(1/L)*int(pw*cos(m*Pi*x/L),x=-L..L);
```

$$a_m := 0$$

```
>   b[m]:=(1/L)*int(pw*sin(m*Pi*x/L),x=-L..L);
```

$$b_m := -\frac{4\,(-1 + (-1)^m)}{\pi\,m^3}$$

She notices that in b_m the coefficients are only non-zero if m is an odd integer. So she substitutes $m = 2n + 1$ into b_m and relabels the coefficients as b_{2n+1}. The new summation index n will take on the values $n = 0, 1, 2, \ldots$.

```
>   b[2*n+1]:=simplify(subs(m=2*n+1,b[m]));
```

$$b_{2n+1} := \frac{8}{\pi\,(2n+1)^3}$$

Since the coefficients a_m are zero for all m (and therefore all n), the Fourier series is then of the form $F = \sum_{n=0}^{\infty} b_{2n+1} \sin((2n+1)\pi x/L)$.

```
>   F:=Sum(b[2*n+1]*sin((2*n+1)*Pi*x/L),n=0..infinity);
```

$$F := \sum_{n=0}^{\infty} \left(\frac{8\sin((2n+1)\,x)}{\pi\,(2n+1)^3} \right)$$

Thus, the original function f in the region $0 < x < L$ can be written as the Fourier sine series F. This is entered in $eq1$.

```
>   eq1:=f=F;
```

$$eq1 := x\,(\pi - x) = \sum_{n=0}^{\infty} \left(\frac{8\sin((2n+1)\,x)}{\pi\,(2n+1)^3} \right)$$

$eq1$ is divided by 8 and evaluated at $x = L/2$.

```
>   eq2:=eval(eq1/8,x=L/2);
```

$$eq2 := \frac{\pi^2}{32} = \frac{1}{8} \left(\sum_{n=0}^{\infty} \left(\frac{8\,(-1)^n}{\pi\,(2n+1)^3} \right) \right)$$

Multiplying $eq2$ by π, and expanding, confirms the series sum.

```
>   eq3:=expand(Pi*eq2);
```

$$eq3 := \frac{\pi^3}{32} = \sum_{n=0}^{\infty} \frac{(-1)^n}{(2n+1)^3}$$

To sum the series directly with Maple, I. M. extracts it from the rhs of *eq3*.

```
>  S:=rhs(eq3);
```

$$S := \sum_{n=0}^{\infty} \frac{(-1)^n}{(2n+1)^3}$$

She expresses the sum S as a hypergeometric function by applying the following `convert` command.

```
>  S:=convert(S,hypergeom);
```

$$S := \mathrm{hypergeom}([\frac{1}{2}, \frac{1}{2}, \frac{1}{2}, 1], [\frac{3}{2}, \frac{3}{2}, \frac{3}{2}], -1)$$

If this function is unfamiliar to you, highlight hypergeom in the computer output with your mouse and open the relevant help window to see its definition. According to Help, it may be possible to convert the hypergeometric function into one of the standard special and elementary functions found in such texts as Handbook of Mathematical Functions by Abramowitz and Stegun ([AS72]). Applying the `convert(StandardFunctions)` command,

```
>  S:=convert(S,StandardFunctions);
```

$$S := -\frac{1}{2} I \,\mathrm{polylog}(3, I) + \frac{1}{2} I \,\mathrm{polylog}(3, -I)$$

yields a combination of polylog functions. Again, if the polylogarithm function is unfamiliar, it may be looked up in Maple's Help. I. M. finally obtains the desired form of the series sum by using `simplify`.

```
>  S:=simplify(S);
```

$$S := \frac{\pi^3}{32}$$

2.4.2   Spiegel's Series Problem

Old age is that time of life when you can feel bad in the morning without having had fun the night before.
Gregarius Nerd, Professor of Mathematics, Erehwon Institute of Technology

In the previous example, we saw that Maple was successful in summing the given series. If Maple is unsuccessful, does that mean that the series cannot be summed? Not necessarily, as you will now see. Sometimes it needs a bit of help in the form of human brain power. This will not be the last time in this book that this is the case.

Let's consider the following infinite series,

$$r\,\sin(\phi) + \frac{1}{3} r^3 \sin(3\,\phi) + \frac{1}{5} r^5 \sin(5\,\phi) + \frac{1}{7} r^7 \sin(7\,\phi) + \cdots, \tag{2.13}$$

which, according to the Schaum Outline Series on Advanced Mathematics by Murray Spiegel ([Spi71]), arises from solving for the steady-state temperature distribution in a thin circular plate of unit radius whose faces are insulated and has each half of its boundary kept at a different constant temperature. In polar coordinates, r is the radial distance from the center of the plate and ϕ the polar angle. Spiegel's Problem 12.52 is to show that the series can be summed and cast into the form $(1/2)\tan^{-1}(2r\sin\phi/(1-r^2))$.

The following recipe solves Spiegel's problem. An operator S is formed to generate the series out to exponent $2N+1$.

```
> restart:
> S:=N->sum(r^(2*m+1)*sin((2*m+1)*phi)/(2*m+1),m=0..N);
```

$$S := N \rightarrow \sum_{m=0}^{N} \frac{r^{(2m+1)}\sin((2m+1)\phi)}{2m+1}$$

Taking $N=3$, then entering S(3) generates the terms displayed in (2.13).

```
> S1:=S(3);
```

$$S1 := r\sin(\phi) + \frac{1}{3}r^3\sin(3\phi) + \frac{1}{5}r^5\sin(5\phi) + \frac{1}{7}r^7\sin(7\phi)$$

Taking $N=\infty$ in S and applying the value command, we find that Maple is unable to directly sum the series, returning the unevaluated sum in $S2$.

```
> S2:=value(S(infinity));
```

$$S2 := \sum_{m=0}^{\infty} \frac{r^{(2m+1)}\sin((2m+1)\phi)}{2m+1}$$

To sum the series, note that the real series can be written as the imaginary part of a *complex series*. To accomplish this, the summand in $S2$ can be taken as the imaginary part of a complex summand, viz., with $I \equiv \sqrt{-1}$,

$$\mathrm{Im}\left(\frac{r^{2m+1}e^{I(2m+1)\phi}}{2m+1}\right) = \mathrm{Im}\left(\frac{(re^{I\phi})^{2m+1}}{2m+1}\right) = \mathrm{Im}\left(\frac{z^{2m+1}}{2m+1}\right)$$

with $z \equiv re^{I\phi}$. Using this result, the complex series, CS, is entered and then successfully summed.

```
> CS:=Sum(z^(2*m+1)/(2*m+1),m=0..infinity);
```

$$CS := \sum_{m=0}^{\infty} \frac{z^{(2m+1)}}{2m+1}$$

```
> CS:=value(CS);
```

$$CS := \frac{1}{2}\ln(\frac{1+z}{1-z})$$

Then, $z=re^{I\phi}$ is substituted into CS,

```
> CS2:=subs(z=r*exp(I*phi),CS);
```

$$CS2 := \frac{1}{2}\ln(\frac{1+re^{(\phi I)}}{1-re^{(\phi I)}})$$

and the complex result $CS2$ broken into real and imaginary parts in $CS3$ with the complex evaluation command and simplified in $CS4$ assuming $r<1$.

```
>   CS3:=evalc(CS2);
>   CS4:=simplify(CS3) assuming r<1;
```

$$CS4 := \frac{1}{4}\ln(\frac{r^2 + 2r\cos(\phi) + 1}{1 - 2r\cos(\phi) + r^2})$$

$$+ \frac{1}{2}I\arctan(\frac{2r\sin(\phi)}{1 - 2r\cos(\phi) + r^2}, -\frac{-1 + r^2}{1 - 2r\cos(\phi) + r^2})$$

The portion of the argument in arctan before the comma is the numerator, the portion after the comma being the denominator. To extract the imaginary part of $CS4$, the `coeff` command is used to pull out the coefficient of I.

```
>   S3:=coeff(CS4,I);
```

$$S3 := \frac{1}{2}\arctan(\frac{2r\sin(\phi)}{1 - 2r\cos(\phi) + r^2}, -\frac{-1 + r^2}{1 - 2r\cos(\phi) + r^2})$$

The operand command, `op` is used to write the series sum in a form which agrees with the result quoted by Spiegel.

```
>   S4:=op(1,S3)*arctan((op([2,1],S3)/op([2,2],S3)));
```

$$S4 := -\frac{1}{2}\arctan(\frac{2r\sin(\phi)}{-1 + r^2})$$

2.5 Supplementary Recipes

02-S01: Euler and Bernoulli Numbers

(a) Taylor expand $\sec(z)$, dropping terms of $O(z^{12})$. The *Euler numbers* E_{2n} are defined by $\sec(z) = \sum_{n=0}^{\infty}(-1)^n E_{2n} z^{2n}/(2n)!$ Using the Euler number command `euler(2*n)`, confirm that the latter expansion of $\sec(z)$ agrees with the Taylor expansion. Generate the Euler numbers $E_0, E_2, E_4,...,E_{10}$.

(b) Taylor expand $z/(e^z - 1)$, dropping terms of $O(z^{12})$. The *Bernoulli numbers* B_n are defined by $z/(e^z - 1) = \sum_{n=0}^{\infty} B_n z^n/n!$ Using the Bernoulli number command `bernoulli(n)`, confirm that the latter expansion agrees with the Taylor expansion. Generate the first 10 Bernoulli numbers.

02-S02: Ms. Curious Approximates an Integral

Ms. Curious has been given the following problem to solve. Consider the integral $f(x) = \int_0^x \sin(t^2)\, dt$. Evaluate this integral analytically and identify the function which occurs. Obtain the smallest polynomial approximation to $f(x)$ which is valid within ± 0.00001 for $0 \le x \le 1$.

02-S03: More Finite Difference Approximations

(a) Confirm the following finite difference approximation,

$$y''''(x) = [y(x + 2h) - 4y(x + h) + 6y(x) - 4y(x - h) + y(x - 2h)]/h^4.$$

Suggest a physical problem for which this FDA might be useful. Taking $y = x^{4.7}$, graphically compare and discuss the FDA (with $h = 0.1$ and 12 digits accuracy) with the exact 4th derivative over the range $x = 0$ to 70.

(b) Show that an FDA to the *Laplacian*, $\nabla^2 f(x,y) \equiv \partial^2 f/\partial x^2 + \partial^2 f/\partial y^2$, is

$$\nabla^2 f(x,y) = (1/12\,h^2)[16\,[f(x+h,y) + f(x,y+h) + f(x-h,y) + f(x,y-h)]$$

$$-[f(x+2\,h,y) + f(x,y+2\,h) + f(x-2\,h,y) + f(x,y-2\,h) + 60\,f(x,y)]].$$

Suggest a physical problem for which this FDA might be useful. Considering $f(x,y) = e^{-x^2 y^2}$ and taking $h=0.1$ and 10 digits accuracy, produce a 3-dimensional color-coded plot of the difference between the exact 2-dimensional Laplacian and the FDA for the range $x=-2..2$, $y=-2..2$.

02-S04: Series Solution

Mimicking a hand calculation, obtain a general series solution, valid near $x=0$, of the following LODE

$$x\,y'' + 2\,y' + x\,y = 0.$$

Show that the series solution may be expressed in a closed form. Confirm this closed-form solution by directly solving the LODE with the `dsolve` command.

02-S05: Chebyshev Polynomials Revisited

Obtain a general series solution of *Chebyshev's equation*,

$$(1 - x^2)\,y'' - x\,y' + p^2\,y = 0$$

valid near $x=0$, by (a) mimicking a hand calculation, (b) using the series option in the `dsolve` command. If p is a positive integer, show that one or the other of the series in the general solution reduces to a finite polynomial. These polynomials are the *Chebyshev polynomials*. The zeroth order Chebyshev polynomial $T_0(x) = 1$. The higher order polynomials $T_m(x)$ are normalized so that the coefficient of the largest power in the mth order polynomial is $2^{(m-1)}$. Derive the Chebyshev polynomials T_1, T_2, ..., T_7 and plot T_1 to T_5 over the range $x=-1$ to 1. Note that the non-finite-polynomial parts of the solution diverge at the end points of the range and are rejected in physical problems.

02-S06: A Fourier Series

Expand $f(\theta) = \theta^2$, $0 < \theta < 2\,\pi$, in a Fourier series F of period $2\,\pi$. Plot f and F (for an upper value $N=20$ of the summation index) in the same figure.

02-S07: Fourier Sine Series

Taking $L=1$, expand each of the following $f(x)$ in a Fourier sine series F of period $2L$, over the interval $(0, L)$:

(a) $f(x) = x\,(L-x)$;

(b) $f(x) = x$, $0 < x < L/2$, and $f(x) = L - x$, $L/2 < x < L$;

(c) $f(x) = 1$, $0 < x < L/2$, and $f(x) = 0$, $L/2 < x < L$.

In each case plot $f(x)$ and F (for an upper value $N = 10$ of the summation index) in the same figure and discuss the goodness of the fit.

02-S08: Fourier Cosine Series

Taking $L=1$, expand each of the $f(x)$ given in **02-S07** in a Fourier cosine series

F of period $2L$, over the interval $(0, L)$. In each case plot $f(x)$ and F (for an upper value $N=10$ of the summation index) in the same figure and discuss the goodness of the fit.

02-S09: Legendre Series

Expand the following $f(x)$ in a series F of Legendre polynomials (up to $n = N = 15$) and plot f and F together in the same figure:

(a) $f(x)=0$, $-1<x<0$, and $f(x)=1$, $0<x<1$;

(b) $f(x)=0$, $-1<x<0$, and $f(x)=x$, $0<x<1$.

02-S10: Directly Evaluating Series Sum

Write the following infinite series out in the summation notation and then use Maple to directly evaluate the series sum in closed form:

(a) $f(x) = 1 + 2x + 3x^2 + 4x^3 + \cdots$

(b) $f(x) = \dfrac{1}{1 \cdot 2} + \dfrac{x}{2 \cdot 3} + \dfrac{x^2}{3 \cdot 4} + \dfrac{x^3}{4 \cdot 5} + \cdots$

(c) $f = \dfrac{1}{1 \cdot 3} + \dfrac{1}{2 \cdot 4} + \dfrac{1}{3 \cdot 5} + \dfrac{1}{4 \cdot 6} + \cdots$

(d) $f(x) = x - \dfrac{4x^3}{3!} + \dfrac{9x^5}{5!} - \dfrac{16x^7}{7!} + \cdots$

02-S11: Another Cosine Series

Expand $f(x) = \sin(x)$, $0 < x < \pi$, in a Fourier cosine series F. Explicitly write out F for the first 5 non-zero terms and plot F and $f(x)$ together. Use F to prove $\sum_{n=1}^{\infty} 1/(4n^2 - 1) = 1/2$. Confirm this by directly summing with Maple.

02-S12: The Complex Series Trick Again

In a certain 2-dimensional electrostatic potential problem, the following infinite series occurs:

$$S = \sin\left(\frac{\pi x}{a}\right) e^{-\frac{\pi y}{a}} + \frac{1}{3}\sin\left(\frac{3\pi x}{a}\right) e^{-\frac{3\pi y}{a}} + \frac{1}{5}\sin\left(\frac{5\pi x}{a}\right) e^{-\frac{5\pi y}{a}} + \cdots$$

Show that the series can be summed and put into the form

$$S = \frac{1}{2} \arctan\left(\frac{\sin(\pi x/a)}{\sinh(\pi y/a)}\right).$$

Chapter 3

Vectors and Matrices

In the first section of this chapter, we see how Maple may be used to deal with vectors in Cartesian coordinates. Examples of vector algebra, the dot and cross products, the gradient operator, and vector identities are presented.

The second section extends the discussion to vectors in orthogonal[1] curvilinear coordinate systems such as spherical polar, cylindrical, and others. The vector operators gradient, divergence, curl, and Laplacian are considered and various important identities and theorems are illustrated.

The third section looks at the manipulation of matrices. Examples of matrix addition and multiplication, calculating the transpose and inverse and eigenvalues and eigenvectors, and diagonalizing and rotating matrices, are provided.

3.1   Vectors: Cartesian Coordinates

For Cartesian coordinates x, y, and z, the unit vectors $\hat{e}_x$, $\hat{e}_y$, and $\hat{e}_z$ always point along the x, y, and z axes for every point in space. A general vector $\vec{A}$ is of the form $\vec{A} = A_x \hat{e}_x + A_y \hat{e}_y + A_z \hat{e}_z$. If $\vec{A}$ is a function of the coordinates, then $\vec{A}$ is called a *vector field*.

The sum of two vectors $\vec{A}$ and $\vec{B}$ is given by
$$\vec{A} + \vec{B} = (A_x + B_x)\hat{e}_x + (A_y + B_y)\hat{e}_y + (A_z + B_z)\hat{e}_z.$$

The dot or scalar product between two vectors $\vec{A}$ and $\vec{B}$ is defined by
$$\vec{A} \cdot \vec{B} = A\,B\,\cos\theta,$$
where $A = \sqrt{A_x^2 + A_y^2 + A_z^2}$ and $B = \sqrt{B_x^2 + B_y^2 + B_z^2}$ are the magnitudes of $\vec{A}$ and $\vec{B}$, respectively.

The cross or vector product of $\vec{A}$ and $\vec{B}$, written as $\vec{A} \times \vec{B}$, is another vector whose magnitude $|\vec{A} \times \vec{B}| = A\,B\,\sin\theta$. The direction of $\vec{A} \times \vec{B}$ is given by the right-hand rule. Put the fingers of the right hand along $\vec{A}$ and curl them towards $\vec{B}$ in the direction of the smaller angle between $\vec{A}$ and $\vec{B}$. The thumb then points in the direction of the new vector.

---

[1]The angle between the unit vectors is 90°.

3.1.1 Bobby Blowfly

When you wanted to go Somewhere, And ended up going Nowhere,
The chances are strong, That you were heading for Erehwon.
An anonymous bard

So that he can better survey the goodies on the various picnic tables below him, Bobby Blowfly is spiraling upwards along a trajectory described by the Cartesian coordinates $x = 2\cos(t)$, $y = \sin(t)$, and $z = 2t/3$, where z (in meters) is measured upwards and t is the time in seconds.

(a) Forming Bobby's position vector $\vec{r}$, calculate his velocity $\vec{v}$, acceleration $\vec{a}$, and speed V at time t.

(b) Determine $\vec{r}$, $\vec{v}$, $\vec{a}$, and V at $t = 3.1$ and 17.3 s.

(c) Calculate the magnitude of his displacement for the interval 3.1 to 17.3 s.

(d) Calculate the distance he travels along the spiral path during the interval.

(e) Calculate the angle in radians and degrees between the velocity vectors at the two times.

(f) Produce a 3-dimensional plot of his trajectory over the time interval with his velocity and acceleration vectors indicated by arrows at the two times.

To solve this vector problem, the VectorCalculus library package is loaded. When combined with the plots package, several warning messages will appear on execution of the following command line which, recall, can be removed by preceding the package commands with `interface(warnlevel=0)`.

>   `restart: with(plots): with(VectorCalculus):`

Bobby's x, y, and z coordinates at time t are entered.

>   `x:=2*cos(t): y:=sin(t): z:=2*t/3:`

Bobby's position vector $\vec{r}$ is entered, using the short-hand[2] syntax `<x,y,z>`, and his velocity $\vec{v}$ and acceleration $\vec{a}$ are calculated. Note that vector symbols are not used in the Maple entries.

>   `r:=<x,y,z>; v:=diff(r,t); a:=diff(r,t,t);`

$$r := 2\cos(t)\, e_x + \sin(t)\, e_y + \frac{2t}{3}\, e_z$$

$$v := -2\sin(t)\, e_x + \cos(t)\, e_y + \frac{2}{3}\, e_z \qquad a := -2\cos(t)\, e_x - \sin(t)\, e_y$$

Since no coordinate system has been specified, the default output is expressed in terms of the Cartesian unit vectors e_x, e_y, and e_z. On the computer screen, these symbols are bold-faced. Using the `DotProduct` command in the VectorCalculus package, Bobby's speed $V = \sqrt{\vec{v} \cdot \vec{v}}$ is calculated. The "long form" of this command is used here. A short-hand syntax will be introduced shortly.

>   `V:=sqrt(DotProduct(v,v));`

---

[2] A longer form is `Vector([x,y,z])`.

$$V := \frac{1}{3}\sqrt{4 + 36\sin(t)^2 + 9\cos(t)^2}$$

The two times, $T1 = 3.1$ s and $T2 = 17.3$ s, are specified and an arrow operator F formed for evaluating an arbitrary specified function f at a time $t = T$.

```
>  T1:=3.1: T2:=17.3: F:=(f,T)->eval(f,t=T):
```

Making use of F, then $\vec{r}$, $\vec{v}$, $\vec{a}$, and V are determined at time $T1$,

```
>  r1:=F(r,T1); v1:=F(v,T1); a1:=F(a,T1); V1:=F(V,T1);
```

$$r1 := (-1.998270301)\, e_x + 0.04158066243\, e_y + 2.066666667\, e_z$$

$$v1 := (-0.08316132486)\, e_x - 0.9991351503\, e_y + \frac{2}{3}\, e_z$$

$$a1 := 1.998270301\, e_x - 0.04158066243\, e_y \qquad V1 := 1.204006353$$

and at time $T2$ (output suppressed here).

```
>  r2:=F(r,T2); v2:=F(v,T2); a2:=F(a,T2); V2:=F(V,T2);
```

Bobby's displacement $\vec{R} = \vec{r}2 - \vec{r}1$ in the time interval $T2 - T1$ is calculated (output suppressed) along with its magnitude $Rmag = \sqrt{\vec{R}\cdot\vec{R}}$. To mimic the hand notation, the short-hand dot syntax is now used to enter the dot product. Inserting spaces before and after the dot help to distinguish the dot product from a decimal point and make for easier readability.

```
>  R:=r2-r1: Rmag:=sqrt(R . R);
```

$$Rmag := 9.739961509$$

The magnitude of Bobby's displacement in the time interval is about 9.7 meters. The distance d that he travels along the path is obtained by calculating the integral $\int_{T1}^{T2} V\, dt$,

```
>  d:=int(V,t=T1..T2);
```

$$d := 23.93269942$$

and is found to be about 24 meters. This distance is considerably more than the magnitude of the displacement.

From the definition of the dot product, the angle θ between the velocities $\vec{v}1$ and $\vec{v}2$ at times $T1$ and $T2$ is obtained by calculating arccos$((\vec{v}1\cdot\vec{v}2)/(V1\,V2))$, where $V1$ and $V2$ are the speeds.

```
>  theta:=arccos((v1 . v2)/(V1*V2));
```

$$\theta := 1.469381108$$

The angle between $\vec{v}1$ and $\vec{v}2$ is about 1.47 radians or, on converting from radians to degrees, about 84°.

```
>  theta2:=convert(theta,units,radian,degree);
```

$$\theta2 := 84.18933595$$

A 3-dimensional plot of Bobby's trajectory over the time interval $T1$ to $T2$ is produced with the **spacecurve** command. To obtain a smooth curve, 500 plotting points are used. The **shading=Z** option is used to vary the color of the trajectory with increasing vertical height z.

```
>  gr||1:=spacecurve([x,y,z],t=T1..T2,numpoints=500,shading=Z):
```

A functional operator f is formed, involving the arrow command, to produce a
cylindrically shaped arrow representing the vector B with its tail at A. The color
c of the arrow must also be provided. The width of the arrow's "body" as well
as its head width and head length are specified. These values are obtained by
trial and error on viewing the final figure.

```
>  f:=(A,B,c)->arrow(A,B,shape=cylindrical_arrow,color=c,
                     width=0.1,head_width=0.3,head_length=0.5):
```

Then f is used to produce a green arrow for the velocity $\vec{v}1$ with its tail at $\vec{r}1$,
a green arrow for $\vec{v}2$ with its tail at $\vec{r}2$, a red arrow for the acceleration $\vec{a}1$
with its tail at $\vec{r}1$, and a red arrow for $\vec{a}2$ with its tail at $\vec{r}2$.

```
>  gr||2:=f(r1,v1,green): gr||3:=f(r2,v2,green):
   gr||4:=f(r1,a1,red): gr||5:=f(r2,a2,red):
```

Using the sequence command, seq, the five graphs are put into a Maple set and
superimposed in Figure 3.1 with the display command.

```
>  display({seq(gr||i,i=1..5)},axes=normal,labels=["x","y","z"]);
```

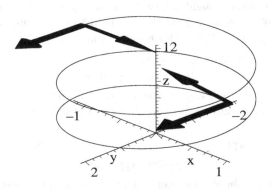

Figure 3.1: Bobby's spiral path with $\vec{v}$ and $\vec{a}$ at $T1$ and $T2$ shown.

As expected, the velocity vectors are tangent to the trajectory, while the acceleration vectors point to the center of the helix. The 3-dimensional plot may
be rotated on the computer screen by dragging with the mouse. It should be
noted that the plot has not been constrained.

3.1.2   Hiking in the Southern Chilkotin

How can you tell that "plop" on the trail is due to a grizzly bear?
It's full of hiker's "bear bells" and pepper spray.
Variation on an anonymous backpacker's saying, R.I.P.

Over the years, I have been fortunate to hike and backpack throughout western
North America, in the mountains of Peru and southern Australia, and in the
jungles of Indonesia. Despite my encounters with a timber wolf, a giant grizzly

bear and many not-so-cuddly black bears, one of my favorite backpacking areas is in the southern Chilkotin area of southwestern British Columbia. This region marks the transition zone from the wet coastal rain forest with jagged glaciated peaks to the west and the drier aspen covered slopes climbing towards smoother reddish brown volcanic tops to the east. This recipe is inspired by my treks in the southern Chilkotin.

After loading the plots and VectorCalculus packages, the height profile h of a representative trekking region is entered. Positive x is to the east, positive y to the north, with all distances in km.

```
>  restart: with(plots): with(VectorCalculus):
>  h:=(x^2+y^2)*exp(-0.5*(x^2+y^2))+1.2*exp(-(x-2.9)^2-(y-2)^2);
```

$$h := \left(x^2 + y^2\right) e^{\left(-0.5\,x^2 - 0.5\,y^2\right)} + 1.2\,e^{\left(-(x-2.9)^2 - (y-2)^2\right)}$$

To visualize the terrain, a 3-dimensional contour plot of h is produced over the horizontal range $x = -4$ to 5 and $y = -4$ to 5, with 30 contours shown. The default is 8 contours. If desired, the heights of the contours can be specified. The option filled=true fills in the surface of the terrain. The color shading is varied in the z direction. A particular angular orientation has been chosen for the plot, but the computer picture may be rotated by dragging with the mouse. The scaling has been left unconstrained to emphasize the vertical features.

```
>  contourplot3d(h,x=-4..5,y=-4..5,contours=30,filled=true,
     shading=z,axes=box,view=[-4..5,-4..5,0..1.3],
     tickmarks=[2,3,3],orientation=[-100,60]);
```

Figure 3.2 shows a volcanic crater with an adjacent hill. Although the location

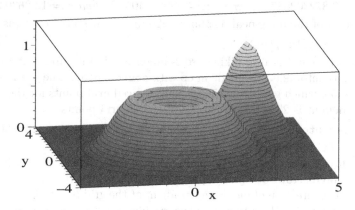

Figure 3.2: Contour plot of trekking region.

and height of the top of the hill can be approximately determined from the picture, more precise values may be found by using the gradient operator, viz.,

$$\operatorname{grad} h \equiv \nabla h = \frac{\partial h}{\partial x}\,\hat{e}_x + \frac{\partial h}{\partial y}\,\hat{e}_y.$$

At the top of the hill, the gradient is zero. The `Gradient` command is used to calculate grad h at an arbitrary point (x, y). Alternately, one can obtain the same result by using `Del` to calculate ∇h.

```
>  G:=Gradient(h,[x,y]); #alternately, G:=Del(h,[x,y]);
```

$$G := \left(2\,x\,e^{(-0.5\,x^2 - 0.5\,y^2)} - 1.0\,(x^2 + y^2)\,x\,e^{(-0.5\,x^2 - 0.5\,y^2)}\right.$$
$$+ 1.2\,(-2\,x + 5.8)\,e^{(-(x-2.9)^2 - (y-2)^2)}\Big)\,\bar{e}_x + \left(2\,y\,e^{(-0.5\,x^2 - 0.5\,y^2)}\right.$$
$$- 1.0\,(x^2 + y^2)\,y\,e^{(-0.5\,x^2 - 0.5\,y^2)} + 1.2\,(-2\,y + 4)\,e^{(-(x-2.9)^2 - (y-2)^2)}\Big)\,\bar{e}_y$$

Notice that overbars appear above the unit (basis) vectors in the output of the gradient operation. This indicates that G is a vector field defined at all points (x, y). It is Maple's way of reminding you that for general curvilinear coordinate systems, such as the spherical polar system, the directions of the unit vectors will vary from point to point in space. Only Cartesian unit vectors are independent of position.

Guided by the figure, a numerical search is made over the range $x = 2$ to 4 and $y = 1$ to 4, using the floating point solve command, to find the zeros of the x and y components (`G[1]` and `G[2]`) of G. The solution *sol* is assigned,

```
>  sol:=fsolve({G[1],G[2]},{x,y},{x=2..4,y=1..4}); assign(sol):
```

$$sol := \{y = 1.980898405,\ x = 2.872302687\}$$

and the x and y coordinates of the peak of the hill are given by *xmax* and *ymax*. The height of the hill, *hmax*, is obtained by evaluating h at *xmax*, *ymax*.

```
>  xmax:=x; ymax:=y; hmax:=eval(h,{x=xmax,y=ymax});
```

$$xmax := 2.872302687 \qquad ymax := 1.980898405 \qquad hmax := 1.226303365$$

The elevation of the hill's peak is about 1226 m. x and y are now unassigned.

```
>  unassign('x','y'):
```

Topographical maps used in hiking are 2-dimensional in nature. A 2-d contour map with 25 contours is now made in `cp` with `contourplot`. The `grid` option is set to 50×50, which generates 2500 equally spaced grid points for the contour plot. The default is 25×25, which produces 625 grid points.

```
>  cp:=contourplot(h,x=-4..5,y=-4..5,contours=25,grid=[50,50],
      filled=true):
```

The `fieldplot` command is used in `fp` to plot the gradient G, placing thick magenta colored arrows at equally spaced grid points (here 15×15), each arrow pointing in the direction of increasing gradient at the grid point. The size of the arrow is a measure of the strength of the gradient, larger arrows corresponding to steeper gradients.

```
>  fp:=fieldplot(G,x=-4..5,y=-4..5,arrows=THICK,grid=[15,15],
      color=magenta):
```

The `textplot` command is used in `tp` to place the word "Top", colored blue, at the location *xmax* $- 0.3$, *ymax* $+ 0.1$, adjacent to the top of the hill.

```
>  tp:=textplot([[xmax-0.3,ymax+0.1,"Top"]],color=blue):
```

After stopping for lunch at the location $(xmax - 2, ymax - 1.6)$ near the lip of the crater, it is desired to hike to the top of the hill. The `pointplot` command is used in `pp` to place size 16 green circles on the contour map at these locations.

```
> pp:=pointplot([[xmax,ymax],[xmax-2,ymax-1.6]],
        symbol=circle,symbolsize=16,color=green):
```

The four graphs, `fp`, `cp`, `tp`, and `pp`, are superimposed with the `display` command, the resulting plot being shown in Figure 3.3.

```
> display([fp,cp,tp,pp],tickmarks=[3,5]);
```

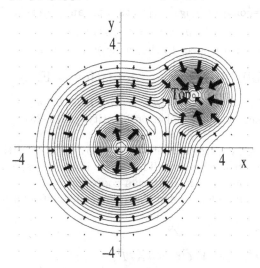

Figure 3.3: Two-dimensional contour plot with gradient arrows.

As expected, the gradient arrows are perpendicular to the contour lines. The gradient at the starting point $xmax - 2$, $ymax - 1.6$ is determined.

```
> G2:=eval(G,{x=xmax-2,y=ymax-1.6});
```

$$G2 := 0.6124645476\,\bar{e}_x + 0.2695348694\,\bar{e}_y$$

The slope at this point is obtained by calculating $\sqrt{\vec{G}2 \cdot \vec{G}2}$.

```
> Slope:=sqrt(G2 . G2);
```

$$Slope := 0.6691501086$$

The slope is about 0.67. It is positive, in agreement with the figure, indicating that if we follow the gradient we will be initially climbing upwards from just inside the lip of the crater. The slope can be expressed in radians by taking the arctangent of the slope, which then can be converted to degrees.

```
> angle:=arctan(Slope);
```

$$angle := 0.5897199394$$

```
> angle:=convert(angle,units,radian,degree);
```

$$angle := 33.78846362$$

Initially we would be climbing at an angle of about 0.6 radians or 34° with the horizontal. The initial direction of travel can be qualitatively deduced from Figure 3.3. Quantitatively, the angle in radians measured with respect to the x-axis (east) can be determined by calculating the arctangent of the ratio of the y component of $G2$ to the x component.

```
>  angle2:=evalf(arctan(G2[2]/G2[1]));
```

$$angle2 := 0.4145759092$$

The angle is about 0.41 radians or, on converting to degrees,

```
>  angle2:=convert(angle2,units,radian,degree);
```

$$angle2 := 23.75344988$$

about 24 degrees with respect to the easterly direction.

3.1.3   Establishing These Identities is Easy

Trying to define yourself is like trying to bite your own teeth.
Alan Watts, American philosopher on identity, (1915–73)

Vector identities play an important role in many areas of mathematical physics, particularly in electromagnetism. This recipe illustrates some examples involving dot and cross products in Cartesian coordinates.

Prove the following identities in Cartesian coordinates for arbitrary $\vec{A}$, $\vec{B}$, $\vec{C}$:

(a) $\vec{A} \cdot (\vec{B} \times \vec{C}) = (\vec{A} \times \vec{B}) \cdot \vec{C}$

(b) $\vec{A} \times (\vec{B} \times \vec{C}) = \vec{B}(\vec{A} \cdot \vec{C}) - \vec{C}(\vec{A} \cdot \vec{B})$

(c) $[\vec{A} \times (\vec{B} \times \vec{C})] + [\vec{B} \times (\vec{C} \times \vec{A})] + [\vec{C} \times (\vec{A} \times \vec{B})] = 0$

(d) $(\vec{B} \times \vec{C}) \times (\vec{C} \times \vec{A}) = \vec{C}(\vec{A} \cdot (\vec{B} \times \vec{C}))$

The VectorCalculus package is loaded and arbitrary vectors $\vec{A}$, $\vec{B}$, $\vec{C}$ entered.

```
>  restart: with(VectorCalculus):
>  A:=<Ax,Ay,Az>; B:=<Bx,By,Bz>; C:=<Cx,Cy,Cz>;
```

$$A := Ax\,e_x + Ay\,e_y + Az\,e_z$$
$$B := Bx\,e_x + By\,e_y + Bz\,e_z$$
$$C := Cx\,e_x + Cy\,e_y + Cz\,e_z$$

Using the long forms of the dot and cross product commands, the difference between the left-hand and right-hand sides in (a) is calculated in *id1*.

```
>  id1:=DotProduct(A,CrossProduct(B,C))
        -DotProduct(CrossProduct(A,B),C);
```

$$id1 := Ax\,(By\,Cz - Bz\,Cy) + Ay\,(Bz\,Cx - Bx\,Cz) + Az\,(Bx\,Cy - By\,Cx)$$
$$- (Ay\,Bz - Az\,By)\,Cx - (Az\,Bx - Ax\,Bz)\,Cy - (Ax\,By - Ay\,Bx)\,Cz$$

Applying **simplify** to *id1* yields zero, thus confirming the first identity.

```
>  id1:=simplify(id1);
```

$$id1 := 0$$

For the remainder of the identities, the short-hand syntax for the dot and crossproducts will be used. We have already used the dot notation for the dot product. The short form for the cross product is &x. To prove the second identity[3] in (b), the left-hand side of the identity is now entered. Again, for clarity, I have left spaces around the cross product symbols.

```
> LHS:= A &x (B &x C);
```

$$
\begin{aligned}
LHS := & \left(Ay \left(Bx\ Cy - By\ Cx \right) - Az \left(Bz\ Cx - Bx\ Cz \right) \right) e_x \\
& + \left(Az \left(By\ Cz - Bz\ Cy \right) - Ax \left(Bx\ Cy - By\ Cx \right) \right) e_y \\
& + \left(Ax \left(Bz\ Cx - Bx\ Cz \right) - Ay \left(By\ Cz - Bz\ Cy \right) \right) e_z
\end{aligned}
$$

Similarly, the right-hand side is entered.

```
> RHS:=B*(A . C) - C*(A . B);
```

$$
\begin{aligned}
RHS := & \left(\left(Ax\ Cx + Ay\ Cy + Az\ Cz \right) Bx - \left(Ax\ Bx + Ay\ By + Az\ Bz \right) Cx \right) e_x \\
& + \left(\left(Ax\ Cx + Ay\ Cy + Az\ Cz \right) By - \left(Ax\ Bx + Ay\ By + Az\ Bz \right) Cy \right) e_y \\
& + \left(\left(Ax\ Cx + Ay\ Cy + Az\ Cz \right) Bz - \left(Ax\ Bx + Ay\ By + Az\ Bz \right) Cz \right) e_z
\end{aligned}
$$

On simplifying $LHS - RHS$, a zero vector results in $id2$, thus confirming (b).

```
> id2:=simplify(LHS - RHS);
```

$$id2 := 0\,e_x$$

The remaining two identities in (c) and (d) are proven in a similar manner as shown in $id3$ and $id4$, respectively. For the latter, the difference between the left-hand and right-hand sides is calculated.

```
> id3:=(A &x (B &x C)) + (B &x (C &x A)) + (C &x (A &x B));
> id3:=simplify(id3);
```

$$id3 := 0\,e_x$$

```
> id4:=(B &x C) &x (C &x A) - C*(A . (B &x C));
> id4:=simplify(id4);
```

$$id4 := 0\,e_x$$

3.1.4 This Task is Not a Chore

The hardest task of a girl's life, nowadays, is to prove
to a man that his intentions are serious.
Helen Rowland, American journalist, *A Guide to Men,Intermezzo (1922)*

In classical mechanics, a standard task is to express the velocity and acceleration of a particle in some other curvilinear coordinate system, in particular in spherical polar and cylindrical coordinates. Curvilinear coordinate systems will be dealt with at length in the next section, but let me show you how easy

---

[3]Often referred to as the BAC–CAB rule, because of the structure of the right-hand side,

it is to obtain the velocity and acceleration of a particle in spherical polar co-ordinates. The relation between the Cartesian coordinates (x, y, z) and the spherical coordinates (r, θ, ϕ) is given by $x = r \sin\theta \cos\phi$, $y = r \sin\theta \sin\phi$, and $z = r \cos\theta$, where r is the radial distance from the origin, θ is the angle between the radius vector and the z-axis, and ϕ is the angle that the projection of the radius vector into the x-y plane makes with the x-axis. The ranges of these coordinates are $0 \le r < \infty$, $0 \le \theta \le \pi$, and $0 \le \phi \le 2\pi$.

After loading the VectorCalculus package,

```
>   restart: with(VectorCalculus):
```

the relations between the two coordinates systems are entered, the spherical variables taken to be time-dependent so that time derivatives can be taken.

```
>   X:=r(t)*sin(theta(t))*cos(phi(t)):

    Y:=r(t)*sin(theta(t))*sin(phi(t)): Z:=r(t)*cos(theta(t)):
```

The position vector $\vec{R} = X\,\hat{e}_x + Y\,\hat{e}_y + Z\,\hat{e}_z$ of a particle is entered as a vector field in Cartesian coordinates, the above relations being automatically substituted.

```
>   R:=VectorField(<X,Y,Z>,'cartesian'[x,y,z]);
```

$$R := r(t)\sin(\theta(t))\cos(\phi(t))\,\overline{e}_x + r(t)\sin(\theta(t))\sin(\phi(t))\,\overline{e}_y + r(t)\cos(\theta(t))\,\overline{e}_z$$

The velocity $\vec{v}$ and acceleration $\vec{a}$ of the particle are calculated by differentiating $\vec{R}$ once and twice, respectively, with respect to t. A line-ending colon has been placed on the acceleration to suppress the very lengthy output. This will be done for all intermediate steps involving the acceleration.

```
>   v:=diff(R,t); a:=diff(R,t,t):
```

$$v := ((\frac{d}{dt}\,r(t))\sin(\theta(t))\cos(\phi(t)) + r(t)\cos(\theta(t))\,(\frac{d}{dt}\,\theta(t))\cos(\phi(t))$$
$$- r(t)\sin(\theta(t))\sin(\phi(t))\,(\frac{d}{dt}\,\phi(t)))\,\overline{e}_x$$
$$+((\frac{d}{dt}\,r(t))\sin(\theta(t))\sin(\phi(t)) + r(t)\cos(\theta(t))\,(\frac{d}{dt}\,\theta(t))\sin(\phi(t))$$
$$+ r(t)\sin(\theta(t))\cos(\phi(t))\,(\frac{d}{dt}\,\phi(t)))\,\overline{e}_y$$
$$+((\frac{d}{dt}\,r(t))\cos(\theta(t)) - r(t)\sin(\theta(t))\,(\frac{d}{dt}\,\theta(t)))\,\overline{e}_z$$

The MapToBasis command can be used to express an arbitrary Cartesian unit vector $\hat{u}$ in terms of the spherical (polar) unit vectors $\hat{e}_r$, $\hat{e}_\theta$, $\hat{e}_\phi$. A functional operator F is formed to do this.

```
>   F:=u->MapToBasis(u,'spherical'[r,theta,phi]):
```

The operator F is applied to the velocity and acceleration. The symbols %1 and %2 which appear in $v2$ indicate sub-expressions. This form of the output is an artifact of exporting Maple into the text as Latex output. Latex is the standard word processing language used in preparing scientific documents, such as this text, which involve mathematical expressions.

```
>   v2:=F(v); a2:=F(a):
```

$$v2 := (\%2\sin(\theta)\cos(\phi) + \%1\sin(\phi)\sin(\theta)$$
$$+ ((\frac{d}{dt}r(t))\cos(\theta(t)) - r(t)\sin(\theta(t))(\frac{d}{dt}\theta(t)))\cos(\theta))\,\overline{e}_r$$
$$+ (\%2\cos(\phi)\cos(\theta) + \%1\sin(\phi)\cos(\theta)$$
$$- ((\frac{d}{dt}r(t))\cos(\theta(t)) - r(t)\sin(\theta(t))(\frac{d}{dt}\theta(t)))\sin(\theta))\,\overline{e}_\theta$$
$$+ (-\%2\sin(\phi) + \%1\cos(\phi))\,\overline{e}_\phi$$

$$\%1 := (\frac{d}{dt}r(t))\sin(\theta(t))\sin(\phi(t)) + r(t)\cos(\theta(t))(\frac{d}{dt}\theta(t))\sin(\phi(t))$$
$$+ r(t)\sin(\theta(t))\cos(\phi(t))(\frac{d}{dt}\phi(t))$$

$$\%2 := (\frac{d}{dt}r(t))\sin(\theta(t))\cos(\phi(t)) + r(t)\cos(\theta(t))(\frac{d}{dt}\theta(t))\cos(\phi(t))$$
$$- r(t)\sin(\theta(t))\sin(\phi(t))(\frac{d}{dt}\phi(t))$$

The `MapToBasis` command has generated trigonometric terms in *v2* (and *a2*) which are not time-dependent. To simplify the velocity and acceleration, an arrow operator G is created to substitute the requisite time-dependence.

```
> G:=u->subs({sin(theta)=sin(theta(t)),cos(theta)=
  cos(theta(t)),cos(phi)=cos(phi(t)),sin(phi)=sin(phi(t))},u):
```

Then G is applied to *v2* and *a2* and the results simplified with the trig option.

```
> v3:=simplify(G(v2),trig); a3:=simplify(G(a2),trig):
```

$$v3 := (\frac{d}{dt}r(t))\,\overline{e}_r + r(t)(\frac{d}{dt}\theta(t))\,\overline{e}_\theta + r(t)\sin(\theta(t))(\frac{d}{dt}\phi(t))\,\overline{e}_\phi$$

The velocity expression given in *v3* is the standard result [FC99] in spherical polar coordinates. The standard result for the acceleration given in *a4* follows on making the algebraic substitution $\cos^2\theta(t) = 1 - \sin^2\theta(t)$.

```
> a4:=algsubs(cos(theta(t))^2=1-sin(theta(t))^2,a3);
```

$$a4 := ((\frac{d^2}{dt^2}r(t)) - r(t)(\frac{d}{dt}\phi(t))^2\sin(\theta(t))^2 - r(t)(\frac{d}{dt}\theta(t))^2)\,\overline{e}_r$$
$$+ (r(t)(\frac{d^2}{dt^2}\theta(t)) + 2(\frac{d}{dt}r(t))(\frac{d}{dt}\theta(t)) - r(t)\sin(\theta(t))\cos(\theta(t))(\frac{d}{dt}\phi(t))^2)\,\overline{e}_\theta$$
$$+ (r(t)\sin(\theta(t))(\frac{d^2}{dt^2}\phi(t)) + 2(\frac{d}{dt}r(t))\sin(\theta(t))(\frac{d}{dt}\phi(t))$$
$$+ 2\,r(t)\cos(\theta(t))(\frac{d}{dt}\theta(t))(\frac{d}{dt}\phi(t)))\,\overline{e}_\phi$$

The above results may also be obtained by making use of the LinearAlgebra package, as illustrated in Supplementary Recipe **03-S04**. The velocity and acceleration are just as easily derived in cylindrical coordinates.

3.2　Vectors: Curvilinear Coordinates

Consider a general 3-dimensional orthogonal coordinate system with coordinates $u = u(x, y, z)$, $v = v(x, y, z)$, $w = w(x, y, z)$. For example, for spherical (polar) coordinates, u is the radial distance r, v is the polar angle θ, and w is the azimuthal angle ϕ. The differential element of length in, say, the u direction is given by $ds_u = h_u\, du$, where the *scale factor* h_u is given by

$$h_u = \sqrt{\left(\frac{\partial x}{\partial u}\right)^2 + \left(\frac{\partial y}{\partial u}\right)^2 + \left(\frac{\partial z}{\partial u}\right)^2}$$

with similar expressions for h_v and h_w.

The unit vectors $\hat{e}_u$, $\hat{e}_v$, $\hat{e}_w$ are related to the Cartesian unit vectors by

$$\hat{e}_u = \frac{\partial \vec{r}}{\partial s_u} = \frac{1}{h_u}\frac{\partial \vec{r}}{\partial u}, \quad \hat{e}_v = \frac{1}{h_v}\frac{\partial \vec{r}}{\partial v}, \quad \hat{e}_w = \frac{1}{h_w}\frac{\partial \vec{r}}{\partial w},$$

where $\vec{r} = x\,\hat{e}_x + y\,\hat{e}_y + z\,\hat{e}_z$ is the position vector with $x = x(u, v, w)$, etc.

The element of area on, say, a surface of constant u is $dA_u = ds_v\, ds_w = h_u h_v\, du\, dv$ while the volume element is $dV = ds_u ds_v ds_w = h_u h_v h_w du\, dv\, dw$.

The gradient (∇), divergence ($\nabla\cdot$), curl ($\nabla\times$), and Laplacian ($\nabla\cdot\nabla \equiv \nabla^2$) operators are calculated as follows [Gri99]:

$$\nabla f = \frac{\hat{e}_u}{h_u}\frac{\partial f}{\partial u} + \frac{\hat{e}_v}{h_v}\frac{\partial f}{\partial v} + \frac{\hat{e}_w}{h_w}\frac{\partial f}{\partial w},$$

$$\nabla\cdot\vec{A} = [\frac{\partial}{\partial u}(A_u\, h_v\, h_w) + \frac{\partial}{\partial v}(A_v\, h_u\, h_w) + \frac{\partial}{\partial w}(A_w\, h_u\, h_v)]/(h_u\, h_v\, h_w),$$

$$\nabla\times\vec{A} = [h_u\hat{e}_u[\frac{\partial}{\partial v}(h_w A_w) - \frac{\partial}{\partial w}(h_v A_v)] + h_v\hat{e}_v[\frac{\partial}{\partial w}(h_u A_u) - \frac{\partial}{\partial u}(h_w A_w)]$$

$$+ h_w\hat{e}_w[\frac{\partial}{\partial u}(h_v A_v) - \frac{\partial}{\partial v}(h_u A_u)]]/(h_u\, h_v\, h_w),$$

$$\nabla^2 f = [\frac{\partial}{\partial u}(\frac{h_v h_w}{h_u}\frac{\partial f}{\partial u}) + \frac{\partial}{\partial v}(\frac{h_u h_w}{h_v}\frac{\partial f}{\partial v}) + \frac{\partial}{\partial w}(\frac{h_u h_v}{h_w}\frac{\partial f}{\partial w})]/(h_u\, h_v\, h_w).$$

3.2.1　From Scale Factors to Vector Operators

Everything that is beautiful and noble is the product
of reason and calculation.
Charles Baudelaire, French poet, (1821–67)

In this recipe, functional operators are created to calculate the scale factors and the vector operators for any 3-dimensional orthogonal coordinate system. As an application, the volume of a sphere, and the gradient, divergence, curl, and Laplacian are calculated for spherical polar coordinates (r, θ, ϕ).

To make the entries easier, the aliases t and p are used for theta and phi.

```
>  restart: alias(theta=t,phi=p):
```

The arrow operator H is introduced to calculate the scale factor and simplify it for a given input coordinate u.

```
>   H:=u->simplify(sqrt(diff(x,u)^2+diff(y,u)^2+diff(z,u)^2),
          symbolic):
```

The relations between the Cartesian coordinates (x, y, z) and (r, θ, ϕ) are entered. Then the spherical polar scale factors h_r, h_θ, and h_ϕ are calculated.

```
>   x:=r*cos(p)*sin(t); y:=r*sin(p)*sin(t); z:=r*cos(t);
```

$$x := r\cos(\phi)\sin(\theta) \qquad y := r\sin(\phi)\sin(\theta) \qquad z := r\cos(\theta)$$

```
>   h[r]:=H(r); h[t]:=H(t); h[p]:=H(p);
```

$$h_r := 1 \qquad h_\theta := r \qquad h_\phi := r\sin(\theta)$$

With all the scale factors now determined, the volume of a sphere of radius R is given by $V = \int_0^{2\pi} \int_0^{\pi} \int_0^{R} h_r\, h_\theta\, h_\phi\, dr\, d\theta\, d\phi$. This integral is now entered,

```
>   V:=Int(Int(Int(h[r]*h[t]*h[p],r=0..R),t=0..Pi),p=0..2*Pi);
```

$$V := \int_0^{2\pi} \int_0^{\pi} \int_0^{R} r^2 \sin(\theta)\, dr\, d\theta\, d\phi$$

and evaluated with the value command.

```
>   Volume:=value(V);
```

$$Volume := \frac{4\,R^3\,\pi}{3}$$

The well-known formula for the volume of a sphere results.

A functional operator g is formed for calculating the gradient of a general function $f(u, v, w)$ for arbitrary input coordinates u, v, and w.

```
>   g:=(u,v,w)->e[u]*diff(f(u,v,w),u)/h[u]
        +e[v]*diff(f(u,v,w),v)/h[v]+e[w]*diff(f(u,v,w),w)/h[w];
```

$$g:=(u,v,w) \rightarrow \frac{e_u \operatorname{diff}(f(u,v,w),u)}{h_u} + \frac{e_v \operatorname{diff}(f(u,v,w),v)}{h_v} + \frac{e_w \operatorname{diff}(f(u,v,w),w)}{h_w}$$

Similarly, arrow operators d and c are created for calculating the divergence and curl of a general vector field $\vec{A} = A_u(u, v, w)\,\hat{e}_u + A_v(u, v, w)\,\hat{e}_v + A_w(u, v, w)\,\hat{e}_w$ for arbitrary coordinates u, v, and w.

```
>   d:=(u,v,w)->(diff(A[u](u,v,w)*h[v]*h[w],u)
        +diff(A[v](u,v,w)*h[u]*h[w],v)
        +diff(A[w](u,v,w)*h[u]*h[v],w))/(h[u]*h[v]*h[w]):
```

```
>   c:=(u,v,w)-> (1/(h[u]*h[v]*h[w]))*
   (h[u]*e[u]*(diff(h[w]*A[w](u,v,w),v)-diff(h[v]*A[v](u,v,w),w))
   +h[v]*e[v]*(diff(h[u]*A[u](u,v,w),w)-diff(h[w]*A[w](u,v,w),u))
   +h[w]*e[w]*(diff(h[v]*A[v](u,v,w),u)-diff(h[u]*A[u](u,v,w),v))):
```

Finally, here's an operator L for calculating the Laplacian of $f(u, v, w)$.

```
>   L:=(u,v,w)->(diff(h[v]*h[w]*diff(f(u,v,w),u)/h[u],u)
        +diff(h[u]*h[w]*diff(f(u,v,w),v)/h[v],v)
        +diff(h[u]*h[v]*diff(f(u,v,w),w)/h[w],w))/(h[u]*h[v]*h[w]):
```

The forms of the gradient and Laplacian are calculated for spherical polar coordinates, the divergence and curl being left for you to do as an exercise.

```
>  gradf:=g(r,t,p); Lapf:=expand(L(r,t,p));
```

$$gradf := e_r\,(\frac{\partial}{\partial r}\,f(r,\theta,\phi)) + \frac{e_\theta\,(\frac{\partial}{\partial \theta}\,f(r,\theta,\phi))}{r} + \frac{e_\phi\,(\frac{\partial}{\partial \phi}\,f(r,\theta,\phi))}{r\sin(\theta)}$$

$$Lapf := \frac{2\,(\frac{\partial}{\partial r}\,f(r,\theta,\phi))}{r} + (\frac{\partial^2}{\partial r^2}\,f(r,\theta,\phi)) + \frac{\cos(\theta)\,(\frac{\partial}{\partial \theta}\,f(r,\theta,\phi))}{r^2\sin(\theta)}$$

$$+ \frac{\frac{\partial^2}{\partial \theta^2}\,f(r,\theta,\phi)}{r^2} + \frac{\frac{\partial^2}{\partial \phi^2}\,f(r,\theta,\phi)}{r^2\sin(\theta)^2}$$

The above expressions agree with those found in standard mathematics texts.

3.2.2   Vector Operators the Easy Way

Civilization is not by any means an easy thing to attain to. There are only two ways by which man can reach it. One is by being cultured, the other by being corrupt.
Oscar Wilde, Anglo-Irish author, *The Picture of Dorian Gray, 1891*

I hope that I am not corrupting you by now demonstrating an easy way of obtaining the vector operators, even for coordinate systems which initially may be unfamiliar to you. It's important that you know how the vector operators are calculated from first principles, so I trust that you did not skip the last recipe. But if you want to solve PDE boundary and initial value problems, the main subject of the **Entrees**, it is desirable, e.g., to obtain the Laplacian operator as quickly as possible.

In this recipe, I will show you how to gain information about any two- or three-dimensional orthogonal coordinate system and illustrate how the gradient, divergence, curl, and Laplacian are easily obtained for paraboloidal coordinates.

To calculate the fore-mentioned vector operators, the VectorCalculus package must first be loaded.

```
>  restart: with(plots): with(VectorCalculus):
```

Executing the following command line will open a help page about the two- and three-dimensional coordinate systems known to Maple.

```
>  ?coords;
```

The majority of coordinate systems that you are likely to encounter in mathematical physics are listed on this help page. The relationship between the given curvilinear coordinate system and Cartesian coordinates is stated. For example, for the paraboloidal coordinates (u,v,w), one has

$$x = u\,v\,\cos(w), \quad y = u\,v\,\sin(w), \quad z = (u^2 - v^2)/2$$

with $0 \le u < \infty$, $0 \le v < \infty$, and $0 \le w \le 2\,\pi$.

A common way of visualizing a coordinate system is to plot the surface corresponding to holding each of the coordinates constant. For example, for spherical polar coordinates, holding the radial coordinate r fixed generates a spherical surface. Although we could create a recipe to do this task for us for paraboloidal coordinates, an easier way is to use the command `coordplot3d` which produces a representative surface corresponding to holding u, v, and w fixed. Executing the following command line results in Figure 3.4.

```
> coordplot3d(paraboloidal,orientation=[-20,40],axes=frame,
  tickmarks=[2,2,2],labels=["x","y","z"]);
```

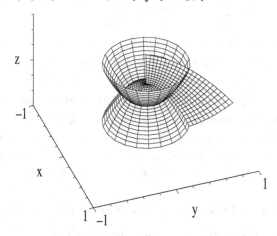

Figure 3.4: Surfaces for fixed u, v, and w in paraboloidal coordinates.

The geometric structure of the three surfaces is easily understood. Let's form $x^2 + y^2 = u^2 v^2 (\cos^2 w + \sin^2 w) = u^2 v^2$, Then, holding, e.g., u fixed, we have $z = (u^2 - v^2)/2 = (u^2 - (x^2 + y^2)/u^2)/2$ which, for fixed u, is the equation of a paraboloid (parabola of revolution about the z-axis). Holding u and v constant produces the two intersecting paraboloids shown in the picture. As you may confirm, the half-plane corresponds to holding w fixed. For a thorough discussion of curvilinear coordinate systems, Morse and Feshbach's *Methods of Theoretical Physics* ([MF53]) is highly recommended.

Entering the assumption that $u > 0$ and $v > 0$ to simplify the results,

```
> assume(u>0,v>0):
```

the coordinates u, v, and w are now set to be paraboloidal.

```
> SetCoordinates('paraboloidal'[u,v,w]):
```

Using the `Gradient` operator, $\nabla f(u, v, w)$ is calculated and simplified.

```
> gradf:=simplify(Gradient(f(u,v,w)));
```

$$gradf := \frac{\frac{\partial}{\partial u} f(u,\, v,\, w)}{\sqrt{u^2 + v^2}}\, \overline{e}_u + \frac{\frac{\partial}{\partial v} f(u,\, v,\, w)}{\sqrt{u^2 + v^2}}\, \overline{e}_v + \frac{\frac{\partial}{\partial w} f(u,\, v,\, w)}{u\, v}\, \overline{e}_w$$

From the structure of the output, one can easily deduce that the scale factors for paraboloidal coordinates are $h_u = h_v = \sqrt{u^2 + v^2}$ and $h_w = u\,v$. Next, $\nabla^2 f(u, v, w)$ is calculated, simplified, and expanded.

```
> Lapf:=expand(simplify(Laplacian(f(u,v,w))));
```

$$Lapf := \frac{\frac{\partial}{\partial u} f(u,\, v,\, w)}{(u^2 + v^2)\, u} + \frac{\frac{\partial^2}{\partial u^2} f(u,\, v,\, w)}{u^2 + v^2} + \frac{\frac{\partial}{\partial v} f(u,\, v,\, w)}{(u^2 + v^2)\, v} + \frac{\frac{\partial^2}{\partial v^2} f(u,\, v,\, w)}{u^2 + v^2}$$

$$+ \frac{\frac{\partial^2}{\partial w^2} f(u,\, v,\, w)}{(u^2 + v^2)\, v^2} + \frac{\frac{\partial^2}{\partial w^2} f(u,\, v,\, w)}{(u^2 + v^2)\, u^2}$$

Entering a general vector field $\vec{A}$ in terms of the coordinates u, v, w,

```
> A:=VectorField(<Au(u,v,w),Av(u,v,w),Aw(u,v,w)>);
```

$$A := \mathrm{Au}(u,\, v,\, w)\,\bar{e}_u + \mathrm{Av}(u,\, v,\, w)\,\bar{e}_v + \mathrm{Aw}(u,\, v,\, w)\,\bar{e}_w$$

$\mathrm{div}\vec{A}$ and $\mathrm{curl}\vec{A}$ are calculated using the Divergence and Curl operators.

```
> divA:=expand(simplify(Divergence(A)));
```

```
> curlA:=expand(simplify(Curl(A)));
```

The lengthy outputs have been suppressed here, so you will have to execute the above command line on the computer to see the results.

3.2.3 These Operators Do Not Have an Identity Crisis

It is always the same: once you are liberated,
you are forced to ask who you are.
Jean Baudrillard, French semiologist, *America, "Astral America" (1986)*

Now that you have learned how to generate vector operators, you should never experience a mathematical crisis in proving a vector operator identity in any orthogonal coordinate system. This recipe provides some examples.

Prove the following identities in the indicated coordinate systems:

(a) $\nabla \cdot (\nabla \times \vec{A}) = 0$, Spherical polar $(\vec{A} = \vec{A}(r, \theta, \phi))$

(b) $\nabla \times (\nabla f) = 0$, Cylindrical $(f = f(\rho, \phi, z))$

(c) $\nabla \times (\nabla \times \vec{A}) = \nabla(\nabla \cdot \vec{A}) - \nabla^2 \vec{A}$, Paraboloidal $(\vec{A} = \vec{A}(u, v, w))$

All three identities involve non-Cartesian coordinate systems, so the coordinates must be specified in each case. For (a), the spherical polar coordinates r, θ, ϕ are required, which are entered and assigned the name sp. The coordinates are then set to be spherical.

```
> restart: with(VectorCalculus):
```

```
> sp:=r,theta,phi: SetCoordinates('spherical'[sp]):
```

A general vector field $\vec{A}$ is entered in spherical coordinates with components labeled Ar, At, and Ap.

```
> A:=VectorField(<Ar(sp),At(sp),Ap(sp)>);
```

$$A := \text{Ar}(r,\ \theta,\ \phi)\,\bar{e}_r + \text{At}(r,\ \theta,\ \phi)\,\bar{e}_\theta + \text{Ap}(r,\ \theta,\ \phi)\,\bar{e}_\phi$$

The left-hand side of **(a)** is just `div(curl`$\vec{A}$`)`. Entering `Divergence(Curl(A))` yields zero, confirming the first vector identity.

```
> id1:=Divergence(Curl(A));
```

$$id1 := 0$$

For **(b)**, the coordinates ρ, ϕ, z are set to be cylindrical.

```
> SetCoordinates('cylindrical'[rho,phi,z]):
```

The left-hand side of the identity in **(b)** is just `curl grad`f, so one could enter `Curl(Gradient(f(rho,phi,z))`. An alternate way is to note that the `Del` operator can be used for calculating ∇. Then using the short-hand syntax for the cross product, we can enter `Del &x Del(f(rho,phi,z))`. The result of executing the command line is zero, confirming the second identity.

```
> id2:=Del&x Del(f(rho,phi,z));
```

$$id2 := 0\,\bar{e}_\rho$$

The identity in **(c)** involves paraboloidal coordinates u, v, w which are set

```
> SetCoordinates('paraboloidal'[u,v,w]):
```

and a general vector field $\vec{A2}$ entered.

```
> A2:=VectorField(<Au(u,v,w),Av(u,v,w),Aw(u,v,w)>);
```

$$A2 := \text{Au}(u,\ v,\ w)\,\bar{e}_u + \text{Av}(u,\ v,\ w)\,\bar{e}_v + \text{Aw}(u,\ v,\ w)\,\bar{e}_w$$

Using `Del` and the short-hand syntax for both the dot and cross products, the right-hand side is subtracted from the left-hand side in **(c)**.

```
> id3:=Del &x (Del &x A2)-Del(Del . A2)+(Del . Del)(A2);
```

$$id3 := 0\,\bar{e}_u$$

The result is zero, confirming the third identity.

3.2.4   Is This Vector Field Conservative?

The most radical revolutionary will become a conservative the day after the revolution.
Hannah Arendt, American political philosopher, (1906–1975)

Conservative vector fields play an important role in physics. A *conservative* field $\vec{A}$ is characterized by having a closed *line integral* $\oint \vec{A} \cdot d\vec{\ell} = 0$ around an arbitrary path, or from Stokes's theorem, $\nabla \times \vec{A} = 0$ everywhere. See Recipe **03-S10**. From the identity $\nabla \times \nabla \phi = 0$ for any function ϕ, one can write $\vec{A} = -\nabla \phi$, where ϕ is the *potential* and the minus sign is inserted by convention. Given a conservative $\vec{A}(\vec{r})$, $\phi(\vec{r})$ may be obtained to within an arbitrary constant by performing the line integral from an arbitrary reference point to $\vec{r}$.

Consider the following vector field,

$$\vec{A} = (2\,x\,\sin y + 2\,y^2 - 12\,x^3 y^3 + 2\,x\,y\,z - 3\,z)\,\hat{e}_x$$
$$+ ((1+x^2)(z + \cos y) - 3\,y^2 + 4\,x\,y - 9\,x^4 y^2)\,\hat{e}_y + ((1+x^2)\,y - 3\,x)\,\hat{e}_z.$$

Show that $\vec{A}$ is conservative and determine the corresponding potential $\phi(x, y, z)$.
Determine the potential $\phi1$ which passes through $x = 0.13$, $y = 1.82$, $z = -1.24$.
Rounding $\phi1$ to the nearest integer Φ, plot the equipotentials corresponding to
$\pm\Phi$ and superimpose the graph on a plot of the vector field.

Here is the solution provided by Ms. I. M. Curious. She loads the plots and
VectorCalculus packages and sets the coordinates to be Cartesian.

```
> restart: with(plots): with(VectorCalculus):
```

```
> SetCoordinates('cartesian'[x,y,z]):
```

The given vector field $\vec{A}$ is entered.

```
> A:=VectorField(<2*x*sin(y)+2*y^2-12*x^3*y^3+2*x*y*z-3*z,
     (1+x^2)*(z+cos(y))-3*y^2+4*x*y-9*x^4*y^2,(1+x^2)*y-3*x>);
```

$$A := (2\,x \sin(y) + 2\,y^2 - 12\,x^3\,y^3 + 2\,x\,y\,z - 3\,z)\,\bar{e}_x +$$
$$((x^2 + 1)\,(z + \cos(y)) - 3\,y^2 + 4\,x\,y - 9\,x^4\,y^2)\,\bar{e}_y + ((x^2 + 1)\,y - 3\,x)\,\bar{e}_z$$

Taking the curl of $\vec{A}$ and simplifying, yields a zero vector in C, confirming that
$\vec{A}$ is a conservative field.

```
> C:=simplify(Curl(A));
```

$$C := 0\,\bar{e}_x$$

Choosing $(0,0,0)$ as the reference point, $\phi(x, y, z)$ is obtained by using the line
integral command to integrate $\vec{A}$ along a straight line from $(0,0,0)$ to (x, y, z).

```
> phi:=-LineInt(A,Line(<0,0,0>,<x,y,z>));
```

$$\phi := 3\,x\,z - y\,x^2\,z + 3\,x^4\,y^3 - y\,z - 2\,x\,y^2 - x^2 \sin(y) + y^3 - \sin(y)$$

The potential is evaluated at $x = 0.13$, $y = 1.82$, $z = -1.24$,

```
> phi1:=eval(phi,{x=0.13,y=1.82,z=-1.24});
```

$$\phi1 := 5.998362305$$

and rounded off to the nearest integer.

```
> Phi:=round(phi1);
```

$$\Phi := 6$$

The `implicitplot3d` command is used in `gr1` to plot the equipotential surfaces
corresponding to $\phi = \Phi = 6$ and $\phi = -\Phi$. The grid is taken to be $20 \times 20 \times 20$,
a `patchcontour` style is used, and the shading is `zhue`.

```
> gr1:=implicitplot3d({phi=Phi,phi=-Phi},x=-2.5..2.5,
     y=-2.5..2.5,z=-2.5..2.5,grid=[20,20,20],
     style=patchcontour,shading=zhue):
```

The `fieldplot3d` command is used in `gr2` to plot the vector field $\vec{A}$, the field
being represented by thick red arrows on a grid $7 \times 7 \times 7$.

```
> gr2:=fieldplot3d(A,x=-2.5..2.5,y=-2.5..2.5,z=-2.5..2.5,
     grid=[7,7,7],arrows=THICK,color=red):
```

I. M. then superimposes the two graphs with the `display` command, choosing
a particular orientation for the 3-dimensional viewing box.

```
>  display({gr1,gr2},axes=boxed,orientation=[40,40],
   tickmarks=[3,3,3],labels=["x","y","z"]);
```

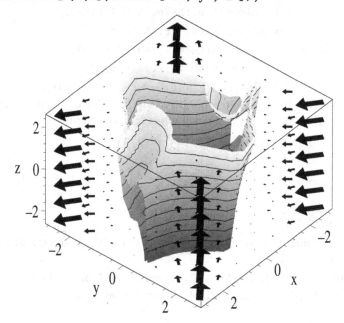

Figure 3.5: Equipotential surfaces and vector field arrows.

The resulting black and white version of the computer plot is shown in Figure 3.5. The equipotential surfaces are quite convoluted in appearance, but it appears that the vector field arrows are perpendicular to them as expected. I. M. suggests that you experiment with the recipe, trying other equipotentials, densities of arrows, etc. You could even change the form of the vector field.

3.2.5 The Divergence Theorem

Nothing leads the scientist so astray as a premature truth.
Jean Rostand, French biologist, *Penses d'un Biologiste, 1939*

The *divergence (Gauss's) theorem* states that for a closed volume V having a bounding surface S, $\int_V (\nabla \cdot \vec{A})\, dv = \oint_S \vec{A} \cdot d\vec{a}$, where $d\vec{a}$ is an element of area on S, dv is an element of volume in V, and $\oint_S$ denotes a closed surface integral.
    Consider the flow of a fluid characterized by the velocity vector field

$$\vec{A} = r\,\cos\theta\,\hat{e}_r + r\,\sin\theta\,\hat{e}_\theta + r\,\sin\theta\cos\phi\,\hat{e}_\phi.$$

in spherical coordinates. Is this a conservative field? Create a 3-dimensional plot of the vector field. Check the divergence theorem for $\vec{A}$, using as your volume V an inverted closed hemispherical bowl of radius R, resting on the x-y plane and centered on the origin.

The plots and VectorCalculus packages are loaded and the coordinates r, θ, and ϕ are set to be spherical. The given vector field $\vec{A}$ is then entered.

```
>   restart: with(plots): with(VectorCalculus):
>   SetCoordinates('spherical'[r,theta,phi]):
>   A:=VectorField(<r*cos(theta),r*sin(theta),
                    r*sin(theta)*cos(phi)>);
```

$$A := r\cos(\theta)\,\overline{e}_r + r\sin(\theta)\,\overline{e}_\theta + r\sin(\theta)\cos(\phi)\,\overline{e}_\phi$$

To plot the vector field, I will temporarily switch to Cartesian coordinates. This can be accomplished with the following MapToBasis command.

```
>   A2:=MapToBasis(A,'cartesian'[x,y,z]);
```

$$A2 := \left(\frac{2\,z\,x}{\sqrt{x^2+y^2+z^2}} - \frac{x\,y}{\sqrt{x^2+y^2}}\right)\overline{e}_x + \left(\frac{2\,z\,y}{\sqrt{x^2+y^2+z^2}} + \frac{x^2}{\sqrt{x^2+y^2}}\right)\overline{e}_y$$

$$+ \left(\frac{z^2}{\sqrt{x^2+y^2+z^2}} + \frac{-x^2-y^2}{\sqrt{x^2+y^2+z^2}}\right)\overline{e}_z$$

Then the fieldplot3d command is used to plot $A2$, producing SLIM red arrows on a $7 \times 7 \times 7$ grid.

```
>   fieldplot3d(A2,x=-2.5..2.5,y=-2.5..2.5,z=-2.5..2.5,
    grid=[7,7,7],arrows=SLIM,color=red,axes=box);
```

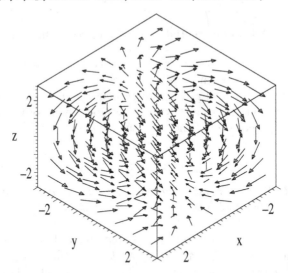

Figure 3.6: Arrows representing the 3-dimensional vector field $\vec{A}$.

The resulting 3-dimensional picture is reproduced in Figure 3.6, but is best viewed on the computer screen where it can be rotated by dragging with the mouse. Looking at the picture, do you think that the vector field is conservative, i.e., has zero curl? Let's calculate the curl as well as the divergence which we will need for confirming the divergence theorem.

```
>  curl:=Del &x A;  div:=simplify(Del . A);
```

$$curl := 2\cos(\phi)\cos(\theta)\,\overline{e}_r - 2\sin(\theta)\cos(\phi)\,\overline{e}_\theta + 3\sin(\theta)\,\overline{e}_\phi$$

$$div := 5\cos(\theta) - \sin(\phi)$$

$\vec{A}$ has a non-zero curl, and therefore is not conservative. Now, let's check the divergence theorem. The left-hand side $\int_0^{2\pi}\int_0^{\pi/2}\int_0^R (\nabla \cdot \vec{A})\,dv$ of the divergence theorem is entered for the volume contained in the hemispherical bowl

```
>  LHS:=Int(Int(Int(div*r^2*sin(theta),r=0..R),
       theta=0..Pi/2),phi=0..2*Pi);
```

$$LHS := \int_0^{2\pi}\int_0^{\frac{\pi}{2}}\int_0^R r^2\sin(\theta)\,(5\cos(\theta) - \sin(\phi))\,dr\,d\theta\,d\phi$$

and then evaluated with the value command.

```
>  LHS:=value(LHS);
```

$$LHS := \frac{5\,R^3\,\pi}{3}$$

On the hemispherical upper surface of the bowl, the unit vector is radial everywhere, i.e., is given by $\hat{e}_r$. This unit vector is entered in n1 as a vector field. On the planar bottom surface of the bowl, the unit vector points in the direction $\hat{e}_\theta$. This unit vector is entered in n2, also as a vector field.

```
>  n1:=VectorField(<1,0,0>);  n2:=VectorField(<0,1,0>);
```

$$n1 := \overline{e}_r \qquad n2 := \overline{e}_\theta$$

The vector field $\vec{A}$ is evaluated on the top (hemisphere) surface in At and on the bottom (planar) surface in Ab.

```
>  At:=eval(A,r=R);  Ab:=eval(A,theta=Pi/2);
```

$$At := R\cos(\theta)\,\overline{e}_r + R\sin(\theta)\,\overline{e}_\theta + R\sin(\theta)\cos(\phi)\,\overline{e}_\phi$$

$$Ab := r\,\overline{e}_\theta + r\cos(\phi)\,\overline{e}_\phi$$

Now, we can evaluate the surface integral on the right-hand side of the divergence theorem. The contribution from the hemispherical top is of the form $\int_0^{2\pi}\int_0^{\pi/2}(\vec{A}t \cdot \hat{n}1)\,R^2\sin\theta\,d\theta\,d\phi$, which is now entered and then evaluated.

```
>  RHS1:=Int(Int((At . n1)*R^2*sin(theta),theta=0..Pi/2),
       phi=0..2*Pi);
```

$$RHS1 := \int_0^{2\pi}\int_0^{\frac{\pi}{2}} R^3\cos(\theta)\sin(\theta)\,d\theta\,d\phi$$

```
>  RHS1:=value(RHS1);
```

$$RHS1 := R^3\,\pi$$

The contribution to the surface integral from the planar bottom is given by $\int_0^{2\pi}\int_0^R (\vec{A}b \cdot \hat{n}2)\,r\sin(\pi/2)\,dr\,d\phi$ which is entered and evaluated.

```
>  RHS2:=Int(Int((Ab . n2)*r*sin(Pi/2),r=0..R),phi=0..2*Pi);
```

$$RHS2 := \int_0^{2\pi}\int_0^R r^2\,dr\,d\phi$$

> ```RHS2:=value(RHS2);```

$$RHS2 := \frac{2\,R^3\,\pi}{3}$$

Adding the two surface integral contributions, $RHS1$ and $RHS2$, completes the evaluation of the right-hand side of the divergence theorem.

> ```RHS:=RHS1+RHS2;```

$$RHS := \frac{5\,R^3\,\pi}{3}$$

The results RHS and LHS are identical, confirming the divergence theorem.

3.3   Matrices

A general *matrix* A of *order* $m \times n$ is of the form

$$A = \begin{pmatrix} a_{11} & a_{12} & a_{13} & \cdots & a_{1n} \\ a_{21} & a_{22} & a_{23} & \cdots & a_{2n} \\ \cdots & \cdots & \cdots & \cdots & \cdots \\ a_{m1} & a_{m2} & a_{m3} & \cdots & a_{mn} \end{pmatrix}$$

the index m labeling the row and n the column. Each number a_{jk} in A is called an *element*. If the number of columns equals the number of rows, the matrix is said to be *square*. A matrix having only one row is called a *row matrix* or a *row vector*, while a matrix having only one column is called a *column matrix* or *column vector*. Some basic properties of matrices are as follows:

(a) Addition and Subtraction: If two matrices $A = (a_{jk})$ and $B = (b_{jk})$ are of the same order, then $A \pm B = (a_{jk} \pm b_{jk})$.

(b) Multiplication: If $A = (a_{jk})$ and λ is any scalar, then $\lambda A = A \lambda = (\lambda a_{jk})$. If $A = (a_{jk})$ is an $m \times n$ matrix and $B = (b_{jk})$ is an $n \times p$ matrix, then the product $A B$ (or $A \cdot B$) is a matrix $C = (c_{jk})$, where $c_{jk} = \sum_{i=1}^{n} a_{ji}\,b_{ik}$. The new matrix C is of order $m \times p$.

(c) Transpose: The transpose A^T of a matrix $A = (a_{jk})$ is $A^T = (a_{kj})$, i.e., interchange rows and columns.

(d) Hermitian matrix: If $A = (a_{jk})$, then the complex conjugate matrix is $A^* = (a_{jk}^*)$. The *Hermitian conjugate* (or *adjoint*) matrix $A^\dagger = (A^T)^*$. A square matrix is *Hermitian* if $A^\dagger = A$.

(e) Inverse of a matrix: If A is a *non-singular* matrix (i.e., its determinant $\det(A) \neq 0$), then the *inverse* matrix A^{-1} is of the form $A^{-1} = (A_{jk})^T / \det(A)$, where (A_{jk}) is the matrix of cofactors A_{jk}. (The *cofactor* A_{jk} is equal to $(-1)^{j+k}$ times the resulting determinant of A obtained by removing all the elements of the jth row and kth column.) It follows that $A A^{-1} = A^{-1} A = I$, where I is the *unit matrix* with each element along its principal diagonal equal to 1 and all off-diagonal elements 0.

Other matrix properties will be introduced in the recipes of this section.

3.3.1 Some Matrix Basics

The basic tool for the manipulation of reality is the manipulation of words. If you can control the meaning of words, you can control the people who must use the words.
Philip K. Dick, American science fiction writer, (1928–82)

In this recipe, the basic Maple syntax for dealing with matrices is introduced.

$$\text{If} \quad A = \begin{pmatrix} 2 & 1 & -1 \\ 1 & -2 & 3 \\ -2 & 1 & 2 \end{pmatrix} \quad \text{and} \quad B = \begin{pmatrix} 1 & -1 & 2 \\ -2 & 1 & 3 \\ 2 & -1 & 1 \end{pmatrix} \quad \text{show that}$$

(a) A and B are non-singular matrices

(b) $(A + B)^2 = A^2 + AB + BA + B^2$

(c) $(AB)^{-1} = B^{-1} A^{-1}$

(d) $(AB)^T = B^T A^T$

The basic Maple library package for dealing with matrices and determinants is the LinearAlgebra package. Replace the colon with a semicolon in with(Linear Algebra) if you wish to see the command structures available in this package.

> `restart: with(LinearAlgebra):`

A matrix can be entered in different ways. For example, A is now entered using the Matrix command with the argument consisting of the elements arranged in order in a list of lists.

> `A:=Matrix([[2,1,-1],[1,-2,3],[-2,1,2]]);`

$$A := \begin{bmatrix} 2 & 1 & -1 \\ 1 & -2 & 3 \\ -2 & 1 & 2 \end{bmatrix}$$

The matrix is clearly a 3×3 (square) matrix. The dimensionality of A can be confirmed using the Dimension command.

> `Dimension(A);`

$$3, 3$$

A matrix such as B can also be entered by using the following syntax.

> `B:=<<1|-1|2>,<-2|1|3>,<2|-1|1>>;`

$$B := \begin{bmatrix} 1 & -1 & 2 \\ -2 & 1 & 3 \\ 2 & -1 & 1 \end{bmatrix}$$

(a) To prove that A and B are non-singular matrices, their determinants are calculated and seen to be non-zero.

> `Determinant(A); Determinant(B);`

$$-19 \qquad -4$$

(b) On the left-hand side, $(A + B)^2$ is calculated.

> `LHS:=(A+B)^2;`

$$LHS := \begin{bmatrix} 9 & 0 & 6 \\ -2 & 1 & 11 \\ 0 & 0 & 9 \end{bmatrix}$$

To multiply two matrices A and B, one can use either the "long-hand" syntax `Multiply(A,B)` or the shorter dot notation `A . B`, again leaving spaces either side of the dot as we did for vector multiplication. The right-hand side $A^2 + AB + BA + B^2$ is now entered using the "short-hand" syntax. For clarity, I have place round brackets around each matrix multiplication.

> `RHS:=A^2+(A . B)+(B . A)+B^2;`

$$RHS := \begin{bmatrix} 9 & 0 & 6 \\ -2 & 1 & 11 \\ 0 & 0 & 9 \end{bmatrix}$$

The matrix result in RHS is the same as in LHS, thus proving the relation.

(c) The inverse matrices A^{-1} and B^{-1} are determined.

> `Ainverse:=MatrixInverse(A); Binverse:=MatrixInverse(B);`

$$Ainverse := \begin{bmatrix} \dfrac{7}{19} & \dfrac{3}{19} & \dfrac{-1}{19} \\ \dfrac{8}{19} & \dfrac{-2}{19} & \dfrac{7}{19} \\ \dfrac{3}{19} & \dfrac{4}{19} & \dfrac{5}{19} \end{bmatrix} \qquad Binverse := \begin{bmatrix} -1 & \dfrac{1}{4} & \dfrac{5}{4} \\ -2 & \dfrac{3}{4} & \dfrac{7}{4} \\ 0 & \dfrac{1}{4} & \dfrac{1}{4} \end{bmatrix}$$

Then $(A B)^{-1} - B^{-1} A^{-1}$ is calculated in $eq1$.

> `eq1:=MatrixInverse(A . B)-(Binverse . Ainverse);`

$$eq1 := \begin{bmatrix} 0 & 0 & 0 \\ 0 & 0 & 0 \\ 0 & 0 & 0 \end{bmatrix}$$

The result is a *zero* or *null matrix*, thus confirming the relation.

(d) The transposes A^T and B^T of A and B are determined.

> `Atranspose:=Transpose(A); Btranspose:=Transpose(B);`

$$Atranspose := \begin{bmatrix} 2 & 1 & -2 \\ 1 & -2 & 1 \\ -1 & 3 & 2 \end{bmatrix} \qquad Btranspose := \begin{bmatrix} 1 & -2 & 2 \\ -1 & 1 & -1 \\ 2 & 3 & 1 \end{bmatrix}$$

Calculating $(A B)^T - B^T A^T$ in $eq2$ yields a null matrix, confirming the relation.

> `eq2:=Transpose(A . B)-(Btranspose . Atranspose);`

$$eq2 := \begin{bmatrix} 0 & 0 & 0 \\ 0 & 0 & 0 \\ 0 & 0 & 0 \end{bmatrix}$$

3.3.2 Eigenvalues and Eigenvectors

A man can become so accustomed to the thought of his own faults that he will begin to cherish them as charming little "personal characteristics".
Helen Rowland, American journalist, *A Guide to Men,* "Brides" *(1922)*

Let A be an $n \times n$ matrix and X a column vector with n elements. The equation $AX = \lambda X$, or $(A - \lambda I)X = 0$, where λ is a number and I the unit or *identity matrix*, has non-trivial solutions if and only if $\det(A - \lambda I) = 0$. When expanded, the determinant yields a polynomial equation of degree n in λ, called the *characteristic polynomial* equation. The roots of this polynomial equation are called the *characteristic values* or *eigenvalues* of the matrix A. Corresponding to each eigenvalue, there will be a non-trivial $(X \neq 0)$ solution X, which is called a *characteristic vector* or *eigenvector*.

Find the eigenvalues and corresponding eigenvectors of the matrix

$$A := \begin{pmatrix} 5 & 7 & -5 \\ 0 & 4 & -1 \\ 2 & 8 & -3 \end{pmatrix}.$$

Determine a set of unit eigenvectors. A *unit eigenvector* has unit length, i.e., the sum of the squares of its components is 1.

To solve this problem, let's load the LinearAlgebra package and enter the given matrix A.

```
>   restart: with(LinearAlgebra):
>   A:=<<5|7|-5>,<0|4|-1>,<2|8|-3>>;
```

$$A := \begin{bmatrix} 5 & 7 & -5 \\ 0 & 4 & -1 \\ 2 & 8 & -3 \end{bmatrix}$$

This problem can be tackled in two equivalent ways, either by mimicking the hand calculation or using specific Maple commands which accomplish the same task. Both ways will be demonstrated. The 3×3 identity matrix is entered.

```
>   IM:=IdentityMatrix(3);
```

$$IM := \begin{bmatrix} 1 & 0 & 0 \\ 0 & 1 & 0 \\ 0 & 0 & 1 \end{bmatrix}$$

The identity matrix is multiplied by λ in S. The simplify command must be applied to actually accomplish the multiplication. The same result could be alternatively achieved by entering ScalarMatrix(lambda,3,3).

```
>   S:=simplify(lambda*IM);
```

$$S := \begin{bmatrix} \lambda & 0 & 0 \\ 0 & \lambda & 0 \\ 0 & 0 & \lambda \end{bmatrix}$$

If we were proceeding by hand, the *characteristic matrix* $A - \lambda I$ would be formed. This is illustrated in CM.

> CM:=A-S;

$$CM := \begin{bmatrix} 5-\lambda & 7 & -5 \\ 0 & 4-\lambda & -1 \\ 2 & 8 & -3-\lambda \end{bmatrix}$$

The characteristic matrix may be more simply obtained directly from A by using the CharacteristicMatrix command as in $CM2$.

> CM2:=CharacteristicMatrix(A,lambda);

$$CM2 := \begin{bmatrix} 5-\lambda & 7 & -5 \\ 0 & 4-\lambda & -1 \\ 2 & 8 & -3-\lambda \end{bmatrix}$$

Proceeding by hand, we then would determine the characteristic polynomial equation by taking the determinant of the characteristic matrix and setting it equal to zero. This is done in CP, a cubic polynomial in λ resulting.

> CP:=Determinant(CM)=0;

$$CP := 6 - 11\lambda + 6\lambda^2 - \lambda^3 = 0$$

The same equation follows on applying the CharacteristicPolynomial command directly to A and setting the result equal to zero.

> CP2:=CharacteristicPolynomial(A,lambda)=0;

$$CP2 := \lambda^3 - 6\lambda^2 + 11\lambda - 6 = 0$$

In the hand calculation, the eigenvalues are obtained by solving the characteristic polynomial equation CP for λ, 3 eigenvalues resulting in L.

> L:=solve(CP,lambda);

$$L := 1,\, 2,\, 3$$

The same three eigenvalues again are more simply obtained by applying the Eigenvalues command to A. Unless otherwise specified, the eigenvalues appear by default in a column format.

> Eigenvalues(A);

$$\begin{bmatrix} 1 \\ 2 \\ 3 \end{bmatrix}$$

Instead of a column format, including the option output=list will produce the eigenvalues in a list format as shown in the following command line.

> Eigenvalues(A,output=list);

$$[1,\, 2,\, 3]$$

Now that the eigenvalues are determined, the corresponding eigenvectors will be calculated. If proceeding by hand, the eigenvalues are extracted separately from L and labeled $L1$, $L2$, and $L3$.

> L1:=L[1]; L2:=L[2]; L3:=L[3];

$$L1 := 1 \qquad L2 := 2 \qquad L3 := 3$$

A general eigenvector X, with three elements, is then entered as a column vector. Note that the elements are separated by commas when inputting a column vector. The goal is to determine the value of $x1$, $x2$, and $x3$ for each of the eigenvalues.

```
>   X:=<<x1,x2,x3>>;
```

$$X := \begin{bmatrix} x1 \\ x2 \\ x3 \end{bmatrix}$$

Then the column vector X is multiplied by the characteristic matrix CM, the latter being evaluated for a specific eigenvalue, e.g., $\lambda = L1$.

```
>   eq1:=eval(CM,lambda=L1) . X;
```

$$eq1 := \begin{bmatrix} 4\,x1 + 7\,x2 - 5\,x3 \\ 3\,x2 - x3 \\ 2\,x1 + 8\,x2 - 4\,x3 \end{bmatrix}$$

Extracting each row entry from *eq1* and setting it equal to 0 yields a set of equations in *system1* for $x1$, $x2$, and $x3$.

```
>   system1:={eq1[1,1]=0,eq1[2,1]=0,eq1[3,1]=0};
```

$$system1 := \{4\,x1 + 7\,x2 - 5\,x3 = 0,\ 3\,x2 - x3 = 0,\ 2\,x1 + 8\,x2 - 4\,x3 = 0\}$$

The system of equations is solved in *sol1* for the unknowns, the answer being expressed in terms of the arbitrary quantity $x2$.

```
>   sol1:=solve(system1,{x1,x2,x3});
```

$$sol1 := \{x3 = 3\,x2,\ x1 = 2\,x2,\ x2 = x2\}$$

The eigenvector $X1$ corresponding to the eigenvalue $L1$ is now obtained by evaluating X with *sol1*.

```
>   X1:=eval(X,sol1);
```

$$X1 := \begin{bmatrix} 2\,x2 \\ x2 \\ 3\,x2 \end{bmatrix}$$

There is an infinite family of eigenvectors $X1$ depending on the value chosen for $x2$. Proceeding by hand, the other two eigenvectors could be similarly obtained. The choice of a specific value for $x2$ will be postponed for the moment.

The multitude of steps in deriving the three eigenvalues and eigenvectors can be completely bypassed by applying the **Eigenvectors** command to A.

```
>   (eiv,V):=Eigenvectors(A);
```

$$eiv, V := \begin{bmatrix} 1 \\ 2 \\ 3 \end{bmatrix}, \begin{bmatrix} \dfrac{2}{3} & \dfrac{1}{2} & -1 \\[2mm] \dfrac{1}{3} & \dfrac{1}{2} & 1 \\[2mm] 1 & 1 & 1 \end{bmatrix}$$

In the output, the eigenvalues are presented in column format (the order of the entries may vary from one execution of the worksheet to the next) before the comma. Corresponding to the first (top) entry, the corresponding eigenvector is given by the first column in the matrix format after the comma. The second column corresponds to the second eigenvalue entry, and so on. So that the eigenvalues and eigenvectors can be extracted separately, the assignment (eiv,V) is made, eiv corresponding to the eigenvalues and V to the eigenvectors.

Choosing $x2=1/3$, the "hand calculated" eigenvector $X1$, corresponding to the eigenvalue 1, is identical to the first column in V.

>   X1:=eval(X1,x2=1/3);

$$X1 := \begin{bmatrix} \dfrac{2}{3} \\ \dfrac{1}{3} \\ 1 \end{bmatrix}$$

The three eigenvectors are now extracted from V by using the Column command. The second argument specifies the column to be extracted.

>   V1:=Column(V,1); V2:=Column(V,2); V3:=Column(V,3);

$$V1 := \begin{bmatrix} \dfrac{2}{3} \\ \dfrac{1}{3} \\ 1 \end{bmatrix} \qquad V2 := \begin{bmatrix} \dfrac{1}{2} \\ \dfrac{1}{2} \\ 1 \end{bmatrix} \qquad V3 := \begin{bmatrix} -1 \\ 1 \\ 1 \end{bmatrix}$$

Finally, it was requested that a set of unit eigenvectors be obtained. The sum of the squares of the elements of each eigenvector can be obtained by applying the Norm(V,2) command. Dividing each of the above eigenvectors by its norm, produces a unit eigenvector.

>   U1:=V1/Norm(V1,2); U2:=V2/Norm(V2,2); U3:=V3/Norm(V3,2);

$$U1 := \begin{bmatrix} \dfrac{\sqrt{14}}{7} \\ \dfrac{\sqrt{14}}{14} \\ \dfrac{3\sqrt{14}}{14} \end{bmatrix} \qquad U2 := \begin{bmatrix} \dfrac{\sqrt{6}}{6} \\ \dfrac{\sqrt{6}}{6} \\ \dfrac{\sqrt{6}}{3} \end{bmatrix} \qquad U3 := \begin{bmatrix} -\dfrac{\sqrt{3}}{3} \\ \dfrac{\sqrt{3}}{3} \\ \dfrac{\sqrt{3}}{3} \end{bmatrix}$$

When first learning how to find eigenvalues and eigenvectors, its important to go through the steps outlined in "our hand calculation". But, after doing this a few times, you will be grateful for Maple's specialized commands.

3.3.3 Diagonalizing a Matrix

An idea is a point of departure and no more. As soon as you elaborate it, it becomes transformed by thought.
Pablo Picasso, Spanish artist, (1881–1973)

Find a matrix C which reduces the matrix

$$A = \begin{pmatrix} 2 & 4 & -6 \\ 4 & 2 & -6 \\ -6 & -6 & -15 \end{pmatrix}$$

to diagonal form (all elements not on the main diagonal are zero) by the transformation $C^{-1}AC$.

After loading the LinearAlgebra package, the matrix A is entered.

```
> restart: with(LinearAlgebra):
> A:=<<2|4|-6>,<4|2|-6>,<-6|-6|-15>>;
```

$$A := \begin{bmatrix} 2 & 4 & -6 \\ 4 & 2 & -6 \\ -6 & -6 & -15 \end{bmatrix}$$

A general matrix C with doubly subscripted elements is entered.

```
> C:=<<c[1,1]|c[1,2]|c[1,3]>,<c[2,1]|c[2,2]|c[2,3]>,
       <c[3,1]|c[3,2]|c[3,3]>>;
```

$$C := \begin{bmatrix} c_{1,1} & c_{1,2} & c_{1,3} \\ c_{2,1} & c_{2,2} & c_{2,3} \\ c_{3,1} & c_{3,2} & c_{3,3} \end{bmatrix}$$

The eigenvalues of A are obtained.

```
> L:=Eigenvalues(A);
```

$$L := \begin{bmatrix} -2 \\ 9 \\ -18 \end{bmatrix}$$

L gives the three distinct eigenvalues. It is desired to find a matrix C which reduces A to a diagonal matrix S with the main diagonal elements given by the eigenvalues and the off-diagonal elements equal to zero. The DiagonalMatrix command is used to form S.

```
> S:=DiagonalMatrix([L[1],L[2],L[3]],3,3);
```

$$S := \begin{bmatrix} -2 & 0 & 0 \\ 0 & 9 & 0 \\ 0 & 0 & -18 \end{bmatrix}$$

Now, we want $C^{-1}AC = S$. Multiplying this matrix equation from the left by C and noting that $CC^{-1} = I$, we have $IAC = AC = CS$, or $AC - CS = 0$. The left-hand side of this last matrix equation is now entered in E.

```
> E:=(A . C)-(C . S);
```

$$E := \left[4\,c_{1,1} + 4\,c_{2,1} - 6\,c_{3,1}\,, \; -7\,c_{1,2} + 4\,c_{2,2} - 6\,c_{3,2}\,, 20\,c_{1,3} + 4\,c_{2,3} - 6\,c_{3,3} \right]$$

$$\left[4\,c_{1,1} + 4\,c_{2,1} - 6\,c_{3,1}\,, \; 4\,c_{1,2} - 7\,c_{2,2} - 6\,c_{3,2}\,, 4\,c_{1,3} + 20\,c_{2,3} - 6\,c_{3,3} \right]$$

$$\left[-6\,c_{1,1} - 6\,c_{2,1} - 13\,c_{3,1}\,, \; -6\,c_{1,2} - 6\,c_{2,2} - 24\,c_{3,2}\,, -6\,c_{1,3} - 6\,c_{2,3} + 3\,c_{3,3} \right]$$

We have to solve the system of equations resulting from setting the 9 entries of E equal to zero. This system of 9 equations is now obtained by using the sequence command twice to sum over the rows and columns of E.

```
> System:={seq(seq(E[i,j]=0,j=1..3),i=1..3)};
```

$$\begin{aligned} System := \{ & 4\,c_{1,1} + 4\,c_{2,1} - 6\,c_{3,1} = 0, \; -7\,c_{1,2} + 4\,c_{2,2} - 6\,c_{3,2} = 0, \\ & 20\,c_{1,3} + 4\,c_{2,3} - 6\,c_{3,3} = 0, \; 4\,c_{1,2} - 7\,c_{2,2} - 6\,c_{3,2} = 0, \\ & 4\,c_{1,3} + 20\,c_{2,3} - 6\,c_{3,3} = 0, \; -6\,c_{1,1} - 6\,c_{2,1} - 13\,c_{3,1} = 0, \\ & -6\,c_{1,2} - 6\,c_{2,2} - 24\,c_{3,2} = 0, \; -6\,c_{1,3} - 6\,c_{2,3} + 3\,c_{3,3} = 0 \} \end{aligned}$$

The double sequence command is used again to enter the 9 unknown matrix elements to be solved for.

```
> Elements:={seq(seq(c[i,j],j=1..3),i=1..3)};
```

$$Elements := \{ c_{1,1},\, c_{1,2},\, c_{1,3},\, c_{2,1},\, c_{2,2},\, c_{2,3},\, c_{3,1},\, c_{3,2},\, c_{3,3} \}$$

The system of 9 equations is solved for the 9 unknown matrix elements.

```
> Solution:=solve(System,Elements);
```

$$\begin{aligned} Solution := \{ & c_{1,3} = c_{2,3},\, c_{3,3} = 4\,c_{2,3},\, c_{1,2} = -2\,c_{3,2},\, c_{1,1} = -c_{2,1}, \\ & c_{2,1} = c_{2,1},\, c_{2,3} = c_{2,3},\, c_{3,2} = c_{3,2},\, c_{3,1} = 0,\, c_{2,2} = -2\,c_{3,2} \} \end{aligned}$$

Then *Solution* is used to evaluate C.

```
> C:=eval(C,Solution);
```

$$C := \begin{bmatrix} -c_{2,1} & -2\,c_{3,2} & c_{2,3} \\ c_{2,1} & -2\,c_{3,2} & c_{2,3} \\ 0 & c_{3,2} & 4\,c_{2,3} \end{bmatrix}$$

The matrix elements $c_{2,1}$, $c_{3,2}$, and $c_{2,3}$ are undetermined. We are free to choose values for these elements. There are an infinity of choices and therefore of C matrices which will diagonalize A with the transformation $C^{-1}\,A\,C$. Let's check that this is so, keeping the three elements unspecified for the moment. The inverse matrix C^{-1} is calculated.

```
> Cinverse:=MatrixInverse(C);
```

$$Cinverse := \begin{bmatrix} -\dfrac{1}{2}\,\dfrac{1}{c_{2,1}} & \dfrac{1}{2}\,\dfrac{1}{c_{2,1}} & 0 \\[2ex] -\dfrac{2}{9}\,\dfrac{1}{c_{3,2}} & -\dfrac{2}{9}\,\dfrac{1}{c_{3,2}} & \dfrac{1}{9}\,\dfrac{1}{c_{3,2}} \\[2ex] \dfrac{1}{18}\,\dfrac{1}{c_{2,3}} & \dfrac{1}{18}\,\dfrac{1}{c_{2,3}} & \dfrac{2}{9}\,\dfrac{1}{c_{2,3}} \end{bmatrix}$$

Provided that non-zero values are chosen for the undetermined elements, the inverse matrix C^{-1} exists. Now we calculate $C^{-1}AC$, obtaining the diagonal matrix S as expected.

> check:=Cinverse . (A . C);

$$check := \begin{bmatrix} -2 & 0 & 0 \\ 0 & 9 & 0 \\ 0 & 0 & -18 \end{bmatrix}$$

A particular matrix C which will accomplish the transformation is now obtained by setting all the undetermined matrix elements equal to 1.

> C:=eval(C,{c[2,1]=1,c[3,2]=1,c[2,3]=1});

$$C := \begin{bmatrix} -1 & -2 & 1 \\ 1 & -2 & 1 \\ 0 & 1 & 4 \end{bmatrix}$$

3.3.4 Orthogonal and Unitary Matrices

No matter how calmly you try to referee, parenting will eventually produce bizarre behavior, and I'm not talking about the kids. Their behavior is always normal.
Bill Cosby, American comedian, *Fatherhood, 1986*

The ideas of orthogonality and normalization which are important in the discussion of vectors can be extended to matrices as illustrated in this simple recipe.

(a) If $A = \begin{pmatrix} a1 \\ a2 \\ a3 \end{pmatrix}$, $B = \begin{pmatrix} b1 \\ b2 \\ b3 \end{pmatrix}$, $C = \begin{pmatrix} c1 \\ c2 \\ c3 \end{pmatrix}$

are real mutually orthogonal unit column vectors, prove that the real matrix

$$ABC = \begin{pmatrix} a1 & b1 & c1 \\ a2 & b2 & c2 \\ a3 & b3 & c3 \end{pmatrix}$$

is an orthogonal matrix. A real matrix M is referred to as an *orthogonal matrix* if its transpose is equal to its inverse, i.e., $M^T = M^{-1}$, or $M^T M = I$.

(b) With $i = \sqrt{-1}$, show that

$$H = \begin{pmatrix} \dfrac{\sqrt{2}}{2} & -i\dfrac{\sqrt{2}}{2} & 0 \\ i\dfrac{\sqrt{2}}{2} & -\dfrac{\sqrt{2}}{2} & 0 \\ 0 & 0 & 1 \end{pmatrix}$$

is a unitary matrix. A complex matrix M is a *unitary matrix* if its complex conjugate transpose is equal to its inverse, i.e., $(M^*)^T = M^{-1}$, or $(M^*)^T M = I$.

After the usual loading of the LinearAlgebra package when dealing with matrices, the column vectors A, B, and C are entered using the Vector command, each argument containing a list of the relevant elements.

```
>  restart: with(LinearAlgebra):
>  A:=Vector([a1,a2,a3]); B:=Vector([b1,b2,b3]);
   C:=Vector([c1,c2,c3]);
```

$$A := \begin{bmatrix} a1 \\ a2 \\ a3 \end{bmatrix} \qquad B := \begin{bmatrix} b1 \\ b2 \\ b3 \end{bmatrix} \qquad C := \begin{bmatrix} c1 \\ c2 \\ c3 \end{bmatrix}$$

A functional operator F is introduced for imposing "orthonormality" on arbitrary column vectors M and N. Normalization follows on setting the parameter $n=1$, orthogonality on setting $n=0$.

```
>  F:=(M,N,n)->(Transpose(M) . N)=n:
```

Using F, normalization is imposed on A, B, and C in $eq1$, $eq2$, and $eq3$, and orthogonality in $eq4$, $eq5$, and $eq6$.

```
>  eq||1:=F(A,A,1); eq||2:=F(B,B,1); eq||3:=F(C,C,1);
```

$$eq1 := a1^2 + a2^2 + a3^2 = 1 \qquad eq2 := b1^2 + b2^2 + b3^2 = 1$$

$$eq3 := c1^2 + c2^2 + c3^2 = 1$$

```
>  eq||4:=F(A,B,0); eq||5:=F(A,C,0); eq||6:=F(B,C,0);
```

$$eq4 := b1\,a1 + b2\,a2 + b3\,a3 = 0 \qquad eq5 := c1\,a1 + c2\,a2 + c3\,a3 = 0$$

$$eq6 := c1\,b1 + c2\,b2 + c3\,b3 = 0$$

The matrix ABC is entered using the Matrix command.

```
>  ABC:=Matrix([[a1,b1,c1],[a2,b2,c2],[a3,b3,c3]]);
```

$$ABC := \begin{bmatrix} a1 & b1 & c1 \\ a2 & b2 & c2 \\ a3 & b3 & c3 \end{bmatrix}$$

To produce an orthogonal matrix OM, the product $(ABC)^T(ABC)$ is formed.

```
>  OM:=Transpose(ABC) . ABC;
```

$$OM := \begin{bmatrix} a1^2 + a2^2 + a3^2\,, & b1\,a1 + b2\,a2 + b3\,a3\,, & c1\,a1 + c2\,a2 + c3\,a3 \\ b1\,a1 + b2\,a2 + b3\,a3\,, & b1^2 + b2^2 + b3^2\,, & c1\,b1 + c2\,b2 + c3\,b3 \\ c1\,a1 + c2\,a2 + c3\,a3\,, & c1\,b1 + c2\,b2 + c3\,b3\,, & c1^2 + c2^2 + c3^2 \end{bmatrix}$$

Evaluating OM with the six orthonormality conditions, yields the identity (unit) matrix, thus confirming that ABC is an orthogonal matrix.

```
>  OM2:=eval(OM,{seq(eq||i,i=1..6)});
```

$$OM2 := \begin{bmatrix} 1 & 0 & 0 \\ 0 & 1 & 0 \\ 0 & 0 & 1 \end{bmatrix}$$

To answer part (b), the interface(imaginaryunit=i) command is used to set $i \equiv \sqrt{-1}$. Then the matrix H is entered.

```
>  interface(imaginaryunit=i):
```

```
> H:=<<sqrt(2)/2|-i*sqrt(2)/2|0>,<i*sqrt(2)/2|-sqrt(2)/2|0>,
    <0|0|1>>;
```

$$H := \begin{bmatrix} \dfrac{\sqrt{2}}{2} & \dfrac{-1}{2}i\sqrt{2} & 0 \\ \dfrac{1}{2}i\sqrt{2} & -\dfrac{\sqrt{2}}{2} & 0 \\ 0 & 0 & 1 \end{bmatrix}$$

The `HermitianTranspose` command is applied to H in T. This command applies the transpose (interchanging rows and columns) and takes the complex conjugate of the elements.

```
> T:=HermitianTranspose(H);
```

$$T := \begin{bmatrix} \dfrac{\sqrt{2}}{2} & \dfrac{-1}{2}i\sqrt{2} & 0 \\ \dfrac{1}{2}i\sqrt{2} & -\dfrac{\sqrt{2}}{2} & 0 \\ 0 & 0 & 1 \end{bmatrix}$$

Since $T = H$, H is a Hermitian matrix. That it is also a unitary matrix is confirmed by forming the product HT, the result R being the unit matrix.

```
> R:=H . T;
```

$$R := \begin{bmatrix} 1 & 0 & 0 \\ 0 & 1 & 0 \\ 0 & 0 & 1 \end{bmatrix}$$

3.3.5 Introducing the Euler Angles

To knock a thing down, especially if it is cocked at an arrogant angle, is a deep delight to the blood.
George Santayana, American philosopher (1863–1952)

The transformation from one 3-dimensional coordinate system (x, y, z) to another (x', y', z') with the same origin can be represented by a matrix equation of the form $X' = RX$, where R is the 3×3 *rotation matrix* and

$$X = \begin{pmatrix} x \\ y \\ z \end{pmatrix}, \quad X' = \begin{pmatrix} x' \\ y' \\ z' \end{pmatrix}.$$

The over-all rotation is made up of three 1-dimensional rotations through the angles ϕ, θ, and ψ, called the *Euler angles*. The choices of these angles is not uniform in the literature[4], so I will adopt the notation of Marion and Thornton [MT95] and Goldstein, Poole, and Safco [GPS02].

---

[4]In fact, Euler angles can be avoided altogether in classical physics by specifying a single plane of rotation and one angle, as discussed for example in [Bay99], [DL03], and [Hes99].

The first rotation is counterclockwise through an angle ϕ about the z axis, the rotation matrix being

$$R_\phi = \begin{pmatrix} \cos\phi & \sin\phi & 0 \\ -\sin\phi & \cos\phi & 0 \\ 0 & 0 & 1 \end{pmatrix}.$$

The new axes are labeled x_1, y_1, and $z_1 = z$. The second rotation is counterclockwise through an angle θ about the x_1 axis, the rotation matrix being

$$R_\theta = \begin{pmatrix} 1 & 0 & 0 \\ 0 & \cos\theta & \sin\theta \\ 0 & -\sin\theta & \cos\theta \end{pmatrix}.$$

The new axes are labeled $x_2 = x_1$, y_2, z_2. The third rotation is counterclockwise through an angle ψ about the z_2 axis, the rotation matrix being

$$R_\phi = \begin{pmatrix} \cos\psi & \sin\psi & 0 \\ -\sin\psi & \cos\psi & 0 \\ 0 & 0 & 1 \end{pmatrix}.$$

The new axes are labeled x', y', $z' = z_2$. The total rotation matrix then is $R = R_\psi\, R_\theta\, R_\phi$.

In this recipe, I will derive the total rotation matrix R and apply its inverse to a triplet of vectors originally pointing along the x, y, z axes. The plots and LinearAlgebra packages are first loaded.

> restart: with(plots): with(LinearAlgebra):

To ease the typing, the aliases p, t, and s are introduced for the input quantities phi, theta, and psi.

> alias(phi=p,theta=t,psi=s):

The three rotation matrices R_ϕ, R_θ, and R_ψ are entered.

> R[p]:=<<cos(p)|sin(p)|0>,<-sin(p)|cos(p)|0>,<0|0|1>>;

$$R_\phi := \begin{bmatrix} \cos(\phi) & \sin(\phi) & 0 \\ -\sin(\phi) & \cos(\phi) & 0 \\ 0 & 0 & 1 \end{bmatrix}$$

> R[t]:=<<1|0|0>,<0|cos(t)|sin(t)>,<0|-sin(t)|cos(t)>>;

$$R_\theta := \begin{bmatrix} 1 & 0 & 0 \\ 0 & \cos(\theta) & \sin(\theta) \\ 0 & -\sin(\theta) & \cos(\theta) \end{bmatrix}$$

> R[s]:=<<cos(s)|sin(s)|0>,<-sin(s)|cos(s)|0>,<0|0|1>>;

$$R_\psi := \begin{bmatrix} \cos(\psi) & \sin(\psi) & 0 \\ -\sin(\psi) & \cos(\psi) & 0 \\ 0 & 0 & 1 \end{bmatrix}$$

The total rotation matrix $R = R_\psi\, R_\theta\, R_\phi$ is then calculated.

> R:=R[s] . (R[t] . R[p]);

$$R := \begin{bmatrix} \cos\psi\cos\phi - \sin\psi\cos\theta\sin\phi, & \cos\psi\sin\phi + \sin\psi\cos\theta\cos\phi, & \sin\psi\sin\theta \\ -\sin\psi\cos\phi - \cos\psi\cos\theta\sin\phi, & -\sin\psi\sin\phi + \cos\psi\cos\theta\cos\phi, & \cos\psi\sin\theta \\ \sin\theta\sin\phi, & -\sin\theta\cos\phi, & \cos\theta \end{bmatrix}$$

The inverse rotation matrix *Rinv* for rotating from the primed coordinates to the unprimed ones is obtained by applying the `MatrixInverse` command to *R* and simplifying with the trigonometric option.

> `Rinv:=simplify(MatrixInverse(R),trig);`

$$Rinv := \begin{bmatrix} \cos\psi\cos\phi - \sin\psi\cos\theta\sin\phi, & -\sin\psi\cos\phi - \cos\psi\cos\theta\sin\phi, & \sin\theta\sin\phi \\ \cos\psi\sin\phi + \sin\psi\cos\theta\cos\phi, & -\sin\psi\sin\phi + \cos\psi\cos\theta\cos\phi, & -\sin\theta\cos\phi \\ \sin\psi\sin\theta, & \cos\psi\sin\theta, & \cos\theta \end{bmatrix}$$

To illustrate vector rotation (the inverse of coordinate rotation), let's evaluate *Rinv* for $\phi = \pi/3$, $\theta = \pi/4$, and $\psi = \pi/6$ radians.

> `Rinv:=eval(Rinv,{p=Pi/3,t=Pi/4,s=Pi/6});`

$$Rinv := \begin{bmatrix} \dfrac{\sqrt{3}}{4} - \dfrac{\sqrt{2}\sqrt{3}}{8} & -\dfrac{1}{4} - \dfrac{3\sqrt{2}}{8} & \dfrac{\sqrt{2}\sqrt{3}}{4} \\[2mm] \dfrac{3}{4} + \dfrac{\sqrt{2}}{8} & -\dfrac{\sqrt{3}}{4} + \dfrac{\sqrt{2}\sqrt{3}}{8} & -\dfrac{\sqrt{2}}{4} \\[2mm] \dfrac{\sqrt{2}}{4} & \dfrac{\sqrt{2}\sqrt{3}}{4} & \dfrac{\sqrt{2}}{2} \end{bmatrix}$$

The following column vectors, each of length 8 units, are entered.

> `v1:=<8,0,0>:  v2:=<0,8,0>:  v3:=<0,0,8>:`

Then the inverse rotation matrix *Rinv* is applied to each vector.

> `w1:=Rinv . v1; w2:=Rinv . v2; w3:=Rinv .   v3;`

$$w1 := \begin{bmatrix} 2\sqrt{3} - \sqrt{2}\sqrt{3} \\ 6 + \sqrt{2} \\ 2\sqrt{2} \end{bmatrix} \qquad w2 := \begin{bmatrix} -2 - 3\sqrt{2} \\ -2\sqrt{3} + \sqrt{2}\sqrt{3} \\ 2\sqrt{2}\sqrt{3} \end{bmatrix} \qquad w3 := \begin{bmatrix} 2\sqrt{2}\sqrt{3} \\ -2\sqrt{2} \\ 4\sqrt{2} \end{bmatrix}$$

The rotated vectors will be of exactly the same length as the original vectors. For example, it is confirmed in the following command line, using the `Norm` command, that *w1* is 8 units long.

> `simplify(Norm(w1,2));`

$$8$$

To plot the original and rotated vectors, an arrow operator `F` is formed to plot an arbitrary vector *v* as a cylindrical arrow with its tail at the origin. The color *c* of the arrow must be specified. Using `F`, then *v1*, *v2*, and *v3* are plotted as red arrows, and *w1*, *w2*, and *w3* as blue arrows.

> `F:=(v,c)->arrow(<0,0,0>,v,shape=cylindrical_arrow,color=c,`
>   `width=0.15,head_width=0.4,head_length=0.5):`
> `a[1]:=F(v1,red): a[2]:=F(v2,red): a[3]:=F(v3,red):`
> `a[4]:=F(w1,blue): a[5]:=F(w2,blue): a[6]:=F(w3,blue):`

The six arrows are superimposed with the `display` command with constrained scaling and a specified orientation of the viewing box, the resulting picture being shown in Figure 3.7. The original vectors *v1*, *v2*, and *v3* are the ones oriented along the coordinate axes.

```
>   display({seq(a[i],i=1..6)},axes=normal,
    orientation=[-36,80],scaling=constrained);
```

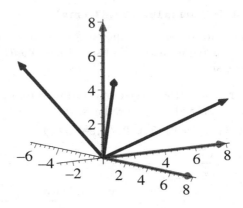

Figure 3.7: Matrix rotation of three vectors.

3.4 Supplementary Recipes

03-S01: Bobby Blowfly Seeks a Warmer Clime

The temperature in a certain region is given by $T = (x-3)^2 + (y-2)^2 + 3(z-1)^2$, with distances in meters and temperatures in degrees Celsius. Bobby Blowfly is currently at the point $x = 1$, $y = 2$, $z = 3$. What is the temperature at this point? Bobby wishes to start flying in a direction which produces the maximum rate of temperature increase. What is the unit vector $\hat{r}$ which points in this direction? What is the maximum rate of temperature change at $(1, 2, 3)$? How much greater is this rate than if he had decided to fly in the direction given by the unit vector $\hat{u} = (\hat{e}_x - \hat{e}_y + \hat{e}_z)/\sqrt{3}$? Plot the 3-dimensional isothermal (constant temperature) surface passing through the point $(1, 2, 3)$ along with the unit vectors $\hat{r}$ and $\hat{u}$ with their tails placed at the point.

03-S02: Jennifer's Vector Assignment

Jennifer asks you to try the following problem. Consider the vectors $\vec{A} = 3\,\hat{e}_x - \hat{e}_y - 7\,\hat{e}_z$, $\vec{B} = 4\,\hat{e}_x + \hat{e}_y + 2\,\hat{e}_z$, and $\vec{C} = \hat{e}_x + 5\,\hat{e}_y + 4\,\hat{e}_z$.

(a) Plot the three vectors as cylindrical arrows in a 3-dimensional picture.

(b) Calculate $\vec{P}_1 = (\vec{A} \times \vec{B}) \times \vec{C}$ and $\vec{P}_2 = \vec{A} \times (\vec{B} \times \vec{C})$. Confirm $\vec{P}_1 \neq \vec{P}_2$.

(c) Calculate the volume $V = |\vec{A} \cdot (\vec{B} \times \vec{C})|$ of a parallelepiped with sides $\vec{A}$, $\vec{B}$, and $\vec{C}$. Show that $V = \begin{vmatrix} A_x & A_y & A_z \\ B_x & B_y & B_z \\ C_x & C_y & C_z \end{vmatrix}$.

03-S03: Another Vector Operator Identity
Using spherical polar coordinates, prove the following vector identity for general 3-dimensional vector fields $\vec{A}$ and $\vec{B}$:

$$\nabla \cdot (\vec{A} \times \vec{B}) = \vec{B} \cdot (\nabla \times \vec{A}) - \vec{A} \cdot (\nabla \times \vec{B}).$$

03-S04: Another Maple Approach
Using the LinearAlgebra package instead of the VectorCalculus package, derive the velocity and acceleration of a particle in spherical polar coordinates.

03-S05: Conservative or Non-conservative?
Consider the vector field
$$\vec{A} = (3x^2 - 6yz)\,\hat{e}_x + (2y + 3xz)\,\hat{e}_y + (1 - 4xyz^2)\,\hat{e}_z.$$

(a) Calculate the line integral of $\vec{A}$ along the straight line joining the points $(0, 0, 0)$ and $(1, 1, 1)$.

(b) Calculate the line integral of $\vec{A}$ along the straight lines from $(0, 0, 0)$ to $(0, 0, 1)$, then to $(0, 1, 1)$, and then to $(1, 1, 1)$.

(c) Calculate the line integral of $\vec{A}$ from $(0, 0, 0)$ to $(1, 1, 1)$ along the path $x=t$, $y=t^2$, $z=t^3$, where t is a parameter.

(d) Having performed the line integrals, is the vector field $\vec{A}$ conservative or non-conservative? Confirm your conclusion by calculating the curl of $\vec{A}$.

(e) Make a 3-dimensional plot of the vector field over an appropriate range of x, y, and z. Is the distribution of arrows consistent with your conclusion?

03-S06: Basic Matrix Operations
Consider the two square matrices, $A = \begin{pmatrix} 1 & 0 & 2 \\ 2 & -1 & 3 \\ 0 & 5 & 6 \end{pmatrix}$, $B = \begin{pmatrix} 2 & -1 & 1 \\ 0 & 1 & 2 \\ 1 & -2 & -1 \end{pmatrix}$.

(a) Use the `Dimension` command to confirm that A and B are 3×3 matrices.

(b) By calculating their determinants, show that A and B are non-singular.

(c) Calculate $A + B$, $A - B$, and $A^2 - B^2$.

(d) Show that $AB \neq BA$, i.e., the two matrices *do not commute*.

(e) Introducing the column vector $C = \begin{pmatrix} 2 \\ 1 \\ 3 \end{pmatrix}$, calculate $(AB)C$.

(f) Show that the product $C(AB)$ is invalid.

(g) Convert C to a row vector $C2$ by using the `Transpose` command.

(h) Calculate the product $C2(AB)$.

03-S07: The Cayley–Hamilton Theorem
Given the 4×4 matrix
$$A = \begin{pmatrix} 2 & 1 & -1 & 1 \\ -2 & 1 & 3 & -2 \\ 1 & -2 & 1 & 1 \\ -3 & 1 & 2 & -1 \end{pmatrix} :$$

(a) Determine the eigenvalues and eigenvectors of A.

(b) Calculate the determinant and trace of A.

(c) Evaluate A^2, A^3, and A^4. Using these results, verify the *Cayley–Hamilton theorem*, which states that a matrix satisfies its own characteristic polynomial equation.

03-S08: Simultaneous Diagonalization

Consider the following 4×4 matrices,

$$A = \begin{pmatrix} 1 & 3 & 0 & 0 \\ 3 & 1 & 0 & 0 \\ 0 & 0 & 1 & 2 \\ 0 & 0 & 2 & 1 \end{pmatrix}, \quad B = \begin{pmatrix} 2 & 3 & 0 & 0 \\ 3 & 2 & 0 & 0 \\ 0 & 0 & 2 & 1 \\ 0 & 0 & 1 & 2 \end{pmatrix}.$$

(a) Show that $AB = BA$. The matrices are said to *commute*.

(b) Determine the eigenvalues of A. Find a general matrix C that reduces A to diagonal form by the transformation $C^{-1}AC$.

(c) Determine the eigenvalues of B. Show that the matrix C of part (b) also reduces B to diagonal form by the transformation $C^{-1}BC$.

03-S09: Orthonormal Vectors

Show that the following vectors form an orthonormal set.

$$A1 = \begin{pmatrix} \cos(\theta) \\ \sin(\theta) \\ 0 \end{pmatrix}, \quad A2 = \begin{pmatrix} -\sin(\theta) \\ \cos(\theta) \\ 0 \end{pmatrix}, \quad A3 = \begin{pmatrix} 0 \\ 0 \\ 1 \end{pmatrix}.$$

03-S10: Stokes's Theorem

Stokes's theorem for a vector field $\vec{A}$ states that $\oint_C \vec{A} \cdot d\vec{\ell} = \int_S (\nabla \times \vec{A}) \cdot d\vec{a}$, where $d\vec{\ell}$ is an element of vector path length along the closed contour C and $d\vec{a}$ is an element of area on the open surface S bounded by C. The direction of the vector area element is related to the direction of the line integral in a right-hand sense. Curling the fingers of the right hand in the direction of performing the line integral, the thumb indicates the sense of the vector area.

Verify Stokes's theorem for $\vec{A} = 3\,y\,\hat{e}_x - x\,z\,\hat{e}_y + y\,z^2\,\hat{e}_z$, where S is the surface of the paraboloid $2\,z = x^2 + y^2$ bounded by $z = 2$ and C is its boundary. I.e., C is the circle $x^2 + y^2 = 4$ lying in the x-y plane at $z = 2$.

03-S11: Solving Linear Equation Systems

The currents I_1, I_2, I_3, I_4 in an electrical network satisfy the system of equations

$$3\,I_1 + 2\,I_2 - I_4 = 65, \quad 2\,I_1 - I_2 + 4\,I_3 + 3\,I_4 = 160,$$
$$-7\,I_1 - 4\,I_2 - 2\,I_4 = 23, \quad 5\,I_1 - I_2 - 2\,I_3 + I_4 = 3.$$

Write the system in matrix form and then determine the currents by (a) using an inverse matrix approach (b) using the direct approach with the `solve` command.

Part II

THE ENTREES

The mathematics is not there till we put it there.
Sir Arthur Eddington, *The Philosophy of Physical Science*, (1882 – 1944)

*If scientific reasoning were limited to the logical
processes of arithmetic, we should not get very far
in our understanding of the physical world.
One might as well attempt to grasp the game of poker
entirely by the use of the mathematics of probability.*
Vannevar Bush, American scientist (1890 – 1974)

*A man ceases to be a beginner in any given science
and becomes a master in that science when he has
learned that...he is going to be a beginner all his life.*
R. G. Collingwood, British philosopher (1889–1943)

*I believe that a scientist looking at nonscientific
problems is just as dumb as the next guy.*
Richard Feynman, American physicist (1918–1988)

Chapter 4

Linear PDEs of Physics

In this chapter, we will solve a wide variety of physical problems involving linear PDEs, first in Cartesian coordinates, then in other curvilinear coordinate systems. The common linear PDEs of physics include:

(1) The *wave equation* (with ψ a continuous scalar function of position $\vec{r}$ and time t, and c the wave velocity),

$$\nabla^2 \psi = \frac{1}{c^2} \frac{\partial^2 \psi}{\partial t^2},\tag{4.1}$$

for vibrating elastic strings and membranes, sound waves in fluids, elastic waves in solids, electromagnetic waves (with ψ a vector), etc.

(2) The *diffusion equation* (with d a positive diffusion constant),

$$\frac{\partial \psi}{\partial t} = d \nabla^2 \psi,\tag{4.2}$$

which applies to time-dependent heat flow, mixing of fluids, neutron diffusion in nuclear reactors, diffusion of impurities in solids, and so on.

(3) *Laplace's equation,*
$$\nabla^2 \psi = 0,\tag{4.3}$$
which applies to steady-state heat flow, irrotational flow of an incompressible fluid, and electro(magneto)statics in charge(current)-free regions.

(4) *Poisson's equation* (with S a source term)
$$\nabla^2 \psi = S(\vec{r}, t),\tag{4.4}$$
which applies, for example, to the electrostatic potential due to a charge distribution and to the steady-state temperature with a heat source present.

(5) The time-independent *Schrödinger equation* (with $\hbar = h/2\pi$ (h being Planck's constant), m the mass, V the potential, and E the total energy),

$$-\frac{\hbar^2}{2m}\nabla^2 \psi + V \psi = E \psi,\tag{4.5}$$

which describes stationary states in quantum mechanics.

4.1   Three Cheers for the String

When I introduce my physics and engineering students to the PDEs of physics, I always begin with the transverse vibrations of the humble string. The reason is quite simple. The vibrating string is familiar and easy to visualize, the basic underlying equation of motion (the 1-dimensional wave equation) easy to derive, and a wide variety of important methods and ideas can be introduced. So this section illustrates a gourmet selection of string recipes. Lest you get ideas of stringing me up after munching on some of the stringy concoctions, I will show you that there is life beyond the string in the second part of this chapter.

4.1.1   Jennifer Finds the General Solution

Friends are like fiddle strings, they must not be screwed too tight.
English Proverb. Collected in: H. G. Bohn, *A Handbook of Proverbs* (1855)

Consider a light, uniform, stretched string of linear density (mass per unit length) ϵ which is horizontal in equilibrium. The goal of this recipe, provided by Jennifer the MIT mathematician, is to obtain the equation of motion for the transverse (vertical) oscillations of the string and then the general solution.

The PDEtools package is loaded, because it contains the **declare** and **dchange** commands. Jennifer needs the former to introduce subscript notation, favored by mathematicians, for the resulting 1-dimensional wave equation and the latter to transform the variables in arriving at the general solution. The plots package is included because a typical solution will be animated.

> `restart: with(PDEtools): with(plots):`

Let the vertical displacement of a point x on the string from equilibrium at time t be $\psi(x,t)$. Consider an infinitesimal element of string of arclength $ds = \sqrt{(dx)^2 + (d\psi)^2} = \sqrt{1 + (\partial\psi/\partial x)^2}\, dx$, located between x and $x + dx$. Since the string is only to move vertically, the horizontal component T of the tension in the string is constant along the string. The vertical component of the tension, which is given by $T\, \partial\psi/\partial x$, will vary along the string. The net vertical force F on the infinitesimal element is equal to the difference between the vertical forces at its ends. This force is entered.

> `F:=T*Diff(psi(x+dx,t),x)-T*Diff(psi(x,t),x);`

$$F := T\left(\frac{\partial}{\partial x}\, \psi(x + dx,\, t)\right) - T\left(\frac{\partial}{\partial x}\, \psi(x,\, t)\right)$$

Since dx is small, F is taylor expanded in powers of dx to order 2 and the order of term removed with the **convert(,polynom)** command.

> `F:=convert(taylor(F,dx=0,2),polynom);`

$$F := T\, \mathrm{D}_1(\psi)(x,\, t) - T\left(\frac{\partial}{\partial x}\, \psi(x,\, t)\right) + T\, \mathrm{D}_{1,\,1}(\psi)(x,\, t)\, dx$$

Newton's second law is applied to the string element in the vertical direction. The net vertical force F is converted to differential form on the lhs of *ode*. This must be equal to the mass of the string element times the acceleration, i.e., to $\epsilon\, ds\, (\partial^2 \psi/\partial t^2)$. Assuming that the vibrations are such that $\partial \psi/\partial x \ll 1$, then $ds \approx dx$. This *linear approximation* is used on the rhs of *ode*.

> `ode:=convert(F,diff)=epsilon*dx*diff(psi(x,t),t,t);`

$$ode := T\left(\frac{\partial^2}{\partial x^2}\,\psi(x,\,t)\right)dx = \varepsilon\, dx\left(\frac{\partial^2}{\partial t^2}\,\psi(x,\,t)\right)$$

Dividing *ode* by $(dx\, T)$ and substituting $\epsilon = T/c^2$, yields the 1-dimensional wave equation *WE*, with c the wave speed. By using the `declare` command, Jennifer has introduced subscript notation for the derivatives in the wave equation.

> `declare(psi(x,t)): WE:=subs(epsilon=T/c^2,ode/(dx*T));`

$$\psi(x,\,t) \text{ will now be displayed as } \psi$$

$$WE := \psi_{x,\,x} = \frac{\psi_{t,\,t}}{c^2}$$

If proceeding by hand, the general solution of *WE* can be obtained by introducing two new independent variables u and v related to x and t by the transformation $u = x + c\,t$, $v = x - c\,t$. Then one would apply the chain rule of differentiation, e.g., calculating,

$$\frac{\partial \psi}{\partial x} = \frac{\partial \psi}{\partial u}\frac{\partial u}{\partial x} + \frac{\partial \psi}{\partial v}\frac{\partial v}{\partial x} = \frac{\partial \psi}{\partial u} + \frac{\partial \psi}{\partial v}, \text{ and then } \frac{\partial^2 \psi}{\partial x^2} = \frac{\partial^2 \psi}{\partial u^2} + 2\frac{\partial^2 \psi}{\partial v \partial u} + \frac{\partial^2 \psi}{\partial v^2}$$

The second time derivative would be similarly calculated. This hand transformation of the wave equation *WE* can be easily accomplished with Maple as follows. Jennifer enters the transformation *tr* and solves for x and t, thus obtaining the inverse transformation *itr*.

> `tr:={u=x+c*t,v=x-c*t}: itr:=solve(tr,{x,t});`

$$itr := \{t = -\frac{v-u}{2\,c},\ x = \frac{u}{2} + \frac{v}{2}\}$$

Then applying the inverse transformation to the wave equation with the `dchange` command, and simplifying, yields the transformed wave equation *WE2*. Note that c is indicated as a parameter (`params=c`) in the command.

> `WE2:=simplify(dchange(itr,WE,[u,v],params=c));`

$$WE2 := \psi_{u,\,u} + 2\,\psi_{u,\,v} + \psi_{v,\,v} = \psi_{u,\,u} - 2\,\psi_{u,\,v} + \psi_{v,\,v}$$

The cross-derivative term is isolated in *WE2*, yielding the greatly simplified wave equation *WE3*.

> `WE3:=isolate(WE2,diff(psi(u,v),u,v));`

$$WE3 := \psi_{u,\,v} = 0$$

Clearly, the solution of *WE3* is a linear combination of an arbitrary function of u and an arbitrary function of v. This general solution can be obtained by applying the partial differential equation solve command, `pdsolve`, to *WE3*.

> `sol:=pdsolve(WE3);`

$$sol := \psi(u,\,v) = \_F2(u) + \_F1(v)$$

The quantities _F1 and _F2 denote arbitrary functions. The general solution
to the original wave equation follows on substituting the transformation *tr* into
the right-hand side of the solution *sol*.

> ```
  sol2:=psi(x,t)=subs(tr,rhs(sol));
  ```

$$sol2 := \psi = \_F2(x+ct) + \_F1(x-ct)$$

Of course, since Jennifer is using Maple, it was not really necessary to mimic
the hand calculation. Applying the `pdsolve` command directly to the original
wave equation *WE* yields a similar general solution.

> ```
  sol3:=pdsolve(WE);
  ```

$$sol3 := \psi = \_F1(x+ct) + \_F2(ct-x)$$

The solution *sol3* is a *linear superposition* of a general wave form traveling in the
negative-x direction (the first term) with speed c and a wave form (the second
term) traveling in the positive-x direction with the same speed. The wave forms
propagate unchanged in shape since dissipative forces have not been included.
As a representative example, Jennifer considers the following specific wave form
ψ which is a linear combination of two Gaussian profiles.

> ```
  psi:=exp(-(x-t)^2)+exp(-(x+t)^2);
  ```

$$\psi := e^{(-(x-t)^2)} + e^{(-(x+t)^2)}$$

On animating $\psi$ with the following command, the initial frame shows a single
pulse of amplitude 2. As the animation progresses, this pulse splits into two
pulses of amplitude 1 propagating in opposite directions with speed $c=1$.

> ```
  animate(psi,x=-10..10,t=0..7,frames=50,numpoints=500);
  ```

4.1.2   Daniel Separates Strings: I Separate Variables

There are strings in the human heart that had better not be vibrated.
Charles Dickens, English novelist, *Barnaby Rudge* (1841)

In contrast to Dickens's quote, there are strings in my heart which vibrate
with pleasure as I watch my young grandson, Daniel, explore his new world.
However, given a multi-stranded string, he would likely try to separate it into
its constituent strands, rather than carrying out the following scenario. But,
who knows? Perhaps, some day, he will separate variables instead of strings.

Young Daniel is playing with a light horizontal string of length L fixed at
$x = 0$ and L, i.e., $\psi(0,t) = \psi(L,t) = 0$. Suppose that he cleverly plucks the
string (which is initially at rest) in such a way that it has an initial profile
$\psi(x,0) = f(x) = h\,x^3\,(L-x)/L^4$. Our task is to use the method of separation of
variables to mathematically determine the subsequent motion of the string and
then animate the string vibrations, taking $L = 1$ m, $h = 5$ m, and $c = 5$ m/s.

After loading the plots package, needed for the animation,

> ```
  restart: with(plots):
  ```
  ```

the 1-dimensional wave equation is entered in *pde*. Unlike Jennifer, I will not bother with the subscript notation.

> pde:=diff(psi(x,t),x,x)=(1/c^2)*diff(psi(x,t),t,t);

$$pde := \frac{\partial^2}{\partial x^2} \psi(x, t) = \frac{\frac{\partial^2}{\partial t^2} \psi(x, t)}{c^2}$$

The separation of variables method assumes that the solution can be written as a product of unknown functions, each function depending on only one of the independent variables. For the present problem, this assumption takes the form $\psi(x,t) = X(x)\,T(t)$. Mentally substituting $\psi(x,t)$ into *pde* and dividing by $\psi(x,t)$ would yield $(d^2 X(x)/dx^2)/X(x) = (d^2 T(t)/dt^2)/(c^2\,T(t))$. The only way this equality can hold is if both sides of the equation are equal to a constant, called the *separation constant*. This then would yield two ODEs, one for determining $X(x)$ and a second for finding $T(t)$. These steps may be achieved by applying the **pdsolve** command to *pde*, but now including the product assumption as a **HINT**.

> pdsolve(pde,psi(x,t),HINT=X(x)*T(t));

$$(\psi(x, t) = X(x)\,T(t))\,\&where\,[\{\frac{d^2}{dt^2} T(t) = \_c_1\,T(t)\,c^2, \frac{d^2}{dx^2} X(x) = \_c_1\,X(x)\}]$$

The resulting ODEs are shown in the above output, with $\_c_1$ the separation constant. The next step is to solve the ODEs for X and T. If, in addition to supplying the **HINT**, the **INTEGRATE** option is included, the separated ODEs are readily solved with the **pdsolve** command.

> pdsolve(pde,psi(x,t),HINT=X(x)*T(t),INTEGRATE);

$$(\psi(x, t) = X(x)\,T(t))\,\&where[\{\{T(t) = \_C3\,e^{(\sqrt{\_c_1}\,c\,t)} + \_C4\,e^{(-\sqrt{\_c_1}\,c\,t)}\},$$
$$\{X(x) = \_C1\,e^{(\sqrt{\_c_1}\,x)} + \_C2\,e^{(-\sqrt{\_c_1}\,x)}\}\}]$$

Finally, the product of X and T must be formed to give $\psi(x,t)$. Including the **build** option in the **pdsolve** command accomplishes this step. In the recipes which follow in this and the following chapter, all three options will be included whenever we wish to separate variables. You will begin to really appreciate the power of the **pdsolve** command when we tackle problems in non-Cartesian coordinate systems.

> sol:=pdsolve(pde,psi(x,t),HINT=X(x)*T(t),INTEGRATE,build);

$$sol := \psi(x, t) = e^{(\sqrt{\_c_1}\,c\,t)}\,\_C3\,\_C1\,e^{(\sqrt{\_c_1}\,x)} + \frac{e^{(\sqrt{\_c_1}\,c\,t)}\,\_C3\,\_C2}{e^{(\sqrt{\_c_1}\,x)}}$$
$$+ \frac{\_C4\,\_C1\,e^{(\sqrt{\_c_1}\,x)}}{e^{(\sqrt{\_c_1}\,c\,t)}} + \frac{\_C4\,\_C2}{e^{(\sqrt{\_c_1}\,c\,t)}\,e^{(\sqrt{\_c_1}\,x)}}$$

To determine the four constants $\_C1$, etc., the two boundary conditions and the two initial conditions must be applied. In order to do this, the form of the solution will be simplified somewhat. First, let's make the substitution $\_c_1 = -k^2$ on the rhs of *sol* and assume that $k > 0$.

```
>  psi:=simplify(subs(_c[1]=-k^2,rhs(sol))) assuming k>0;
```

$$\psi := \_C3 \_C1\, e^{(k\,(c\,t+x)\,I)} + \_C3 \_C2\, e^{(k\,(c\,t-x)\,I)}$$
$$+ \_C4 \_C1\, e^{(-I\,k\,(c\,t-x))} + \_C4 \_C2\, e^{(-I\,k\,(c\,t+x))}$$

Then ψ is converted to trigonometric form and the result expanded.

```
>  psi2:=expand(convert(psi,trig));
```

The lengthy output, which has been suppressed here in the text, involves the terms $\cos(k\,x)$, $\sin(k\,x)$, $\cos(c\,k\,t)$, and $\sin(c\,k\,t)$. To satisfy the boundary condition at $x = 0$, the $\cos(k\,x)$ terms are removed from $\psi 2$ by substituting $\cos(k\,x) = 0$. At $x = L$, the boundary condition yields $\sin(k\,L) = 0$ or $k = n\,\pi/L$, with $n = 1,\ 2,\ 3,\ ....$. This is also substituted. Since the transverse velocity of the string is initially zero, we must also set $\sin(c\,k, t) = 0$.

```
>  psi3:=subs({cos(k*x)=0,sin(c*k*t)=0,k=n*Pi/L},psi2);
```

On then factoring $\psi 3$, the result shown in $\psi 4$ results.

```
>  psi4:=factor(psi3);
```

$$\psi 4 := (-\_C2 + \_C1)\,(\_C3 + \_C4)\cos(\frac{n\,\pi\,c\,t}{L})\sin(\frac{n\,\pi\,x}{L})\,I$$

The following **select** command is used to replace the "ugly" coefficient combination in $\psi 4$ with the symbol A_n. The resulting term, labeled $\psi 5$ here, is the nth term in the infinite series which will represent $\psi(x,t)$.

```
>  psi5:=A[n]*select(has,psi4,{sin,cos});
```

$$\psi 5 := A_n \cos(\frac{n\,\pi\,c\,t}{L})\sin(\frac{n\,\pi\,x}{L})$$

The initial shape of the string is entered.

```
>  f:=h*x^3*(L-x)/L^4:
```

Using linear superposition, $\psi(x,t) = \sum_{n=1}^{\infty} A_n \sin(n\,\pi\,x/L)\cos(n\,\pi\,c\,t/L)$. At $t = 0$, one has $f = \sum_{n=1}^{\infty} A_n \sin(n\,\pi\,x/L)$. Multiplying this equation by $\sin(m\,\pi\,x/L)$ and integrating from $x = 0$ to L, the only term which will survive in the series due to orthogonality of the sine functions is the term corresponding to $n = m$. This procedure is carried out in the following equation.

```
>  eq:=int(f*sin(n*Pi*x/L),x=0..L)
        =int(subs(t=0,psi5)*sin(n*Pi*x/L),x=0..L):
```

The equation is then solved for A_n, assuming that n is an integer.

```
>  A[n]:=solve(eq,A[n]) assuming n::integer;
```

$$A_n := -\frac{12\,h\,(4 - 4\,(-1)^n + n^2\,\pi^2\,(-1)^n)}{n^5\,\pi^5}$$

The parameter values $L = 1$, $h = 5$, and $c = 5$ are entered and a formal sum performed with the first 25 terms in the series being retained.

```
>  L:=1: h:=5: c:=5:
```

```
>  psi6:=Sum(psi5,n=1..25);
```

$$\psi 6 := \sum_{n=1}^{25} \left(-\frac{60\,(4-4\,(-1)^n + n^2\,\pi^2\,(-1)^n)\cos(5\,n\,\pi\,t)\sin(n\,\pi\,x)}{n^5\,\pi^5}\right)$$

Applying the `value` command to $\psi 6$, the motion of the string is now animated over the time range $t=0$ to 160 seconds. The opening frame of the animation is shown in Figure 4.1. By executing the command line, you will be able to observe the vibrations of the string.

```
>   animate(value(psi6),x=0..L,t=0..160,frames=100,
        thickness=2,scaling=constrained,tickmarks=[4,3]);
```

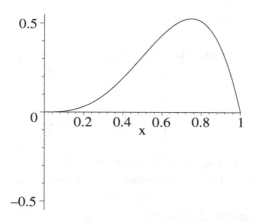

Figure 4.1: The initial shape of the plucked string.

4.1.3 Daniel Strikes Again: Mr. Fourier Reappears

We are all instruments endowed with feeling and memory.
Our senses are so many strings that are struck by surrounding
objects and that also frequently strike themselves.
Denis Diderot, French philosopher, (1713–84)

If Daniel didn't separate the strands or otherwise destroy the string, another possible scenario is that he would strike it in some manner. Consider a string fixed at $x = 0$ and L to be initially horizontal, but is struck in such a way that it is given the initial piecewise velocity profile $g(x)=4\,v\,x/L$ for $x \leq L/4$, $g(x)=4\,(v/L)\,(L/2-x)$ for $L/4 \leq x \leq L/2$, and $g(x)=0$ for $x \geq L/2$. Using the Fourier series approach, determine the subsequent motion of the string and animate it for $L=20$ cm, $v=5$ cm/s, and $c=1$ cm/s.

The plots package is needed for the animation. To perform the integration involving the piece-wise velocity profile, it is necessary to assume that both v and L are positive.

```
>   restart: with(plots): assume(v>0,L>0):
```

A functional operator ψ is introduced to generate the necessary Fourier series. From the previous recipe, it is clear that the spatial part should be built up of terms of the form $\sin(n\pi x/L)$, with $n = 1, 2, 3, ...$, in order to satisfy the boundary conditions at $x = 0$ and L. The time part is left quite general, so that the recipe will run even if the initial conditions are changed.

```
> psi:=N->Sum(sin(n*Pi*x/L)*(a[n]*cos(n*Pi*c*t/L)
     +b[n]*sin(n*Pi*c*t/L)),n=1..N);
```

$$\psi := N \rightarrow \sum_{n=1}^{N} \sin(\frac{n\pi x}{L})\left(a_n \cos(\frac{n\pi ct}{L}) + b_n \sin(\frac{n\pi ct}{L})\right)$$

The given initial spatial $(f(x))$ and velocity $(g(x))$ profiles are entered.

```
> f(x):=0:
  g(x):=piecewise(x<L/4,4*v*x/L,x<L/2,(4*v/L)*(L/2-x),x<L,0);
```

$$g(x) := \begin{cases} \dfrac{4vx}{L} & x < \dfrac{L}{4} \\[2ex] \dfrac{4v(\dfrac{L}{2} - x)}{L} & x < \dfrac{L}{2} \\[2ex] 0 & x < L \end{cases}$$

The coefficients a_n and b_n are explicitly evaluated.

```
> a[n]:=(2/L)*int(f(x)*sin(n*Pi*x/L),x=0..L);
```

$$a_n := 0$$

```
> b[n]:=simplify((2/(n*Pi*c))*int(g(x)*sin(n*Pi*x/L),x=0..L));
```

$$b_n := -\frac{8vL\left(\sin(\frac{n\pi}{2}) - 2\sin(\frac{n\pi}{4})\right)}{n^3\pi^3 c}$$

The first 25 terms in the Fourier series are calculated, being left as formal sum.

```
> sol:=psi(25);
```

$$sol := \sum_{n=1}^{25} \left(-\frac{8\sin(\frac{n\pi x}{L})vL\left(\sin(\frac{n\pi}{2}) - 2\sin(\frac{n\pi}{4})\right)\sin(\frac{n\pi ct}{L})}{n^3\pi^3 c}\right)$$

For animation purposes, the parameter values are substituted into the solution. The transverse velocity of the string is then calculated so that we can decide if 25 terms is sufficient to fit the initial velocity profile and thus ensure an accurate animation of the string.

```
> sol2:=subs({L=20,v=5,c=1},sol): vel:=diff(sol2,t):
```

Substituting the parameter values into $g(x)$ and evaluating the velocity at $t = 0$, $g(x)$ and *vel* are now plotted in the same figure, the former curve being colored blue, the latter colored red. The scaling is constrained. The resulting picture is reproduced in Figure 4.2.

```
> plot([subs({L=20,v=5,c=1},g(x)),eval(vel,t=0)],x=0..20,color
    =[blue,red],thickness=2,scaling=constrained,tickmarks=[4,4]);
```

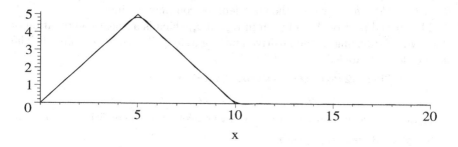

Figure 4.2: Fourier series superimposed on initial velocity profile.

Clearly, 25 terms produces a very good fit to the initial velocity profile. The
motion of the string, described by *sol2*, is then animated.

```
>   animate(sol2,x=0..20,t=0..50,frames=100,thickness=2,
        scaling=constrained,tickmarks=[4,3]);
```
What do you think happens? You will have to execute the recipe to find out.

4.1.4   The 3-Piece String

What is called an acute knowledge of human nature is mostly noth-
ing but the observer's own weaknesses reflected back from others.
G. C. Lichtenberg, German physicist, philosopher, (1742–99)

In this recipe, we will derive the energy reflection and transmission coefficients
for a monochromatic (frequency ω) plane wave traveling from $x = -\infty$ in an
infinitely long string which has a different constant linear density in the region
$x = 0$ to L than in the rest of the string. If the linear density in region 1 ($x < 0$)
is ϵ, in region 2 ($0 < x < L$) is ϵ_2, and in region 3 ($x > L$) is ϵ again, the
corresponding wave numbers are $k_1 = k = \omega \sqrt{\epsilon/T}$, $k_2 = \omega \sqrt{\epsilon_2/T} = r k$ with
$r \equiv \sqrt{(\epsilon_2/\epsilon)}$, and $k_3 = k$, respectively, where T is the tension in the string.

Using complex notation, in region 1 the wave form is $\psi1 = e^{i(kx-\omega t)} +$
$b1\, e^{-i(kx+\omega t)}$. (The physical solution is the real part of $\psi1$.) The spatial part
is now entered, $e^{-i\omega t}$ being omitted since it will cancel.

```
>   restart:
>   psi||1:=exp(I*k*x)+b1*exp(-I*k*x);
```

$$\psi1 := e^{(k\,x\,I)} + b1\, e^{(-I\,k\,x)}$$

The first term is the "incident" wave with unit amplitude[1] traveling in the

---

[1]Since the reflection and transmission coefficients involve ratios, the incident wave amplitude can be chosen to be 1 without any loss of generality.

positive x-direction, the second the "reflected" wave with amplitude $b1$ traveling in the negative x-direction. The energy reflection coefficient, which measures the fraction of the incident wave energy in region 1 which is reflected, then is $R = |b1|^2 = (b1)(b1^*)$, where the star denotes complex conjugate.

    The spatial part of the solution in region 2, which is a linear combination of plane waves traveling in the positive and negative x directions, is entered. The amplitudes are labeled $a2$ and $b2$.

>   `psi||2:=a2*exp(I*r*k*x)+b2*exp(-I*r*k*x);`

$$\psi 2 := a2\, e^{(r\,k\,x\,I)} + b2\, e^{(-I\,r\,k\,x)}$$

In region 3, there will only be a plane wave traveling in the positive x-direction.

>   `psi||3:=a3*exp(I*k*x);`

$$\psi 3 := a3\, e^{(k\,x\,I)}$$

The energy transmission coefficient, which measures the fraction of the incident energy in region 1 which is transmitted into region 3, will be given by $T = |a3|^2$. Clearly, since energy must be conserved, one must have $R + T = 1$. To determine R and T, the four unknown coefficients must be determined. The string is continuous, so $\psi 1 = \psi 2$ at $x = 0$ and $\psi 2 = \psi 3$ at $x = L$. These boundary conditions are entered in *eq1* and *eq2*.

>   `eq||1:=eval(psi||1=psi||2,x=0);`

$$eq1 := 1 + b1 = a2 + b2$$

>   `eq||2:=eval(psi||2=psi||3,x=L);`

$$eq2 := a2\, e^{(r\,k\,L\,I)} + b2\, e^{(-I\,r\,k\,L)} = a3\, e^{(k\,L\,I)}$$

The continuity of the slopes at $x = 0$ and L is entered in *eq3* and *eq4*.

>   `eq||3:=eval(diff(psi||1=psi||2,x),x=0);`

$$eq3 := k\,I - b1\,k\,I = a2\,r\,k\,I - b2\,r\,k\,I$$

>   `eq||4:=eval(diff(psi||2=psi||3,x),x=L);`

$$eq4 := a2\,r\,k\,e^{(r\,k\,L\,I)}\,I - b2\,r\,k\,e^{(-I\,r\,k\,L)}\,I = a3\,k\,e^{(k\,L\,I)}\,I$$

The sequence of four equations is solved for $b1$, $a2$, $b2$, and $a3$.

>   `sol:=solve({seq(eq||i,i=1..4)},{b1,a2,b2,a3}); assign(sol):`

On assigning the solution, R is calculated by multiplying $b1$ by its complex conjugate, employing the complex evaluation command, and simplifying.

>   `R:=simplify(evalc(b||1*conjugate(b||1)));`

$$R := \frac{(-1 + \cos(r\,k\,L)^2)\,(r^2 - 1)^2}{-2\,r^2 \cos(r\,k\,L)^2 - r^4 + r^4 \cos(r\,k\,L)^2 - 2\,r^2 - 1 + \cos(r\,k\,L)^2}$$

The energy transmission coefficient $T = (a3)(a3^*)$ is similarly calculated.

>   `T:=simplify(evalc(a||3*conjugate(a||3)));`

$$T := -\frac{4\,r^2}{-1 - 2\,r^2 + \cos(r\,k\,L)^2 - 2\,r^2 \cos(r\,k\,L)^2 + r^4 \cos(r\,k\,L)^2 - r^4}$$

As a check, let's confirm that the sum $S \equiv R + T = 1$.

```
>  S:=simplify(R+T);
```

$$S := 1$$

To get a feeling for the behavior of R and T, let's choose, say, $r=2$, i.e., region 2 has a density 4 times that of regions 1 and 3. Setting $kL=z$ in R and T, the reflection and energy coefficients are given by $T1$ and $R1$,

```
>  r:=2: T1:=algsubs(k*L=z,T); R1:=algsubs(k*L=z,R);
```

$$T1 := -\frac{16}{-25 + 9\cos(2z)^2} \qquad R1 := \frac{-9 + 9\cos(2z)^2}{-25 + 9\cos(2z)^2}$$

which are then plotted along with $S=1$ over the range $z=0$ to 6.

```
>  plot([R1,T1,S],z=0..6,color=[red,blue,green],thickness=2,
   labels=["kL","R,T"],numpoints=100);
```

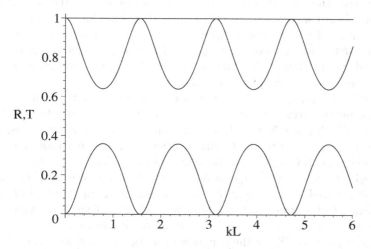

Figure 4.3: R and T for the 3-piece string as a function of kL.

The bottom oscillatory curve is R, the top oscillatory curve is T, and the horizontal line is the sum S of the two. Whenever $rz \equiv rkL = n\pi$, with $n = 0, 1, 2, ...$, the reflection coefficient R is 0 and the transmission coefficient T is 1 (100% transmission).

4.1.5 Encore?

Applause is a receipt, not a bill.
Artur Schnabel, American pianist, on why he didn't give encores, (1882–1951)

Early in my academic career, I was assigned the task of teaching the freshman physics class of several hundred in a huge tiered lecture hall. Being a theoretician, I approached this task with some trepidation, knowing that in order to demonstrate some basic ideas of physics, and to keep the class awake,

I would have to intersperse the lecture material with demonstrations. In one demo, a heavy iron ball was suspended from a long vertical rope attached to a hidden catwalk far above the lecture podium. Standing with my back against a 4 × 8 plywood sheet, I would pull the heavy ball away from the vertical a sizeable distance, bring it up to my chin, and release it. I would stand there bravely, lecturing on the conservation of energy, as the ball completed a few oscillations without, of course, hitting my chin.

This seemed a bit tame, so I decided to add a humorous twist. A departmental assistant was placed out of sight on the catwalk with instructions to give the rope a good heave on the third swing. I would step away from the plywood sheet after the second swing and the iron ball would crash into the plywood. I would then express my relief at having avoided injury by a clear violation of energy conservation. Unfortunately, either the assistant couldn't count or was paid off by the students. He pushed on the second swing! As the ball hurtled towards my chin at an alarming speed, I knew that something was wrong, but I reacted slowly. I got my hands up in time to avoid serious injury, but the momentum of the ball knocked me against the plywood sheet which toppled with a crash to the floor. The students loved it, whistling and cheering and demanding an encore. Somewhat dazed, I declined!

However, I did repeat the demo the next year with a better trained assistant, replacing the plywood with a large plate of old glass painted black so the students didn't know it was glass. This time it was a dazzling success as I stepped away in time and the ball shattered the glass. The down side was that I had to sweep up all the glass, as the janitorial staff refused to do so.

Although the following recipe is not identical to the situation described above, it is inspired by that early exciting demo. A small iron ball of mass M is attached to the lower end of a long vertical rope of length L which has a uniform linear density ϵ. Derive the equation of motion for small transverse oscillations of the rope. Solve the equation of motion for the normal modes of oscillation if the initially vertical rope is given a non-zero transverse velocity. Taking $M = 10$ kg, $L = 10$ m, $\epsilon = 0.1$ kg/m, and the gravitational acceleration $g = 10$ m/s^2, animate the normal mode with the lowest frequency.

The plots library package is needed for the animation and the plottools package required in order to rotate the animation by 90°.

>   `restart: with(plots): with(plottools):`

Taking the origin at the top of the rope and measuring the vertical distance y downwards, the tension T in the rope will be given by $T = M g + \epsilon (L - y) g$. At the bottom of the rope $(y = L)$, the upward tension has to only balance the weight of the ball, but at the top of the rope $(y = 0)$ it has to balance both the weight of the ball and the weight of the entire rope.

>   `T:=M*g+epsilon*(L-y)*g;`

$$T := M g + \varepsilon (L - y) g$$

The transverse (horizontal) oscillations of the rope will be described by the 1-dimensional wave equation $\partial(T\, \partial\psi/\partial y)/\partial y = \epsilon\, \partial^2\psi/\partial t^2$. On entering this PDE, the tension is automatically substituted, yielding the equation of motion *pde*.

```
>  pde:=diff(T*diff(psi(y,t),y),y)=epsilon*diff(psi(y,t),t,t);
```

$$pde := -\varepsilon\, g\left(\frac{\partial}{\partial y}\,\psi(y,\,t)\right) + (M\,g + \varepsilon\,(L-y)\,g)\left(\frac{\partial^2}{\partial y^2}\,\psi(y,\,t)\right) = \varepsilon\left(\frac{\partial^2}{\partial t^2}\,\psi(y,\,t)\right)$$

Since the rope initially has zero displacement, but a non-zero velocity, a normal-mode solution of the form $\psi(y,t) = X(y)\sin(\omega t)$ is sought. The `pdsolve` command is applied, with the assumptions that ω, ϵ, M, L, and g are positive, and $y < L$.

```
>  sol:=pdsolve(pde,psi(y,t),HINT=X(y)*sin(omega*t),INTEGRATE,
   build) assuming omega::positive, epsilon::positive,
   M::positive,L::positive, g::positive, y<L;
```

$$sol := \psi(y,\,t) = \sin(\omega\,t)\,\_C1\,\mathrm{BesselJ}\!\left(0,\,\frac{2\,\omega\,\sqrt{M+\varepsilon\,L-\varepsilon\,y}}{\sqrt{\varepsilon}\,\sqrt{g}}\right)$$

$$+\,\sin(\omega\,t)\,\_C2\,\mathrm{BesselY}\!\left(0,\,\frac{2\,\omega\,\sqrt{M+\varepsilon\,L-\varepsilon\,y}}{\sqrt{\varepsilon}\,\sqrt{g}}\right)$$

The spatial part of the solution is a linear combination of zeroth-order Bessel functions of the first and second kinds. The spatial part X is now extracted.

```
>  X:=simplify(rhs(sol)/sin(omega*t),symbolic);
```

$$X := \_C1\,\mathrm{BesselJ}\!\left(0,\,\frac{2\,\omega\,\sqrt{M+\varepsilon\,L-\varepsilon\,y}}{\sqrt{\varepsilon}\,\sqrt{g}}\right) + \_C2\,\mathrm{BesselY}\!\left(0,\,\frac{2\,\omega\,\sqrt{M+\varepsilon\,L-\varepsilon\,y}}{\sqrt{\varepsilon}\,\sqrt{g}}\right)$$

Assuming that the free-end boundary condition $(dX/dy)|_{y=L} = 0$ applies, we solve for $\_C1$.

```
>  _C1:=solve(subs(y=L,diff(X,y))=0,_C1);
```

$$\_C1 := -\frac{\_C2\,\mathrm{BesselY}\!\left(1,\,\dfrac{2\,\omega\,\sqrt{M}}{\sqrt{\varepsilon}\,\sqrt{g}}\right)}{\mathrm{BesselJ}\!\left(1,\,\dfrac{2\,\omega\,\sqrt{M}}{\sqrt{\varepsilon}\,\sqrt{g}}\right)}$$

The parameter values are entered and X divided by the arbitrary constant $\_C2$ and expanded to yield $X2$.

```
>  L:=10: M:=10: g:=10: epsilon:=1/10: X2:=expand(X/_C2);
```

$$X2 := -\frac{\mathrm{BesselY}(1,\,2\,\omega\,\sqrt{10})\,\mathrm{BesselJ}\!\left(0,\,2\,\omega\,\sqrt{-\dfrac{y}{10}+11}\right)}{\mathrm{BesselJ}(1,\,2\,\omega\,\sqrt{10})}$$

$$+\,\mathrm{BesselY}\!\left(0,\,2\,\omega\,\sqrt{-\dfrac{y}{10}+11}\right)$$

The fixed-end boundary condition at $y=0$ is applied to the numerator of $X2$.

```
>  bc:=eval(numer(X2),y=0)=0;
```

$$bc := -\text{BesselY}(1, 2\omega\sqrt{10})\,\text{BesselJ}(0, 2\omega\sqrt{11})$$
$$+ \text{BesselY}(0, 2\omega\sqrt{11})\,\text{BesselJ}(1, 2\omega\sqrt{10}) = 0$$

The left-hand side of bc is plotted in Figure 4.4 over the frequency range $\omega = 0.1$ to 20 and the vertical view limited so as to be able to clearly see the zeros.

```
>  plot(lhs(bc),omega=0.1..20,view=[0..20,-0.2..0.1]);
```

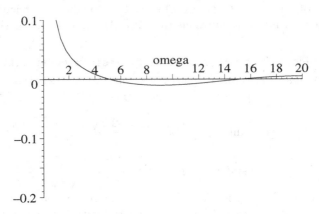

Figure 4.4: Plot of lhs of the boundary condition at $y=0$ versus frequency.

Guided by the plot, the lowest zero is numerically sought between $\omega = 4$ and 6.

```
>  omega:=fsolve(bc,omega=4..6);
```

$$\omega := 5.137764180$$

The lowest possible frequency for a normal mode is $\omega = 5.14$ Hz. The mode with the lowest frequency then is given by ψ. Higher frequency normal modes can be obtained by selecting other zeros in the fixed-end boundary condition.

```
>  psi:=X2*sin(omega*t);
```

$$\psi := \left(-\frac{\text{BesselY}(1, 10.27552836\sqrt{10})\,\text{BesselJ}(0, 10.27552836\sqrt{-\frac{y}{10}+11})}{\text{BesselJ}(1, 10.27552836\sqrt{10})}\right.$$
$$\left. + \text{BesselY}(0, 10.27552836\sqrt{-\frac{y}{10}+11})\right)\sin(5.137764180\,t)$$

The solution ψ is animated, the scaling being left unconstrained for better viewing. Using the **rotate** command, the figure is rotated by $-90°$ so that the string is initially vertical and the subsequent time-dependent displacement is horizontal.

```
>  rotate(animate(psi,y=0..L,t=0..5,color=red,
      thickness=2,frames=50,numpoints=100),-Pi/2);
```

You will have to execute the worksheet on your computer to see the animation.

4.2   Beyond the String

Many more string examples are given in the **Supplementary Recipes**, but it's time to move on to other physical examples and also look at two- and three-dimensional situations. All the recipes in this section involve Cartesian coordinates. The next section is devoted entirely to non-Cartesian examples.

4.2.1   Heaviside's Telegraph Equation

Get your facts first,
and then you can distort them as much as you please.
Mark Twain, American author on propaganda, (1835–1910)

Consider an electrical transmission line carrying a current $I(x,t)$ that has a uniform inductance L, capacitance C to ground (zero potential), and resistance R (all per unit length) and has a leakage (leakage coefficient G per unit length) proportional to the potential V from the line to ground.

(a) Show that $V(x,t)$ satisfies the *telegraph equation*,

$$\frac{\partial^2 V}{\partial x^2} = LC\,\frac{\partial^2 V}{\partial t^2} + (RC + LG)\frac{\partial V}{\partial t} + RGV. \qquad (4.6)$$

(b) Under what conditions does the telegraph equation reduce to a wave equation? a diffusion equation? Identify the wave velocity for the former, the diffusion coefficient for the latter.

(c) Show that a general solution of the form $V(x,t) = e^{-kt} f(\alpha x + \beta t)$, where f is an arbitrary function, will satisfy (4.6), provided that $RC = LG$ and k, α, and β take on certain forms which are to be determined. Briefly discuss what this solution means.

This problem was first solved by the English electrical engineer Oliver Heaviside in 1887. Heaviside also reduced Maxwell's equations into the simpler form that we now know, invented the differential operator (D) notation for solving differential equations, predicted the existence of an ionized reflective layer (the Heaviside layer) in the ionosphere which bounces radio signals back to earth, and predicted that the mass of an electric charge would increase with velocity.

   Now let's answer the question. In order to use the symbol I for the current, the imaginary unit $\sqrt{-1}$ is set equal to j with the following interface command.

```
> restart: interface(imaginaryunit=j):
```

The rate of change of the voltage V with distance x is given by *eq1*. The first term $-RI$ on the rhs is the potential drop per unit length due to the resistance, which is given by Ohm's law. The second term $-L\,\partial I/\partial t$ is the back emf per unit length due to the inductance.

```
> eq1:=diff(V(x,t),x)=-R*I(x,t)-L*diff(I(x,t),t);
```

$$eq1 := \frac{\partial}{\partial x}V(x,t) = -RI(x,t) - L\left(\frac{\partial}{\partial t}I(x,t)\right)$$

The rate of change of the current I with x is given by *eq2*. The first term on the rhs is the leakage of current per unit length from the line to ground, the second term the capacitive loss per unit length to ground.

> `eq2:=diff(I(x,t),x)=-G*V(x,t)-C*diff(V(x,t),t);`

$$eq2 := \frac{\partial}{\partial x} I(x,\, t) = -G\, V(x,\, t) - C\left(\frac{\partial}{\partial t} V(x,\, t)\right)$$

eq1 and *eq2* are differentiated with respect to x, t, respectively, in *eq3* and *eq4*.

> `eq3:=diff(eq1,x); eq4:=diff(eq2,t);`

$$eq3 := \frac{\partial^2}{\partial x^2} V(x,\, t) = -R\left(\frac{\partial}{\partial x} I(x,\, t)\right) - L\left(\frac{\partial^2}{\partial x\, \partial t} I(x,\, t)\right)$$

$$eq4 := \frac{\partial^2}{\partial x\, \partial t} I(x,\, t) = -G\left(\frac{\partial}{\partial t} V(x,\, t)\right) - C\left(\frac{\partial^2}{\partial t^2} V(x,\, t)\right)$$

A PDE for $V(x,t)$ alone is obtained by substituting *eq2* and *eq4* into *eq3*. Collecting the $\partial V/\partial t$ terms then yields the telegraph equation *TE*.

> `eq5:=subs({eq2,eq4},eq3);`

> `TE:=collect(eq5,diff(V(x,t),t));`

$$TE := \frac{\partial^2}{\partial x^2} V(x,\, t) = (RC + LG)\left(\frac{\partial}{\partial t} V(x,\, t)\right) + RG\, V(x,\, t) + LC\left(\frac{\partial^2}{\partial t^2} V(x,\, t)\right)$$

The wave equation *WE* follows on setting $R=0$ and $G=0$ in *TE*, the wave velocity being $c=1/\sqrt{LC}$.

> `WE:=eval(TE,{R=0,G=0});`

$$WE := \frac{\partial^2}{\partial x^2} V(x,\, t) = LC\left(\frac{\partial^2}{\partial t^2} V(x,\, t)\right)$$

The diffusion equation *DE* results on setting $L = 0$ and $G = 0$ in *TE*, the diffusion coefficient being $d = 1/(RC)$.

> `DE:=eval(TE,{L=0,G=0});`

$$DE := \frac{\partial^2}{\partial x^2} V(x,\, t) = RC\left(\frac{\partial}{\partial t} V(x,\, t)\right)$$

To answer the last part of the question, the given ansatz is entered.

> `ansatz:=V(x,t)=exp(-k*t)*f(alpha*x+beta*t);`

$$ansatz := V(x,\, t) = e^{(-kt)} f(\alpha x + \beta t)$$

The **pdetest** command is used to test whether the proposed solution satisfies the telegraph equation *TE*. If it does, the answer would be zero, but instead it yields an algebraic equation, appropriately in terms of differential operators, the notation introduced by Heaviside.

> `eq6:=pdetest(ansatz,TE);`

$$eq6 := -e^{(-kt)}(-(D^{(2)})(f)(\alpha x + \beta t)\alpha^2 - RCkf(\alpha x + \beta t)$$
$$+ RCD(f)(\alpha x + \beta t)\beta - LGkf(\alpha x + \beta t) + LGD(f)(\alpha x + \beta t)\beta$$
$$+ RGf(\alpha x + \beta t) + LCk^2 f(\alpha x + \beta t) - 2LCkD(f)(\alpha x + \beta t)\beta$$
$$+ LC(D^{(2)})(f)(\alpha x + \beta t)\beta^2)$$

By inspection of the output in *eq6*, the second-order differential terms will cancel if we choose $\beta = \alpha/\sqrt{LC}$. Assuming that $G = RC/L$, all remaining terms will cancel provided that $k = R/L$. To confirm that this is so, these relations are entered and the new ansatz, *ansatz2* displayed, and entered as the argument in pdetest.

```
> beta:=alpha/sqrt(L*C): G:=R*C/L: k:=R/L: ansatz2:=ansatz;
```

$$ansatz2 := V(x, t) = e^{\left(-\frac{Rt}{L}\right)} f\left(\alpha x + \frac{\alpha t}{\sqrt{LC}}\right)$$

```
> pdetest(ansatz2,TE);
```
$$0$$

The answer is zero, indicating that a solution to the telegraph equation exists for $G = RC/L$ of the form $V = e^{-Rt/L} f(\alpha(x + t/\sqrt{LC}))$. This solution implies "distortionless" propagation of the input wave form with velocity $1/\sqrt{LC}$, the wave form decreasing in amplitude with time.

4.2.2 Spiegel's Diffusion Problems

Any solution to a problem changes the problem.
R. W. Johnson, American journalist, *Washingtonian, Nov. 1979*

Murray Spiegel's *Advanced Mathematics* [Spi71] is probably the classic source book of solved problems relevant to mathematical physics. The following recipe is based on two diffusion examples (Problems **12.16** and **12.17**) taken from that text. Of course, we shall use computer algebra here, instead of performing a hand calculation. I shall take each example one step further by animating the solutions. Simply staring at the series solutions is no substitute for observing what actually happens as time progresses. So here are the problems.

(a) Using the method of separation of variables, solve the 1-dimensional heat diffusion equation in a thin bar 3 m in length whose surface is insulated and has a diffusion coefficient of 2 m^2/s. Its ends are kept at $0\,°C$ $(T(0,t) = T(3,t) = 0)$ and it has the initial temperature profile $T(x,0) = 5\sin(4\pi x) - 3\sin(8\pi x) + 2\sin(10\pi x)$. Animate the temperature profile over the time interval $t = 0$ to 0.025 seconds.

(b) If the bar in part (a) has an initial internal constant temperature of $25\,°C$ with the ends held at $0\,°C$, use the Fourier series method to derive the temperature distribution for arbitrary time $t > 0$. Animate the solution over the time interval $t = 0$ to 2 seconds, keeping 20 terms in the series.

After loading the plot package, needed for the animations, the 1-dimensional heat diffusion equation for the bar of length $L=3$ and diffusion coefficient $d=2$ is entered in *pde*.

```
> restart: with(plots): d:=2: L:=3:
> pde:=diff(T(x,t),t)=d*diff(T(x,t),x,x);
```

$$pde := \frac{\partial}{\partial t}\, T(x,\, t) = 2\,(\frac{\partial^2}{\partial x^2}\, T(x,\, t))$$

The `pdsolve` command is applied to *pde*. Since the separation of variables method is requested, the hint that $T(x,t) = X(x)\,Y(t)$ is provided. The separated ODEs are integrated and the product solution built in *sol*.

```
> sol:=pdsolve(pde,HINT=X(x)*Y(t),INTEGRATE,build);
```

$$sol := T(x,\, t) = \_C3\,(e^{(-c_1\, t)})^2\, \_C1\, e^{(\sqrt{-c_1}\, x)} + \frac{\_C3\,(e^{(-c_1\, t)})^2\, \_C2}{e^{(\sqrt{-c_1}\, x)}}$$

Without loss of generality the integration constant $\_C3$ can be set equal to 1 on the rhs of *sol*. For later convenience, $\_C1$ and $\_C2$ are relabeled as A and B, respectively, and the separation constant $\_c_1 = -k^2$.

```
> T:=subs({_C3=1,_C1=A,_C2=B,_c[1]=-k^2},rhs(sol));
```

$$T := (e^{(-k^2\, t)})^2\, A\, e^{(\sqrt{-k^2}\, x)} + \frac{(e^{(-k^2\, t)})^2\, B}{e^{(\sqrt{-k^2}\, x)}}$$

T is simplified with the symbolic option in $T2$.

```
> T2:=simplify(T,symbolic);
```

$$T2 := A\, e^{(-k\,(2\, k\, t - x\, I))} + B\, e^{(-k\,(2\, k\, t + x\, I))}$$

To satisfy the boundary condition $T(0,t) = 0$, we must set $B = -A$ in $T2$. The complex evaluation command is then applied so as to convert the complex exponential into a sine function.

```
> T3:=evalc(subs(B=-A,T2));
```

$$T3 := 2\, I\, A\, e^{(-2\, k^2\, t)} \sin(k\, x)$$

To satisfy the boundary condition $T(3,t)=0$, we must have $\sin(3\,k)=0$, so $k = m\,\pi/3$, with $m=1,\, 2,\, 3,\, ...$. The coefficient is removed from $T3$ by substituting $A=1/(2\,I)$.

```
> k:=m*Pi/L: T4:=subs(A=1/(2*I),T3);
```

$$T4 := e^{(-\frac{2\, m^2\, \pi^2\, t}{9})} \sin(\frac{m\, \pi\, x}{3})$$

The general solution is of the form $T(x,t) = \sum_{m=1}^{\infty} C_m\, e^{(-\frac{2\, m^2\, \pi^2\, t}{9})} \sin(\frac{m\, \pi\, x}{3})$. The initial temperature profile will be satisfied by retaining the three terms corresponding to $m = 12,\, 24$, and 30, with $C_{12} = 5$, $C_{24} = -3$, and $C_{30} = 2$. Using $T4$, this is done in $T5$.

```
> T5:=5*eval(T4,m=12)-3*eval(T4,m=24)+2*eval(T4,m=30);
```

$$T5 := 5\, e^{(-32\, \pi^2\, t)} \sin(4\, \pi\, x) - 3\, e^{(-128\, \pi^2\, t)} \sin(8\, \pi\, x) + 2\, e^{(-200\, \pi^2\, t)} \sin(10\, \pi\, x)$$

The temperature profile given in *T5* is animated over the time interval $t=0$ to 0.025 s, the opening frame of the animation (initial temperature profile) being shown in Figure 4.5. As t increases, $T(x,t)$ decays to zero inside the bar.

```
>   animate(T5,x=0..L,t=0..0.025,frames=100,numpoints=500,
    thickness=2,labels=["x","T"]);
```

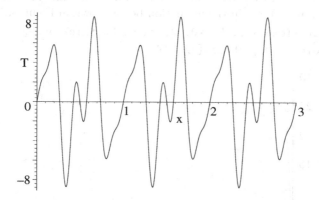

Figure 4.5: The initial temperature profile in the bar for part **(a)**.

Now let's tackle part **(b)** of the problem. The initial temperature profile is a constant 25 °C everywhere inside the bar. Using the Fourier series approach, the general Fourier coefficient is given by $C_m = (2/L) \int_0^L 25 \sin(m \pi x/L)\, dx$ with $L = 3$ and m a positive integer. This coefficient is now calculated.

```
>   C[m]:=(2/L)*int(25*sin(m*Pi*x/L),x=0..L) assuming m::integer;
```

$$C_m := -\frac{50\,(-1+(-1)^m)}{m\,\pi}$$

The formal Fourier series representation (out to $m=20$) of the temperature is displayed in *Temp*, and then explicitly calculated with the **value** command.

```
>   Temp:=Sum(C[m]*T4,m=1..20);
```

$$Temp := \sum_{m=1}^{20}\left(-\frac{50\,(-1+(-1)^m)\,e^{\left(-\frac{2m^2\pi^2 t}{9}\right)}\sin(\frac{m\pi x}{3})}{m\,\pi}\right)$$

```
>   Temp:=value(Temp);
```

$$Temp:=\frac{100\,e^{\left(-\frac{2\pi^2 t}{9}\right)}\sin(\frac{\pi x}{3})}{\pi}+\frac{100\,e^{(-2\pi^2 t)}\sin(\pi x)}{3\,\pi}$$

$$+\frac{20\,e^{\left(-\frac{50\pi^2 t}{9}\right)}\sin(\frac{5\pi x}{3})}{\pi}+\frac{100}{7}\frac{e^{\left(-\frac{98\pi^2 t}{9}\right)}\sin(\frac{7\pi x}{3})}{\pi}+\frac{100\,e^{(-18\pi^2 t)}\sin(3\pi x)}{9\,\pi}$$

$$+\frac{100}{11}\frac{e^{\left(-\frac{242\pi^2 t}{9}\right)}\sin(\frac{11\pi x}{3})}{\pi}+\frac{100}{13}\frac{e^{\left(-\frac{338\pi^2 t}{9}\right)}\sin(\frac{13\pi x}{3})}{\pi}+\frac{20\,e^{(-50\pi^2 t)}\sin(5\pi x)}{3\,\pi}$$

$$+\frac{100}{17}\,\frac{e^{\left(-\frac{578\,\pi^2 t}{9}\right)}\sin(\frac{17\pi x}{3})}{\pi}+\frac{100}{19}\,\frac{e^{\left(-\frac{722\,\pi^2 t}{9}\right)}\sin(\frac{19\pi x}{3})}{\pi}$$

The temperature distribution is now animated over the time interval $t=0$ to 2 seconds, the initial frame in the animation being shown in Figure 4.6.

```
>   animate(Temp,x=0..L,t=0..2,frames=50,thickness=2,
        numpoints=500,labels=["x","T"]);
```

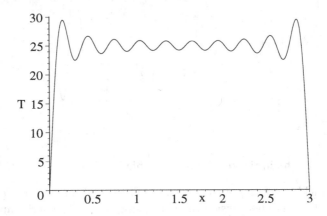

Figure 4.6: The initial temperature profile for part **(b)** with 20 terms.

Because of the step function nature of the initial temperature profile at the ends of the bar, Gibb's oscillations appear in the initial series representation, oscillating around the exact profile. The oscillations quickly disappear as the animation progresses and once again $T(x,t)$ decreases to zero inside the bar.

4.2.3   Introducing Laplace's Equation

An editor is someone who separates the wheat from the chaff and then prints the chaff.
Adlai Stevenson, American politician, referring to news reporting (1900–65)

In electrostatics, the electric field $\vec{E}$ is related to the electric potential V by $\vec{E}=-\nabla V$. In SI units, Maxwell's equations are $\nabla\times\vec{E}=0$ and $\nabla\cdot\vec{E}=\rho/\epsilon_0$, where ρ is the electric charge density and ϵ_0 the permittivity of free space. Substituting $\vec{E}$ into the first relation yields the vector identity $\nabla\times\nabla V=0$, while the second relation produces Poisson's equation, $\nabla^2 V=-\rho/\epsilon_0$. In charge-free regions of space, $\rho=0$ and Poisson's equation reduces to Laplace's equation $\nabla^2 V=0$. This recipe involves solving a two-dimensional electrostatic boundary-value problem, using Laplace's equation in Cartesian coordinates.

Determine the potential $V(x,y)$ in the charge-free region $x \geq 0$, $0 \leq y \leq a$, given that the boundary at $x=0$ is held at the constant potential V_0, while the boundaries at $y=0$ and a are held at zero potential. The asymptotic boundary condition is $V \rightarrow 0$ for $x \rightarrow \infty$. Taking $a=1$ meter and $V_0 = 2$ volts, produce two- and three-dimensional contour plots of $V(x,y)$ for $x \geq 0$, $0 \leq y \leq a$.

After loading the plots library package, needed for the contour plots, the two-dimensional form of Laplace's equation is entered for the potential $V(x,y)$.

```
>  restart: with(plots):
>  pde:=diff(V(x,y),x,x)+diff(V(x,y),y,y)=0;
```

$$pde := (\frac{\partial^2}{\partial x^2} V(x,\,y)) + (\frac{\partial^2}{\partial y^2} V(x,\,y)) = 0$$

pde is analytically solved, assuming that $V(x,y)=X(x)\,Y(y)$.

```
>  sol:=pdsolve(pde,HINT=X(x)*Y(y),INTEGRATE,build);
```

$$sol := V(x,\,y) = \_C3 \sin(\sqrt{-c_1}\,y)\,\_C1\,e^{(\sqrt{-c_1}\,x)} + \frac{\_C3 \sin(\sqrt{-c_1}\,y)\,\_C2}{e^{(\sqrt{-c_1}\,x)}}$$

$$+ \_C4 \cos(\sqrt{-c_1}\,y)\,\_C1\,e^{(\sqrt{-c_1}\,x)} + \frac{\_C4 \cos(\sqrt{-c_1}\,y)\,\_C2}{e^{(\sqrt{-c_1}\,x)}}$$

The substitution $\_c_1 = k$ is made on the rhs of *sol* and the result simplified.

```
>  sol:=simplify(subs(sqrt(_c[1])=k,rhs(sol)));
```

$$sol := (\,\_C3 \sin(k\,y)\,\_C1\,e^{(2\,k\,x)} + \_C3 \sin(k\,y)\,\_C2$$

$$+ \_C4 \cos(k\,y)\,\_C1\,e^{(2\,k\,x)} + \_C4 \cos(k\,y)\,\_C2\,)\,e^{(-k\,x)}$$

To satisfy the boundary condition $V=0$ at $y=0$, the $\cos(k\,y)$ term is removed from *sol* by setting it equal to zero. To satisfy $V=0$ at $y=a$ for arbitrary $x>0$, we must have $\sin(k\,a)=0$, so $k=n\,\pi/a$, where n is a positive integer. The term $e^{2\,k\,x}$ must also be removed so that $V \rightarrow 0$ as $x \rightarrow \infty$.

```
>  sol2:=subs({cos(k*y)=0,exp(2*k*x)=0,k=n*Pi/a},sol);
```

$$sol2 := \_C3 \sin(\frac{n\,\pi\,y}{a})\,\_C2\,e^{(-\frac{n\,\pi\,x}{a})}$$

Using the operand command, op, the coefficient combination is replaced with *A*. The arguments may have to be changed if the terms are ordered differently.

```
>  sol3:=A*op(2,sol2)*op(4,sol2);
```

$$sol3 := A \sin(\frac{n\,\pi\,y}{a})\,e^{(-\frac{n\,\pi\,x}{a})}$$

Making use of orthogonality of the sine functions over the interval $y=0$ to a, the coefficient A is calculated, assuming that n is an integer.

```
>  A:=(2/a)*int(V0*sin(n*Pi*y/a),y=0..a) assuming n::integer;
```

$$A := -\frac{2\,V0\,(-1+(-1)^n)}{n\,\pi}$$

Entering the given values of the parameters, the result in *sol3* is summed. A large number of terms (100) is kept to obtain smooth contour plots.

```
>   a:=1: V0:=2: V:=sum(sol3,n=1..100):
```

Using the `contourplot` command, V is plotted over the interval $x = 0$ to 2
and $y = 0$ to $a = 2$. Constant potential curves (equipotentials) are plotted for
0.2, 0.4,...,1.6, 1.8 volts. The grid is taken to be 50×50 in order to obtain
reasonably smooth curves. The resulting picture is shown in Figure 4.7, the 0.2
volt curve being furthest to the right.

```
>   contourplot(V,x=0..2,y=0..a,contours=[seq(i*V0/10,i=1..9)],
      thickness=2,grid=[50,50]);
```

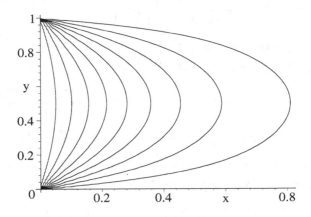

Figure 4.7: Equipotentials corresponding to 0.2 (far right), 0.4,...,1.8 volts.

The `contourplot3d` command is used to produce a three-dimensional contour plot, the equipotential curves again being spaced 0.2 volts apart.

```
>   contourplot3d(V,x=0..1,y=0..a,contours=[seq(i*V0/10,i=1..9)],
      grid=[60,60],shading=zhue,filled=true,axes=boxed,
      tickmarks=[2,2,2]);
```

The resulting picture can be viewed by executing the recipe on your computer.

4.2.4   Grandpa's "Trampoline"

In the first place God made idiots. This was for practice.
Then He made School Boards.
Mark Twain, *Following the Equator, (1897)*

From his sundeck, my grandson Daniel can peek over the fence and see his
neighbor's children happily bouncing up and down on their trampoline after
coming home from school. He is too young to join in the fun, so must be content (not very!) to watch. On the other hand, I am getting too old to engage in
trampoline antics, being more content to substitute the following recipe instead.

A trampoline, consisting of a square horizontal elastic membrane fixed on its four edges ($x=0$, b and $y=0$, b), has an initial transverse profile $\psi(x,y,0) \equiv f = A x^2 y \, (b-x)(b-y)^3$ and is released from rest. Determine $\psi(x,y,t)$ for arbitrary time $t > 0$. Taking $A = 1/5$ m^{-7}, $b=2$ m, and the wave speed $c=1$ m/s, animate the trampoline motion.

The plots package is loaded, because we shall be animating the motion of the trampoline. A functional operator F is created to calculate the second derivative of $\psi(x,y,t)$ with respect to an arbitrary variable v.

```
>  restart: with(plots): F:=v->diff(psi(x,y,t),v,v):
```

The transverse vibrations of the trampoline are governed by the two-dimensional wave equation which is entered using F.

```
>  pde:=F(x)+F(y)=F(t)/c^2;
```

$$pde := \left(\frac{\partial^2}{\partial x^2} \psi(x,\, y,\, t)\right) + \left(\frac{\partial^2}{\partial y^2} \psi(x,\, y,\, t)\right) = \frac{\frac{\partial^2}{\partial t^2} \psi(x,\, y,\, t)}{c^2}$$

The wave equation is solved by the method of separation of variables, the assumed form $\psi(x,y,t) = X(x)\,Y(y)\,T(t)$ being supplied as a hint.

```
>  sol:=pdsolve(pde,HINT=X(x)*Y(y)*T(t),INTEGRATE,build);
```

The output (not displayed here) has two separation constants, $\_c_1$ and $\_c_2$. To simplify the form of the solution, the substitution $\_c_1 = -\alpha^2$ and $\_c_2 = -\beta^2$ is made on the rhs of *sol* and the result simplified with the symbolic option.

```
>  psi1:=simplify(subs({_c[1]=-alpha^2,_c[2]=-beta^2},
         rhs(sol)),symbolic);
```

$$\psi1 := \_C5 \sin(c\sqrt{\alpha^2 + \beta^2}\, t) \_C3 \_C1 \, e^{((\beta y + \alpha x)\, I)} + \cdots$$
$$+ \_C6 \cos(c\sqrt{\alpha^2 + \beta^2}\, t) \_C3 \_C1 \, e^{((\beta y + \alpha x)\, I)} + \cdots$$

The initial transverse velocity is zero, so the time-dependent sine terms must be removed from $\psi1$.

```
>  psi2:=remove(has,psi1,sin);
```

The exponential terms in $\psi2$ are converted to a trig form in $\psi3$ and the result expanded. The terms $\cos(\beta y)$, $\sin(\beta y)$, $\cos(\alpha x)$, and $\sin(\alpha x)$ appear.

```
>  psi3:=expand(convert(psi2,trig));
```

To satisfy the fixed edge conditions along $y=0$ and $x=0$, the cosine terms are removed from the solution by setting them equal to zero, while one must have $\alpha = m\pi/b$ and $\beta = n\pi/b$, with $m, n = 1, 2, 3, \ldots$, to satisfy the conditions along $x=b$ and $y=b$. Making these substitutions and factoring yields $\psi4$.

```
>  psi4:=factor(subs({cos(beta*y)=0,cos(alpha*x)=0,
         alpha=m*Pi/b,beta=n*Pi/b},psi3));
```

$$\psi4 := -\cos\!\left(c\pi\sqrt{\frac{m^2 + n^2}{b^2}}\, t\right) \sin\!\left(\frac{n\pi y}{b}\right) \sin\!\left(\frac{m\pi x}{b}\right) \_C6 \, (\_C1 - \_C2)(\_C3 - \_C4)$$

The following `select` command is used to replace the cumbersome coefficient combination in $\psi 4$ with the symbol $B_{m,n}$.

> psi5:=B[m,n]*select(has,psi4,{sin,cos});

$$\psi 5 := B_{m,n} \cos(c\,\pi\,\sqrt{\frac{m^2+n^2}{b^2}}\,t)\,\sin(\frac{n\,\pi\,y}{b})\,\sin(\frac{m\,\pi\,x}{b})$$

The initial transverse profile of the trampoline is entered.

> f:=A*x^2*y*(b-x)*(b-y)^3;

$$f := A\,x^2\,y\,(b-x)\,(b-y)^3$$

Using orthogonality of the sine functions, the coefficient $B_{m,n}$ is evaluated by performing the double integration

$$B_{m,n} = (2/b)^2 \int_0^b \int_0^b f\,\sin(m\,\pi\,x/b)\,\sin(n\,\pi\,y/b)\,dx\,dy,$$

and assuming that m and n are integers.

> B[m,n]:=(2/b)^2*int(int(f*sin(m*Pi*x/b)*sin(n*Pi*y/b),
>   x=0..b),y=0..b) assuming m::integer,n::integer;

$$B_{m,\,n} := -48\,b^7\,A(-4 + 8\,(-1)^{(1+m)} + n^2\,\pi^2 + 2\,(-1)^m\,\pi^2\,n^2 + 4\,(-1)^n$$
$$+8\,(-1)^{(n+m)}\big)\Big/(m^3\,\pi^8\,n^5)$$

The solution then will be of the form

$$\psi(x,y,t) = \sum_{m=1}^{\infty} \sum_{n=1}^{\infty} B_{m,n}\,\sin(m\,\pi\,x/b)\,\sin(n\,\pi\,y/b)\,\cos(c\,\pi\,\sqrt{m^2+n^2}\,t/b).$$

A functional operator G is formed, using $\psi 5$, for explicitly calculating this double Fourier series out to N terms in each sum,

> G:=N->sum(sum(psi5,m=1..N),n=1..N):

The given parameters are entered, and the double series calculated for $N=10$. You will have to execute the recipe on your own computer to see the $10 \times 10 = 100$ terms in ψ.

> A:=1/5: b:=2: c:=1: N:=10: psi:=G(N);

Using the animate command, the answer ψ is animated over the time interval $t=0$ to 5 seconds, 50 frames being used. Execute the recipe and enjoy!

> animate(plot3d,[psi,x=0..b,y=0..b],t=0..5,frames=50,
>   axes=boxed,shading=zhue,tickmarks=[3,3,3]);

4.2.5   Irma Insect's Isotherm

A man thinks he amounts to a great deal but to a flea or a mosquito a human being is merely something good to eat.
Don Marquis, American humorist, (1878–1937)

In the Erehwon zoo, Irma insect lives inside a rectangular enclosure with walls at $x=0$, a and $y=0$, b, floor at $z=0$ and ceiling at $z=c$. In the winter, the floor and walls have a temperature of zero degrees while the ceiling has a steady temperature profile $T(x, y, c) = 1600\,x\,(a - x)\,y\,(b - y)$ degrees. Determine the temperature inside Irma's enclosure. Irma is most comfortable when the temperature is $20°$. If $a = b = c = 1$, plot the 3-dimensional isothermal surface corresponding to this temperature. What is T at the center of the enclosure?

The steady-state temperature inside the enclosure is given by the solution of Laplace's equation, $\nabla^2 T(x, y, z) = 0$. For variety, let's load the VectorCalculus package and use the Laplacian command in Cartesian coordinates.

```
>  restart: with(plots): with(VectorCalculus):
>  pde:=Laplacian(T(x,y,z),'cartesian'[x,y,z])=0;
```

$$pde := (\frac{\partial^2}{\partial x^2}\,T(x,\,y,\,z)) + (\frac{\partial^2}{\partial y^2}\,T(x,\,y,\,z)) + (\frac{\partial^2}{\partial z^2}\,T(x,\,y,\,z)) = 0$$

The temperature of the walls at $x = 0$ and a is held at zero degrees. On separating variables, the x part of the solution will be a linear combination of a sine and a cosine. To satisfy $T = 0$ at $x = 0$ and a, this part of the solution must involve $\sin(m\,\pi\,x/a)$, with m a positive integer. Similarly, the y part of the solution must be $\sin(n\,\pi\,y/b)$, with n a positive integer. So, let's apply the pdsolve command with the hint that the solution is of the form $\sin(m\,\pi\,x/a)\,\sin(n\,\pi\,y/b)\,Z(z)$, with the structure of $Z(z)$ to be determined.

```
>  sol:=pdsolve(pde,T(x,y,z),HINT=sin(m*Pi*x/a)*
       sin(n*Pi*y/b)*Z(z),INTEGRATE,build);
```

$$sol := T(x,\,y,\,z) = \sin(\frac{m\,\pi\,x}{a})\sin(\frac{n\,\pi\,y}{b})\_C1\sin(\frac{\pi\,\sqrt{-m^2\,b^2 - n^2\,a^2}\,z}{a\,b})$$
$$+ \sin(\frac{m\,\pi\,x}{a})\sin(\frac{n\,\pi\,y}{b})\_C2\cos(\frac{\pi\,\sqrt{-m^2\,b^2 - n^2\,a^2}\,z}{a\,b})$$

Since the floor at $z = 0$ is held at zero degrees, the cosine term must be removed from the rhs of the solution sol. The result is the general term $T_{m,n}$ in a double Fourier series, with the coefficient $\_C1$ to be determined.

```
>  T[m,n]:=remove(has,rhs(sol),cos);
```

$$T_{m,n} := \sin(\frac{m\,\pi\,x}{a})\sin(\frac{n\,\pi\,y}{b})\_C1\sin(\frac{\pi\,\sqrt{-m^2\,b^2 - n^2\,a^2}\,z}{a\,b})$$

The temperature profile at $z = c$ is entered in f. To determine the form of $\_C1$, the combination $\sin(m\,\pi\,x/a)\,\sin(n\,\pi\,y/b)$ is first entered in g.

```
>  f:=1600*x*(a-x)*y*(b-y): g:=sin(m*Pi*x/a)*sin(n*Pi*y/b):
```

Making use of orthogonality of the sine functions, f is multiplied by g and the double integral over x and y carried out. This must be equal to $T_{m,n}$, evaluated at $z = c$, multiplied by g, with the same double integration performed.

```
>  eq:=int(int(f*g,x=0..a),y=0..b)
        =int(int(eval(T[m,n],z=c)*g,x=0..a),y=0..b):
```

Then eq is solved for $\_C1$, and simplified assuming that m and n are integers.

```
>   _C1:=simplify(solve(eq,_C1)) assuming m::integer,n::integer;
```

$$\_C1 := \frac{25600\, a^2\, b^2\, (1 - (-1)^m + (-1)^m\, (-1)^n - (-1)^n)}{\sin\!\left(\dfrac{\pi\, \sqrt{-m^2\, b^2 - n^2\, a^2}\, c}{a\, b}\right) m^3\, \pi^6\, n^3}$$

The dimensions of the enclosure are entered, and the series representation of the temperature T calculated, keeping $50 \times 50 = 2500$ terms to ensure accurate numerical results.

```
>  a:=1.0: b:=1.0: c:=1.0: T:=sum(sum(T[m,n],m=1..50),n=1..50):
```

Setting $T = 20$, the `implicitplot3d` command is used to plot the 20 degree isotherm, the result being the bowl-shaped surface shown in Figure 4.8.

```
>  implicitplot3d(T=20,x=0..a,y=0..b,z=0..c,axes=box,
   orientation=[10,70],style=PATCHCONTOUR,tickmarks=[2,2,2]);
```

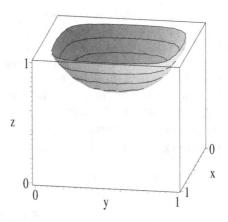

Figure 4.8: The $20\,^\circ$C isotherm.

The temperature is now evaluated at the center of the enclosure,

```
>  T2:=evalf(eval(T,{x=a/2,y=b/2,z=c/2}));
```

$$T2 := 11.36308962$$

and found to be about 11 degrees.

4.2.6 Daniel Hits Middle C

The notes I handle no better than many pianists. But the pauses between the notes – ah, that is where the art resides.
Artur Schnabel, American pianist, *Chicago Daily News, 11 June 1958*

A horizontal bar of nickel-iron (Young's modulus, $Y = 2.1 \times 10^{11}$ N/m^2, and density, $\rho = 7.8 \times 10^3$ kg/m^3) of length $L = 1$ m, whose cross-section is rectangular with width $W = 0.1$ m and height $H = 0.049$ m, is clamped (i.e., $\psi = \psi' = 0$) at $x = 0$ and L. Daniel strikes the bar sharply at the midpoint of one of its wider sides in such a way that the initial transverse velocity is (approximately) $v(x,0) = v_0\,\delta(x - L/2)$, with $v_0 = 1$ m^2/s and δ the Dirac delta function. Determine the transverse displacement $\psi(x,t)$ of the bar at arbitrary time $t > 0$. Show that the fundamental frequency corresponds to approximately middle C (f=261.63 Hz). Animate the motion of the bar over the interval t=0 to $10\,T$, where $T = 1/f$ is the fundamental period.

After loading the plots library package, needed for the animation,

```
> restart: with(plots):
```

the equation of motion for transverse oscillations of the bar is entered.

```
> pde:=a^4*diff(psi(x,t),x$4)+diff(psi(x,t),t,t)=0;
```

$$pde := a^4 \left(\frac{\partial^4}{\partial x^4}\,\psi(x,\,t)\right) + \left(\frac{\partial^2}{\partial t^2}\,\psi(x,\,t)\right) = 0$$

The parameter $a = (S\,\kappa^2\,Y/\epsilon)^{1/4}$, where $S = WH$ is the cross-sectional area of the bar, $\kappa = H/\sqrt{12}$ is the radius of gyration about the horizontal midplane through the bar, and $\epsilon = \rho\,S$ is the linear density.

Using pdsolve, *pde* is solved with the hint $\psi(x,t) = X(x)\,T(t)$.

```
> sol:=pdsolve(pde,HINT=X(x)*T(t),INTEGRATE,build);
```

Setting the separation constant $\_c_1 = k^4$ in *sol*, the frequency is $\omega = a^2\,k^2$.

```
> sol:=simplify(subs(_c[1]=k^4,sol),symbolic);
```

$$sol := \psi(x,\,t) = e^{(-I\,k\,x)}\,\_C5\,\sin(a^2\,k^2\,t)\,\_C1 + \cdots + \_C6\,\cos(a^2\,k^2\,t)\,\_C4\,e^{(k\,x)}$$

Since the bar is initially horizontal, i.e., $\psi(x,0) = 0$, the cosine term must be removed from the rhs of *sol*. The result is then factored.

```
> psi:=factor(remove(has,rhs(sol),cos));
```

$$\psi := \_C5\,\sin(a^2\,k^2\,t)(\_C1\,e^{(-I\,k\,x)} + e^{(-k\,x)}\,\_C2 + \_C3\,e^{(k\,x\,I)} + e^{(k\,x)}\,\_C4)$$

To apply the boundary conditions at the end of the bar, the spatial part is extracted with the select command. The time part is also selected.

```
> X:=select(has,psi,x); T:=select(has,psi,t);
```

$$X := \_C1\,e^{(-I\,k\,x)} + e^{(-k\,x)}\,\_C2 + \_C3\,e^{(k\,x\,I)} + e^{(k\,x)}\,\_C4$$

$$T := \sin(a^2\,k^2\,t)$$

X is converted to trig form, and the cosine, sine, cosh, and sinh terms collected.

```
> X:=collect(convert(X,trig),[cos,sin,cosh,sinh]);
```

$$X := (\_C3 + \_C1)\cos(k\,x) + (-\_C1\,I + \_C3\,I)\sin(k\,x)$$
$$+ (\_C4 + \_C2)\cosh(k\,x) + (\_C4 - \_C2)\sinh(k\,x)$$

Using the op command to extract the relevant operands, new coefficients $A1$, etc, are introduced into X in the following line.

```
> X:=add(A||i*op([i,2],X),i=1..4);
```

$$X := A1\,\cos(k\,x) + A2\,\sin(k\,x) + A3\,\cosh(k\,x) + A4\,\sinh(k\,x)$$

The clamped-end boundary conditions are applied at $x=0$ in *bc1* and *bc2* and at $x=L$ in *bc3* and *bc4*.

```
>  bc1:=eval(X,x=0)=0;  bc2:=eval(diff(X,x),x=0)=0;
```
$$bc1 := A1 + A3 = 0 \qquad bc2 := A2\,k + A4\,k = 0$$

```
>  bc3:=eval(X,x=L)=0;  bc4:=eval(diff(X,x),x=L)=0;
```
$$bc3 := A1\cos(k\,L) + A2\sin(k\,L) + A3\cosh(k\,L) + A4\sinh(k\,L) = 0$$
$$bc4 := -A1\sin(k\,L)\,k + A2\cos(k\,L)\,k + A3\sinh(k\,L)\,k + A4\cosh(k\,L)\,k = 0$$

On attempting to solve the four boundary conditions for the four unknown coefficients, one of the coefficients will be undetermined and a transcendental equation for k will result. Let's choose the undetermined coefficient to be $A4$. Then, $bc1$, $bc2$, and $bc3$ are solved for $A1$, $A2$, and $A3$ and $sol2$ assigned.

```
>  sol2:=solve({bc1,bc2,bc3},{A||1,A||2,A||3}); assign(sol2):
```
The coefficient $A4$ is temporarily set equal to 1 and the fourth boundary condition divided by k^3 and simplified with the symbolic option.

```
>  A||4:=1: bc4:=simplify(bc4/k^3,symbolic);
```
$$bc4 := \frac{2\,(-1 + \cos(k\,L)\cosh(k\,L))}{k^2\,(\cos(k\,L) - \cosh(k\,L))} = 0$$

We set $k\,L = K$ in $bc4$ and isolate the $\cos(K)$ term to the left of the equation.

```
>  eq:=isolate(subs(k*L=K,bc4),cos(K));
```
$$eq := \cos(K) = \frac{1}{\cosh(K)}$$

On comparing the transcendental equation eq with the corresponding equation for the clamped-free end situation in Recipe **01-1-3**, we see that they differ by a minus sign on the right-hand side. As before, eq must be solved numerically for the allowed K values. For large K, $\cosh(K)$ becomes very large and $\cos K$ approaches zero. The allowed K values, therefore, are approximately the zeros of the cosine function. A functional operator f is introduced to determine the zeros in the range $3(n-1)$ to $3n$, where n will take on the values 1, 2, etc.

```
>  f:=n->fsolve(eq,K,3*(n-1)..3*n):
```
Then forming $f(n)$, dividing by L, and using the sequence command, the first four zeros of k are given in $sol3$ which is assigned.

```
>  sol3:=seq(k||n=f(n)/L,n=1..4); assign(sol3):
```
$$sol3 := k1 = 0., \; k2 = \frac{4.730040745}{L}, \; k3 = \frac{7.853204624}{L}, \; k4 = \frac{10.99560784}{L}$$

The first zero, $k1$, must be rejected, because for $k = 0$ the lhs of $bc4$ is finite. The parameters L, v_0, W, H, Y, and ρ are entered, and the radius of gyration κ, cross-sectional area S, linear density ϵ, and the parameter a calculated.

```
>  L:=1: v0:=1: W:=0.1: H:=0.049: Y:=2.1*10^11: rho:=7.8*10^3:
>  kappa:=evalf(H/sqrt(12)); S:=W*H; epsilon:=rho*S;
   a:=(S*kappa^2*Y/epsilon)^(1/4);
```
$$\kappa := 0.01414508160 \quad S := 0.0049 \quad \varepsilon := 38.22000 \quad a := 8.567101289$$

The coefficient $A4$, which had been temporarily set equal to 1, is relabeled

B, and will now be calculated. The initial transverse velocity is of the form $v(x, 0) = v_0 \, \delta(x - L/2) = \sum_{n=1}^{\infty} a^2 \, k_n^2 \, B_n \, X_n$, where X_n is the nth spatial mode. Multiplying $v(x, 0)$ by X_n, integrating x over the range 0 to L, and using the orthogonality of the spatial modes, yields $v_0 \, X_n(x = L/2) = B_n \, a^2 \, k_n^2 \, \int_0^L X_n^2 \, dx$, which is easily solved for B_n. The coefficient B is evaluated in the following line for a general k value.

```
>  B:=v0*eval(X,x=L/2)/(a^2*k^2*int(X^2,x=0..L)):
```

The product $B \, X \, T$ is evaluated at $k = kn$ and the first three terms of the complete solution ψ added. Only the fundamental ($n=2$) contribution is shown here, the remaining terms having rapidly decreasing amplitudes.

```
>  psi:=simplify(add(eval(B*X*T,k=k||n),n=2..4));
```

$$\psi := -0.0009671479296 \sin(1642.092308 \, t) \cos(4.730040745 \, x)$$
$$+ 0.0009502249823 \sin(1642.092308 \, t) \sin(4.730040745 \, x)$$
$$+ 0.0009671479296 \sin(1642.092308 \, t) \cosh(4.730040745 \, x)$$
$$- 0.0009502249823 \sin(1642.092308 \, t) \sinh(4.730040745 \, x) + \cdots$$

The fundamental frequency $\omega = a^2 \, k2^2$ is 1642 rads/s or $f = \omega/(2\pi) \approx 261.3$ Hz, which is close to middle C. The fundamental period $T = 1/f$ is also calculated.

```
>  omega:=a^2*k||2^2; f:=evalf(omega/(2*Pi)); T:=1/f;
```

$$\omega := 1642.092308 \quad f := 261.3471078 \quad T := 0.003818536004$$

Finally, the vibrations of the bar are animated over the interval $t = 0$ to $10 \, T$.

```
>  animate(psi,x=0..L,t=0..10*T,frames=100,thickness=2);
```

Execute the recipe on your computer and click on the plot and on the start arrow to see the vibrations.

4.2.7 A Poisson Recipe

Life is good for only two things, discovering mathematics and teaching mathematics
Simon Poisson, French mathematician, (1781–1840)

In this day and age, being overweight and having high blood pressure and/or high cholesterol cause great concern among an aging population. To address these issues, all types of diets are proposed and cookbooks written. Amongst the latter are the *HeartSmart* [Ste94] cook books, sponsored by the Heart and Stroke Foundation of Canada. Each book presents "over 200 healthful & delicious recipes", with a section devoted to fish and seafood. So, in the spirit of keeping us intellectually healthy, here's a delicious Poisson (equation) recipe.

The simplest non-trivial Poisson equation problems involve point or line charges in the vicinity of one or more "grounded" (zero potential) conducting surfaces. Since the charge densities for these sources may be characterized by Dirac δ-functions, the solutions of these problems are electrostatic Green's functions. As a simple 2-dimensional example, let us consider an infinitely long line charge characterized by the charge density $\rho = 4\pi\,\epsilon_0\,\delta(x-a)\,\delta(y-b)$, with ϵ_0 the permittivity of free space. The line charge is located between two infinite grounded conducting plates at $y = 0$ and at $L > b > 0$. To firmly establish the geometry in our minds, a finite portion of the two infinite conducting planes is plotted in the x-y plane as green lines in gr1, with $L = 1$.

```
>   restart: with(plots):
>   gr1:=plot([[[-1/2,0],[3/2,0]],[[-1/2,1],[3/2,1]]],color=
         green,thickness=2,tickmarks=[4,3],labels=["x","y"]):
```

Taking $a = 0.5$ and $b = 0.8$, a point is plotted in gr2, representing an end-on view of the line charge. The geometry for the Green's function problem is then displayed by superimposing the two graphs in Figure 4.9.

```
>   gr2:=pointplot([[0.5,0.8]],symbol=circle,symbolsize=12):
>   display({gr1,gr2},scaling=constrained);
```

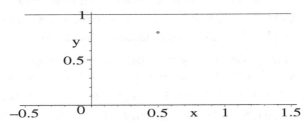

Figure 4.9: End-on view of line charge between two grounded conducting plates.

The relevant Poisson equation for the Green's function potential G is

$$\nabla^2 G = \frac{\partial^2 G}{\partial x^2} + \frac{\partial^2 G}{\partial y^2} = -\frac{\rho}{\epsilon_0} = -4\pi\,\delta(x-a)\,\delta(y-b). \tag{4.7}$$

For the region outside the line charge, Equation (4.7) becomes homogeneous and a solution is easily constructed. In the y direction G is zero at $y = 0$ and L, so G must include terms of the structure $\sin(n\pi y/L)$, with $n = 1, 2, 3, \ldots$. As $x \to \pm\infty$, we must have $G \to 0$. For $x < a$, the Green's function GL to the "left" of the line charge will be built up of terms of the structure $e^{n\pi(x-a)/L}$. We have used the fact that our final result should ultimately depend only on the difference $x - a$. The factor $n\pi/L$ is included so that the homogeneous form of Poisson's equation will be satisfied. The nth term in the infinite series then takes the following form for GL, the coefficients A_n still to be determined.

```
>   GL:=A[n]*sin(n*Pi*y/L)*exp(n*Pi*(x-a)/L);
```

$$GL := A_n \sin\left(\frac{n\pi y}{L}\right) e^{\left(\frac{n\pi(x-a)}{L}\right)}$$

For $x > a$, the Green's function GR to the "right" of the line charge is built up of terms of the structure $e^{-n\pi(x-a)/L}$. The nth term in the infinite series then takes the following form for GR, the coefficients B_n still to be determined.

```
>   GR:=B[n]*sin(n*Pi*y/L)*exp(-n*Pi*(x-a)/L);
```

$$GR := B_n \sin(\frac{n\pi y}{L}) e^{(-\frac{n\pi(x-a)}{L})}$$

As a check on, e.g., GL, we confirm that $\partial^2(GL)/\partial x^2 + \partial^2(GL)/\partial y^2 = 0$.

```
> check:=diff(GL,x,x)+diff(GL,y,y);
```

$$check := 0$$

The Green's function must be continuous, i.e., $GL = GR$, at $x = a$ for arbitrary y between 0 and L. The continuity condition is implemented in *eq1*.

```
> eq1:=expand(eval(GL=GR,x=a)/sin(n*Pi*y/L));
```

$$eq1 := A_n = B_n$$

To determine the discontinuity in $\partial G/\partial x$ at $x = a$, we appeal to the divergence (Gauss's) theorem, viz., $\oint_V (\nabla \cdot \vec{A}) \, dv = \oint_S (\hat{n} \cdot \vec{A}) \, ds$ for a vector field $\vec{A}$. Here $\hat{n}$ is the outward unit normal to the closed surface S, enclosing a volume V. Let's take V to be a thin (thickness 2ϵ, where ultimately $\epsilon \to 0$) slice between the plates of unit length in the z direction, with faces at $x - \epsilon$ and $x + \epsilon$, and $\vec{A} = \nabla G$. Then, making use of Poisson's equation, the divergence theorem yields

$$\int\int \nabla^2 G \, dx \, dy = \int (\hat{n} \cdot \nabla G) \, d\ell = \int \frac{\partial G}{\partial n} \, d\ell = -4\pi \int\int \delta(x-a) \, \delta(y-b) \, dx \, dy = -4\pi.$$

Here $\partial G/\partial n$ is the normal derivative of G and the line integral ($\int d\ell$) is carried out along the perimeter of the slice in the x-y plane. As $\epsilon \to 0$, the "ends" of the slice at $y = 0$ and L will make no contribution to the line integral, so that $\lim_{\epsilon\to 0} [\int_{\text{along } x=a+\epsilon}(\partial G/\partial x) \, dy - \int_{\text{along } x=a-\epsilon}(\partial G/\partial x) \, dy] = -4\pi$, which implies (in the limit $\epsilon \to 0$) that $(\partial G/\partial x)_{a+\epsilon} - (\partial G/\partial x)_{a-\epsilon} = -4\pi\delta(y-b)$. Multiplying this result by $\sin(n\pi y/L)$, integrating from $y = 0$ to L, and using orthogonality, yields the desired second relation for the coefficients. This calculation is implemented in *eq2*, the assumptions that $b > 0$, $L > b$, and n is an integer being included to accomplish the integration over the δ-function.

```
> eq2:=int(eval(diff(GL-GR,x),x=a)*sin(n*Pi*y/L),y=0..L)
       =4*Pi*int(Dirac(y-b)*sin(n*Pi*y/L),y=0..L)
       assuming b>0,L>b,n::integer;
```

$$eq2 := \frac{1}{2} n\pi A_n + \frac{1}{2} n\pi B_n = 4\pi \sin(\frac{n\pi b}{L})$$

eq1 and *eq2* are solved for A_n and B_n, and the solution assigned.

```
> sol:=solve({eq1,eq2},{A[n],B[n]}): assign(sol):
```

Keeping 30 terms in the sum, the complete Green's function to the left and right of $x = a$ are given in *GL2* and *GR2*, respectively.

```
> GL2:=Sum(GL,n=1..30); GR2:=Sum(GR,n=1..30);
```

$$GL2 := \sum_{n=1}^{30} \left(\frac{4\sin(\frac{n\pi b}{L}) \sin(\frac{n\pi y}{L}) e^{(\frac{n\pi(x-a)}{L})}}{n} \right)$$

$$GR2 := \sum_{n=1}^{30} \left(\frac{4\sin(\frac{n\pi b}{L}) \sin(\frac{n\pi y}{L}) e^{\left(-\frac{n\pi(x-a)}{L}\right)}}{n} \right)$$

The parameter values $a=0.5$, $b=0.8$, and $L=1$ are entered, and the piecewise Green's function, $G=GL2$ for $x < a$ and $GR2$ for $x > a$, formed.

> `a:=0.5: b:=0.8: L:=1:`

> `G:=piecewise(x<a,value(GL2),x>a,value(GR2)):`

Using `contourplot`, equipotentials are drawn in cp for $G=1/8$, $1/4$, $1/2$, 1, 2, 3 and then superimposed on `gr1` and `gr2`, the result being shown in Fig. 4.10. The curve closest to the line charge is for $G=3$, the furthest for $G=1/8$.

> `cp:=contourplot(G,x=-2..2,y=0..L,contours=[1/8,1/4,1/2,`
> `    1,2,3],grid=[70,70],color=blue,thickness=2):`

> `display({gr1,gr2,cp},scaling=constrained);`

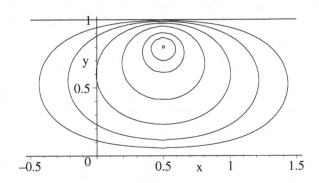

Figure 4.10: Equipotentials around line charge between grounded plates.

4.3   Beyond Cartesian Coordinates

Some representative non-Cartesian examples are now presented. Many more are included in the **Supplementary Recipes** at the end of the chapter.

4.3.1   Is It Separable?

Alliance. In international politics, the union of two thieves who have their hands so deeply inserted in each other's pockets that they cannot separately plunder a third.
Ambrose Bierce, American author, *The Devil's Dictionary (1881–1906)*

The method of separation of variables can be applied to other orthogonal curvilinear coordinate systems besides the Cartesian system. I have found that, at first, some students are surprised that the variable separation method works at all. After a while, they assume that it always works. It turns out that the *scalar Helmholtz equation*, $\nabla^2 S + k^2 S = 0$, which is the spatial part of either the wave or diffusion equations with k a constant, is separable [MF53] in 11, and only 11, 3-dimensional orthogonal curvilinear coordinate systems. Fortunately, these include spherical polar and cylindrical coordinates, which are the two most commonly used non-Cartesian systems. An example of a 3-dimensional coordinate system for which the Helmholtz equation is not separable are the *bispherical* coordinates u, v, w, which are related to x, y, z by

$$x = \frac{a \sin u \cos v}{\cosh w - \cos u}, \quad y = \frac{a \sin u \sin v}{\cosh w - \cos u}, \quad z = \frac{a \sinh w}{\cosh w - \cos u}. \tag{4.8}$$

Here a is a scale factor and $0 \le u < \pi$, $0 \le v \le 2\pi$, $-\infty < w < \infty$.

As the following recipe illustrates, Laplace's equation is not separable in bispherical coordinates either, but can be separated into three ODEs by a modified separation assumption. This is useful, e.g., in determining the potential outside two spheres of equal diameters, held at different potentials, and with their centers separated by a distance greater than the sphere diameter.[2]

Although, the bispherical system is known (with $a = 1$) to Maple, it is instructive to tackle the following problem from first principles.

(a) Plot the contours in the x-z plane corresponding to holding u and w fixed. What surfaces are generated if v is constant?

(b) Calculate the scale factors and the Laplacian operator.

(c) Show that Laplace's equation is not completely separable if one makes the "standard" ansatz, $S(u, v, w) = U(u) V(v) W(w)$.

(d) Show that Laplace's equation is completely separable if one assumes that $S(u, v, w) = \sqrt{(\cosh w - \cos u)} \, U(u) V(v) W(w)$. Assuming that $\cosh w > \cos u$, identify any special functions which occur in the separated ODEs.

It is assumed that $u \ge 0$, $u < \pi$, $v \ge 0$, $v \le 2\pi$, and $\cosh w > \cos u$. The coordinate relations are then entered.

```
> restart: with(plots): assume(u>=0,u<Pi,v>=0,v<=2*Pi,
    cosh(w)>cos(u)):
> x:=a*sin(u)*cos(v)/(cosh(w)-cos(u));
> y:=a*sin(u)*sin(v)/(cosh(w)-cos(u));
> z:=a*sinh(w)/(cosh(w)-cos(u));
```

To plot the surfaces corresponding to holding w fixed, let's form $X^2 + Y^2 + (Z - a \coth w)^2 = x^2 + y^2 + (z - a \coth w)^2$ and simplify the right-hand side.

```
> eq1:=X^2+Y^2+(Z-a*coth(w))^2
        =simplify(x^2+y^2+(z-a*coth(w))^2);
```

$$eq1 := X^2 + Y^2 + (Z - a \coth(w))^2 = \frac{a^2}{\sinh(w)^2}$$

The result is the equation of a sphere of radius $a/\sinh w$ centered at $X = 0$,

---

[2] An excellent source of electrostatic problems in bispherical and other coordinate systems is *Problems in Mathematical Physics* by Lebedev, Skal'skaya, and Uflyand (Pergamon, 1966).

$Y = 0$, and $Z = a \coth w$. Different choices of w will generate different size spheres with centers located at different Z values.

Similarly, $(\sqrt{X^2 + Y^2} - a \cot u)^2 + Z^2 = (\sqrt{x^2 + y^2} - a \cot u)^2 + z^2$ is entered in *eq2* and simplified.

```
> eq2:=(sqrt(X^2+Y^2)-a*cot(u))^2+Z^2
    =simplify((sqrt(x^2+y^2)-a*cot(u))^2+z^2,symbolic);
```

$$eq2 := (\sqrt{X^2 + Y^2} - a\cot(u))^2 + Z^2 = \frac{a^2}{\sin(u)^2}$$

To see what type of surface is generated by holding u fixed, let's set $a = 1$ and $Y = 0$ so as to generate plots in the X-Z plane. The `unapply` command is used to free up w and u in *eq1* and *eq2* for plotting purposes.

```
> a:=1: Y:=0: A:=unapply(eq1,w): B:=unapply(eq2,u):
```

Using A and B and the `implicitplot` command, representative contours are drawn for fixed w (solid circles) in `gr1` and fixed u (dashed circles) in `gr2`

```
> gr1:=implicitplot({seq(A(n/2),n=-4..-1),seq(A(n/2),n=1..4)},
    X=-5*a..5*a,Z=-5*a..5*a,grid=[70,70],color=red,thickness=2):
```

```
> gr2:=implicitplot({seq(B(0.2*Pi*n),n=1..4)},X=-5*a..5*a,
    Z=-5*a..5*a,grid=[70,70],color=blue,thickness=2,linestyle=2):
```

and superimposed to produce Fig. 4.11.

```
> display({gr1,gr2},scaling=constrained);
```

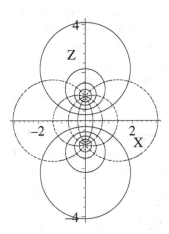

Figure 4.11: Contours for fixed w (solid circles) and fixed u (dashed).

The corresponding 3-dimensional surfaces result on rotating the figure about the vertical (Z) axis, the circles becoming spheres, hence the name "bispherical" for the coordinate system. From the defining relations between the bispherical and Cartesian systems, one has $y/x = \sin v / \cos v = \tan v$, so holding v fixed will produce half-planes passing through the z-axis.

A functional operator L is created for generating the Laplacian of $S(u, v, w)$ on specifying the coordinates u, v, w and the scale factors h_u, h_v, h_w.

```
>  L:=(u,v,w,hu,hv,hw)->(diff(hv*hw*diff(S(u,v,w),u)/hu,u)
       +diff(hu*hw*diff(S(u,v,w),v)/hv,v)
       +diff(hu*hv*diff(S(u,v,w),w)/hw,w))/(hu*hv*hw):
```

An operator H is formed for producing and simplifying the scale factors.

```
>  H:=u->simplify(sqrt(diff(x,u)^2+diff(y,u)^2+diff(z,u)^2),
         symbolic):
```

Using H the three scale factors are explicitly calculated.

```
>  h[u]:=H(u); h[v]:=H(v); h[w]:=H(w);
```

$$h_u := \frac{1}{\cosh(w) - \cos(u)} \qquad h_v := \frac{\sin(u)}{\cosh(w) - \cos(u)} \qquad h_w := \frac{1}{\cosh(w) - \cos(u)}$$

Then employing L, Laplace's equation, $\nabla^2 S(u, v, w) = 0$, is generated.

```
>  Lap:=L(u,v,w,h[u],h[v],h[w])=0;
```

$$Lap := \left(\frac{\cos(u)(\frac{\partial}{\partial u}S(u,v,w))}{\cosh(w) - \cos(u)} + \frac{\sin(u)^2(\frac{\partial}{\partial u}S(u,v,w))}{(\cosh(w) - \cos(u))^2} + \frac{\sin(u)(\frac{\partial^2}{\partial u^2}S(u,v,w))}{\cosh(w) - \cos(u)} \right.$$

$$+ \frac{\frac{\partial^2}{\partial v^2}S(u,v,w)}{(\cosh(w) - \cos(u))\sin(u)} + \frac{\sin(u)(\frac{\partial}{\partial w}S(u,v,w))\sinh(w)}{(\cosh(w) - \cos(u))^2}$$

$$+ \left. \frac{\sin(u)(\frac{\partial^2}{\partial w^2}S(u,v,w))}{\cosh(w) - \cos(u)} \right) (\cosh(w) - \cos(u))^3 / \sin(u) = 0$$

An unsuccessful attempt is made to separate Laplace's equation by assuming that $S(u, v, w) = U(u) V(v) W(w)$.

```
>  pdsolve(Lap,HINT=U(u)*V(v)*W(w));
```

$$Warning: \quad Incomplete \ separation.$$

$$(S(u,\ v,\ w) = V(v) \ \_F1(u,\ w)) \ \&where[\{ \frac{d^2}{dv^2} V(v) = \_c2 \ V(v), \cdots \cdots$$

The separation is incomplete, an ODE resulting for $V(v)$, but the u and w dependence remaining coupled in a PDE, which is not displayed here in the text. Supplying the modified separation assumption as a hint, Laplace's equation is now completely separated.

```
>  pdsolve(Lap,HINT=sqrt((cosh(w)-cos(u)))*U(u)*V(v)*W(w),
   INTEGRATE);
```

$(S(u, v, w) = \sqrt{\cosh(w) - \cos(u)}\, U(u)\, V(v)\, W(w))$ &where

$[\{\{W(w) = \_C5\, e^{(\sqrt{\_c3}\, w)} + \_C6\, e^{(-\sqrt{\_c3}\, w)}\},$

$\{V(v) = \_C3\, e^{(\sqrt{\_c2}\, v)} + \_C4\, e^{(-\sqrt{\_c2}\, v)}\},$

$\{U(u) = \_C1\, (\frac{1}{2}\cos(2u) - \frac{1}{2})^{(1/2\, I\, \sqrt{\_c2})} \sin(2u) \text{hypergeom}([\frac{1}{2} I \sqrt{\_c2}$

$+\frac{1}{2}\sqrt{\_c3} + \frac{3}{4}, \frac{1}{2} I \sqrt{\_c2} - \frac{1}{2}\sqrt{\_c3} + \frac{3}{4}], [\frac{3}{2}], \frac{1}{2}\cos(2u) + \frac{1}{2}) /$

$\sqrt{1 - \cos(2u)} + \cdots\}\}]$

$W(w)$ and $V(v)$ are both expressed in terms of exponentials, but U is given in terms of *hypergeometric* functions. The hypergeometric function $F(a, b; c; z)$ is given by the following infinite series [AS72], where Γ is the Gamma function,

$$F(a, b; c; z) = \frac{\Gamma(c)}{\Gamma(a)\,\Gamma(b)} \sum_{n=0}^{\infty} \frac{\Gamma(a+n)\,\Gamma(b+n)}{\Gamma(c+n)} \frac{z^n}{n!}.$$

4.3.2   A Shell Problem, Not a Shell Game

Insurance. An ingenious modern game of chance in which the player is permitted to enjoy the comfortable conviction that he is beating the man who keeps the table.
Ambrose Bierce, American author, *The Devil's Dictionary (1881–1906)*

In creating physics exams at the freshman level, I often play a bit of a shell game, presenting problems similar to those that the students have solved for homework, but in new guises and combinations. The hope is that they really understand the underlying principles and methods and haven't merely memorized the solutions to the homework problems. At the senior level, I rely less on "disguise" and more on having students explore challenging problems, even "standard" ones, in some depth. A computer algebra approach is encouraged as an auxiliary tool. The following recipe, submitted by Ms. I. M. Curious, is based on a standard problem appearing on an exam given to my senior electromagnetic theory class.

A very long circular cylindrical shell of dielectric constant ϵ and inner and outer radii a and b, respectively, is placed in a previously uniform electric field $\vec{E}_0$ with the cylinder axis perpendicular to the field. The medium inside ($r < a$) and outside ($r > b$) the cylindrical shell has a dielectric constant of unity.

(a) Determine the potential and electric field in the three regions.

(b) Taking $\epsilon = 3$, $a = 1$, $b = 2$, and $E_0 = 1$, plot the equipotentials and electric field vectors in all three regions in a single figure. Discuss the results and explore the effect of changing the parameter values.

In addition to the plots library package, I. M. loads the plottools and VectorCalculus packages. Plottools contains the `circle` command which she will use for drawing the inner and outer radii of the cylindrical shell. The VectorCalculus package is needed for the `Laplacian` and `Gradient` commands.

```
> restart: with(plots): with(plottools): with(VectorCalculus):
```
Neglecting end effects, the cylindrical shell is taken to be infinitely long in the z direction, thus reducing the problem to 2 dimensions in the x-y plane. Noting the circular symmetry, I. M. introduces the polar coordinates (r, θ) with $x = r \cos \theta$ and $y = r \sin \theta$, r being measured from the cylinder axis and θ from the x-axis. Laplace's equation is entered in polar coordinates and expanded.

```
> pde:=expand(Laplacian(phi(r,theta),'polar'[r,theta]))=0;
```

$$pde := \frac{\frac{\partial}{\partial r} \phi(r, \theta)}{r} + \left(\frac{\partial^2}{\partial r^2} \phi(r, \theta)\right) + \frac{\frac{\partial^2}{\partial \theta^2} \phi(r, \theta)}{r^2} = 0$$

Using the separation of variables method, a general solution is sought of the form $\phi(r, \theta) = R(r) \Theta(\theta)$. For convenience, I. M. replaces the separation constant $\sqrt{\_c_1}$ that appears on the rhs of *sol* with the symbol k.

```
> sol:=pdsolve(pde,HINT=R(r)*Theta(theta),INTEGRATE,build);
> sol2:=subs(sqrt(_c[1])=k,rhs(sol));
```

$$sol2 := \_C3 \sin(k\,\theta) \_C1\, r^k + \frac{\_C3 \sin(k\,\theta) \_C2}{r^k} + \_C4 \cos(k\,\theta) \_C1\, r^k$$
$$+ \frac{\_C4 \cos(k\,\theta) \_C2}{r^k}$$

Taking the electric field to be in the x direction, I. M. notes that the solution must have reflection symmetry (is unchanged if $\theta \to -\theta$) around the x axis. So she removes the sine terms, which are odd functions of θ, from *sol2*. As $r \to \infty$, the electric field must remain uniform and is given by $\vec{E}_0 = E_0\, \hat{e}_x = -(\partial \phi/\partial x)\, \hat{e}_x$, so the asymptotic potential is $\phi = -E_0\, x = -E_0\, r \cos \theta$, the arbitrary constant in the potential being set equal to zero. This immediately implies that $k = 1$, which must hold in every region to satisfy the boundary conditions. I. M. substitutes $k = 1$ and, without loss of generality, also sets the redundant coefficient $\_C4$ equal to 1 as well.

```
> sol3:=subs({_C4=1,k=1},remove(has,sol2,sin));
```

$$sol3 := \cos(\theta) \_C1\, r + \frac{\cos(\theta) \_C2}{r}$$

I. M. labels ϕ in the regions $r < a$, $a < r < b$, and $r > b$ as $\phi 1$, $\phi 2$, and $\phi 3$. An operator P is formed for relabeling the coefficients $\_C1$ and $\_C2$ for each ϕ.

```
> P:=(u,v)->subs({_C1=u,_C2=v},sol3):
```
For $r < a$, the $1/r$ term must be removed from $\phi 1$ for it to remain finite at $r = 0$. So I. M. forms $\phi 1$ by setting $v = 0$ in P and $u = A_1$. For $\phi 2$, she chooses $u = A_2$, $v = B_2$. For $\phi 3$, she takes $u = -E0$, in order to match the asymptotic boundary condition as $r \to \infty$, and $v = B_3$.

```
> phi1:=P(A[1],0); phi2:=P(A[2],B[2]); phi3:=P(-E0,B[3]);
```

$$\phi 1 := \cos(\theta)\, A_1\, r$$

$$\phi 2 := \cos(\theta)\, A_2\, r + \frac{\cos(\theta)\, B_2}{r}$$

$$\phi 3 := -\cos(\theta)\, E0\, r + \frac{\cos(\theta)\, B_3}{r}$$

With A_1, A_2, B_2, and A_3 unknown, 4 boundary conditions are required. The potentials $\phi 1 = \phi 2$ at $r = a$ and $\phi 2 = \phi 3$ at b for arbitrary θ. The following operator F is created to match the potentials u and v at a radius $r = R$.

```
> F:=(u,v,R)-> expand(eval((u=v)/cos(theta),r=R)):
```

Using F, the above boundary conditions are applied in *eq1* and *eq2*.

```
> eq1:=F(phi1,phi2,a); eq2:=F(phi2,phi3,b);
```

$$eq1 := A_1\, a = A_2\, a + \frac{B_2}{a}$$

$$eq2 := A_2\, b + \frac{B_2}{b} = -E0\, b + \frac{B_3}{b}$$

The radial component of the *displacement vector* $\vec{D} = \epsilon\,\vec{E}$ is continuous at the boundaries, so $\partial\phi 1/\partial r = \epsilon\,(\partial\phi 2/\partial r)$ at $r = a$ and $\epsilon\,(\partial\phi 2/\partial r) = \partial\phi 3/\partial r$ at $r = b$. An operator G is created for equating the radial derivative of u and v at a radius R. Using G, the two boundary conditions are applied in *eq3* and *eq4*.

```
> G:=(u,v,R)->expand(eval(diff(u=v,r),r=R)/cos(theta)):
```

```
> eq3:=G(phi1,epsilon*phi2,a); eq4:=G(epsilon*phi2,phi3,b);
```

$$eq3 := A_1 = \varepsilon\, A_2 - \frac{\varepsilon\, B_2}{a^2}$$

$$eq4 := \varepsilon\, A_2 - \frac{\varepsilon\, B_2}{b^2} = -E0 - \frac{B_3}{b^2}$$

The four equations are solved for the four coefficients and *sol4* assigned.

```
> sol4:=solve({eq1,eq2,eq3,eq4},{A[1],A[2],B[2],B[3]});
  assign(sol4):
```

The potentials $\phi 1$, $\phi 2$, and $\phi 3$ are now determined, the coefficients being automatically substituted.

```
> phi1:=phi1; phi2:=simplify(phi2); phi3:=phi3;
```

$$\phi 1 := -\frac{4\cos(\theta)\,\varepsilon\, E0\, b^2\, r}{2\,\varepsilon\, a^2 - a^2 + 2\,\varepsilon\, b^2 + b^2 - \varepsilon^2\, a^2 + b^2\, \varepsilon^2}$$

$$\phi 2 := -\frac{2\cos(\theta)\, E0\, b^2\,(r^2\,\varepsilon + r^2 + \varepsilon\, a^2 - a^2)}{(2\,\varepsilon\, a^2 - a^2 + 2\,\varepsilon\, b^2 + b^2 - \varepsilon^2\, a^2 + b^2\, \varepsilon^2)\, r}$$

$$\phi 3 := -\cos(\theta)\, E0\, r + \frac{\cos(\theta)\, E0\, b^2\,(-\varepsilon^2\, a^2 + b^2\, \varepsilon^2 + a^2 - b^2)}{(2\,\varepsilon\, a^2 - a^2 + 2\,\varepsilon\, b^2 + b^2 - \varepsilon^2\, a^2 + b^2\, \varepsilon^2)\, r}$$

An operator EF is formed for calculating the electric field in polar coordinates, given some potential f.

> `EF:=f->-Gradient(f,'polar'[r,theta]):`

The electric field is then explicitly calculated in each region, but only the field *EF1* in region 1 is displayed here.

> `EF1:=EF(phi1); EF2:=EF(phi2); EF3:=EF(phi3);`

$$EF1 := \frac{4\cos(\theta)\,\varepsilon\,E0\,b^2}{2\,\varepsilon\,a^2 - a^2 + 2\,\varepsilon\,b^2 + b^2 - \varepsilon^2\,a^2 + \varepsilon^2\,b^2}\,\overline{e}_r$$
$$-\frac{4\sin(\theta)\,\varepsilon\,E0\,b^2}{2\,\varepsilon\,a^2 - a^2 + 2\,\varepsilon\,b^2 + b^2 - \varepsilon^2\,a^2 + \varepsilon^2\,b^2}\,\overline{e}_\theta$$

The parameter values $\epsilon = 3$, $a = 1$, $b = 2$, and $E_0 = 1$ are now entered.

> `epsilon:=3: a:=1: b:=2: E0:=1:`

For plotting purposes, I. M. changes to Cartesian coordinates by entering $r = \sqrt{x^2 + y^2}$, $\cos\theta = x/r$ and $\sin\theta = y/r$.

> `r:=sqrt(x^2+y^2): cos(theta):=x/r: sin(theta):=y/r:`

The potential V for all three regions is formed with the `piecewise` command and the radical expressions simplified with the `radsimp` command. The electric field *Ef* is then calculated from V using the `Gradient` command and again the radicals are simplified.

> `V:=radsimp(piecewise(r<a,phi1,r<b,phi2,phi3));`
> `Ef:=-radsimp(Gradient(V,[x,y]));`

$$V := \begin{cases} -\dfrac{4\,x}{5} & \sqrt{x^2+y^2} < 1 \\[2mm] -\dfrac{4\,(2\,x^2 + 2\,y^2 + 1)\,x}{15\,(x^2+y^2)} & \sqrt{x^2+y^2} < 2 \\[2mm] -\dfrac{(5\,x^2 + 5\,y^2 - 8)\,x}{5\,(x^2+y^2)} & \textit{otherwise} \end{cases}$$

The equipotentials of V are produced in `gr1` over the range $x = -4..4$, $y = -4..4$ with the `contourplot` command, 25 contours being requested. I. M. colors the plot by including `filled=true` as an option.

> `gr1:=contourplot(V,x=-4..4,y=-4..4,contours=25,filled=true):`

An operator `C` is formed to plot a thick red circle of radius r, centered at the origin. Then `C` is used in `gr2` and `gr3` to produce circles of radius a and b, representing the inner and outer radii of the cylindrical shell.

> `C:=r->circle([0,0],r,color=red,thickness=3):`
> `gr2:=C(a): gr3:=C(b):`

The `fieldplot` command is used in `gr4` to plot the electric field vectors as medium sized blue arrows. The density of the arrows is controlled.

> `gr4:=fieldplot([Ef[1],Ef[2]],x=-4..4,y=-4..4,color=blue,`
> `        arrows=MEDIUM,grid=[20,20]):`

The four graphs are superimposed with the `display` command, the scaling being constrained. The resulting picture is shown in Figure 4.12.

> `display({gr1,gr2,gr3,gr4},scaling=constrained);`

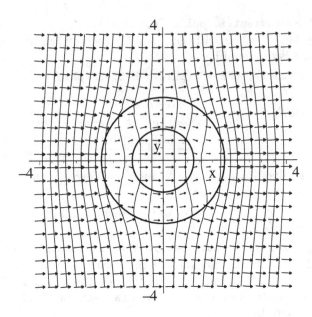

Figure 4.12: The electric field arrows and equipotentials for a cylindrical dielectric shell inserted in a previously uniform (horizontal) electric field.

Referring to the figure, I. M. notes that the electric field for $r < a$ is completely horizontal and therefore parallel to the asymptotic field. The arrows are slightly shorter however. Quantitatively, the ratio of the electric field for $r < a$ to the asymptotic field is $A_1/E_0 = 4/5$. Unlike the situation for a conductor, the surfaces of the dielectric shell are not equipotentials so the electric field vectors are not perpendicular to the inner and outer surfaces of the shell.

I. M. leaves it to you, the reader, to explore her recipe. For example, you might take ϵ to be much larger, or alter b with a fixed. She reminds you that in interpreting any result to remember that the arrows are not field lines.

4.3.3   The Little Drummer Boy

Shall I play for you! pa rum pum pum on my drum.
From the Christmas carol, *Little Drummer Boy*

Little Daniel loves to bang on large pots and pans with a wooden spoon, creating his own version of music, which to untrained adult ears such as mine sounds a lot like noise. Here's a quieter version of "drum playing".

A large circular drumhead of radius $r = a = 1$ m has its perimeter fixed. If the drumhead has an initial shape $U(r, \theta, 0) = r(1 - r^2/a^2) \sin(2\theta)/20$ and is released from rest, determine the shape of the drumhead at arbitrary time

$t > 0$. Take the wave speed to be $c = 1$ m/s. Then, animate the motion of the drumhead in time steps of 0.1 s over the interval $t = 0$ to 2 s.

After loading the plots and VectorCalculus packages, the wave equation *pde* is entered in polar coordinates (r, θ), making use of the Laplacian command.

```
> restart: with(plots): with(VectorCalculus):
> pde:=expand(Laplacian(U(r,theta,t),'polar'[r,theta]))
    =(1/c^2)*diff(U(r,theta,t),t,t);
```

$$pde := \frac{\frac{\partial}{\partial r} U(r, \theta, t)}{r} + \left(\frac{\partial^2}{\partial r^2} U(r, \theta, t)\right) + \frac{\frac{\partial^2}{\partial \theta^2} U(r, \theta, t)}{r^2} = \frac{\frac{\partial^2}{\partial t^2} U(r, \theta, t)}{c^2}$$

Then *pde* is analytically solved, assuming that $U(r, \theta, t) = R(r)\,\Theta(\theta)\,T(t)$.

```
> sol:=pdsolve(pde,HINT=R(r)*Theta(theta)*T(t),INTEGRATE,
    build);
```

$$sol := U(r, \theta, t) = e^{(\sqrt{-c_3}\,t)}\, e^{(\sqrt{-c_2}\,\theta)}\, \_C5\,\_C3\,\_C1\, \mathrm{BesselJ}\!\left(\sqrt{-\_c_2}, \frac{\sqrt{-\_c_3}\,r}{c}\right)$$
$$+ e^{(\sqrt{-c_3}\,t)}\, e^{(\sqrt{-c_2}\,\theta)}\, \_C5\,\_C3\,\_C2\, \mathrm{BesselY}\!\left(\sqrt{-\_c_2}, \frac{\sqrt{-\_c_3}\,r}{c}\right) + \cdots$$

The answer involves exponentials and Bessel functions of the first and second kinds. The separation constants $\_c_2$ and $\_c_3$ on the rhs of *sol* are replaced with $-p^2$ and $-c^2\,k^2$, and the result simplified with the symbolic option.

```
> U1:=simplify(subs({_c[2]=-p^2,_c[3]=-c^2*k^2},rhs(sol)),
    symbolic);
```

$$U1 := \_C5\,\_C3\,\_C1\, \mathrm{BesselJ}(p,\, k\,r)\, e^{((c\,k\,t + p\,\theta)\,I)}$$
$$+ \_C5\,\_C3\,\_C2\, \mathrm{BesselY}(p,\, k\,r)\, e^{((c\,k\,t + p\,\theta)\,I)} + \cdots$$

The Bessel functions $Y_p(k\,r)$ of the second kind diverge at $r = 0$, so are removed from *U1*. Then *U2* is converted to trig form and expanded in *U3*.

```
> U2:=remove(has,U1,BesselY);
> U3:=expand(convert(U2,trig));
```

$$U3 := \_C5\,\_C3\,\_C1\, \mathrm{BesselJ}(p,\, k\,r)\, \cos(c\,k\,t)\, \cos(p\,\theta)$$
$$- \_C5\,\_C3\,\_C1\, \mathrm{BesselJ}(p,\, k\,r)\, \sin(c\,k\,t)\, \sin(p\,\theta)$$
$$+ \_C5\,\_C4\,\_C1\, \mathrm{BesselJ}(p,\, k\,r)\, \sin(c\,k\,t)\, \cos(p\,\theta)\, I$$
$$+ \_C5\,\_C3\,\_C1\, \mathrm{BesselJ}(p,\, k\,r)\, \cos(c\,k\,t)\, \sin(p\,\theta)\, I + \cdots$$

The initial transverse velocity of the drumhead is zero, so the $\sin(c\,k\,t)$ terms must be removed from *U3*. Since the initial shape involves a sine function only, there can be no cosine terms present in the solution. The $\cos(p\,\theta)$ terms are therefore also removed in *U4* and the result then factored.

```
> U4:=factor(remove(has,U3,{sin(c*k*t),cos(p*theta)}));
```

$$U4 := \_C1\,(\_C3 - \_C4)(\_C5 + \_C6)\sin(p\,\theta)\cos(c\,k\,t)\,\mathrm{BesselJ}(p,\, k\,r)\,I$$

The solution must vanish at the perimeter, so the Bessel functions $J_p(k\,r)$ must be zero at $r = a$. Thus, $k\,a$ must equal the mth zero of J_p, with $m = 1,\ 2,\ ....$ The allowed k values are now entered.

```
>  k:=BesselJZeros(p,m)/a:
```

The parameter values $a = 1$ and $c = 1$ are given. To match the initial angular dependence, we must have $p = 2$, i.e., only the Bessel function J_2 will be present in the solution. The radial portion of the initial shape is entered in f. The form of the Bessel functions is given in g.

```
>  a:=1: c:=1: p:=2: f:=r*(1-r^2/a^2)/20; g:=BesselJ(p,k*r);
```

$$f := \frac{r\,(-r^2 + 1)}{20} \qquad g := \mathrm{BesselJ}(2,\ \mathrm{BesselJZeros}(2,\ m)\,r)$$

The messy constants are removed from $U4$ with the following select command.

```
>  U[2,m]:=select(has,U4,{cos,sin,BesselJ});
```

$$U_{2,\,m} := \cos(\mathrm{BesselJZeros}(2,\ m)\,t)\,\sin(2\,\theta)\,\mathrm{BesselJ}(2,\ \mathrm{BesselJZeros}(2,\ m)\,r)$$

Making use of orthogonality and noting that the weight function for the Bessel functions is r, the mth coefficient is given by $A_{2,m} = \int_0^a f\,r\,g\,dr / \int_0^a r\,g^2\,dr$.

```
>  A[2,m]:=int(f*r*g,r=0..a)/int(r*g^2,r=0..a):
```

The first 4 terms of the series solution U are now displayed in decimal form.

```
>  U:=evalf(sum(A[2,m]*U[2,m],m=1..4));
```

$$\begin{aligned}
U := \ & 0.04223579864\ \cos(5.135622302\,t)\,\sin(2.\,\theta)\,\mathrm{BesselJ}(2.,\ 5.135622302\,r) \\
& + 0.002159293112\ \cos(8.417244140\,t)\,\sin(2.\,\theta)\,\mathrm{BesselJ}(2.,\ 8.417244140\,r) \\
& + 0.005912588156\ \cos(11.61984117\,t)\,\sin(2.\,\theta)\,\mathrm{BesselJ}(2.,\ 11.61984117\,r) \\
& + 0.001008506062\ \cos(14.79595178\,t)\,\sin(2.\,\theta)\,\mathrm{BesselJ}(2.,\ 14.79595178\,r)
\end{aligned}$$

For animation purposes, we convert from polar coordinates to Cartesian coordinates, setting $r = \sqrt{x^2 + y^2}$ and using the trig identity $\sin(2\,\theta) = 2\,\sin\theta\,\cos\theta = 2\,x\,y/r^2$. Note that a floating point evaluation is used in entering the latter so that the substitution will actually occur. This is necessary because a floating point evaluation was used in expressing U.

```
>  r:=sqrt(x^2+y^2): sin(evalf(2*theta)):=2*x*y/r^2:
```

The solution is then expressed as the piecewise function $UU = U$ for $r < a$ and 0 for $r > a$.

```
>  UU:=evalf(piecewise(r<a,U,r>a,0)):
```

To animate UU, a functional operator gr is created to make a 3-dimensional plot of the drumhead shape on the ith time step, the stepsize being 0.1 s.

```
>  gr:=i->plot3d(eval(UU,t=0.1*i),x=-a..a,y=-a..a,
      style=patchcontour,shading=zhue):
```

Then using the sequence command, the profiles on 20 consecutive time steps are displayed. The insequence=true option is included to produce the animation.

```
>  display([seq(gr(i),i=1..20)],insequence=true,axes=framed);
```

If you wish to see the drumhead animation, execute the recipe on your computer, then click on the computer plot and on the start arrow in the tool bar.

4.3.4 The Cannon Ball

The sound of a kiss is not so loud as that of a cannon,
but its echo lasts a great deal longer.
Oliver Wendell Holmes Sr., American writer, physician, (1809–94)

In the re-enactment of a Civil war battle, a cannon is fired and a hot spherical iron cannon ball plunges into an icy lake whose temperature is very close
to freezing (0°C). If the cannon ball has a radius $R = 20$ cm and is initially
100°C throughout on entering the lake, determine the temperature distribution inside the cooling cannon ball as a function of time. Plot the temperature
distribution in 1 minute intervals up to 15 minutes. What is the temperature
at the center of the cannon ball 15 minutes after plunging into the lake? For
iron, $K/(\rho C) = 0.185$ in cgs units, where K is the thermal conductivity, ρ the
density, and C the specific heat.

After loading the plots and VectorCalculus packages,

```
>  restart: with(plots): with(VectorCalculus):
```
the heat diffusion equation $\nabla^2 T = (1/a^2)\,(\partial T/\partial t)$, with T the temperature and
$a^2 \equiv K/(\rho C)$, is entered in spherical polar coordinates (r, θ, ϕ). r is the radial
distance from the center of the cannon ball, θ is the angle that the radial vector
makes with the z-axis, and ϕ is the angle that the projection of the radial vector
into the x-y plane makes with the x-axis.

```
>  pde:=expand(Laplacian(T(r,theta,phi,t),'spherical'))
      [r,theta,phi]=diff(T(r,theta,phi,t),t)/a^2;
```

$$pde := \frac{2\left(\frac{\partial}{\partial r} T(r, \theta, \phi, t)\right)}{r} + \left(\frac{\partial^2}{\partial r^2} T(r, \theta, \phi, t)\right) + \frac{\cos(\theta)\left(\frac{\partial}{\partial \theta} T(r, \theta, \phi, t)\right)}{r^2 \sin(\theta)}$$

$$+ \frac{\frac{\partial^2}{\partial \theta^2} T(r, \theta, \phi, t)}{r^2} + \frac{\frac{\partial^2}{\partial \phi^2} T(r, \theta, \phi, t)}{r^2 \sin(\theta)^2} = \frac{\frac{\partial}{\partial t} T(r, \theta, \phi, t)}{a^2}$$

Since the initial temperature of the cannon ball is uniform throughout, the solution will have no angular dependence. Assuming that $T(r, \theta, \phi, t) = R(r) F(t)$,
the heat flow equation *pde* is analytically solved using the pdsolve command.

```
>  sol:=pdsolve(pde,HINT=R(r)*F(t),INTEGRATE,build);
```

$$sol := T(r, \theta, \phi, t) = \frac{\_C3\, e^{(a^2\, \_c_1\, t)}\, \_C1 \sinh(\sqrt{\_c_1}\, r)}{r}$$

$$+ \frac{\_C3\, e^{(a^2\, \_c_1\, t)}\, \_C2 \cosh(\sqrt{\_c_1}\, r)}{r}$$

The hyperbolic sine (sinh) term remains finite at the origin $(r=0)$, but the cosh
term diverges to ∞ and must be removed from the rhs of *sol*.

```
>  T1:=remove(has,rhs(sol),cosh);
```

$$T1 := \frac{-C3\, e^{(a^2 - c_1 t)} \_C1 \sinh(\sqrt{-c_1}\, r)}{r}$$

The separation constant $\_c_1$ is replaced with $-k^2$ in $T1$ and the result simplified with the `symbolic` option.

>   T2:=simplify(subs(_c[1]=-k^2,T1),symbolic);

$$T2 := \frac{-C3\, e^{(-a^2 k^2 t)} \_C1 \sin(k\,r)\, I}{r}$$

It should be noted that the term $\sin(k\,r)/r$ is ([AS72]) just the zeroth order *spherical Bessel function* of the first kind.[3] Making use of the `select` command, the coefficient combination $\_C3\_C1\, I$ in $T2$ is replaced with the symbol A.

>   T3:=A*select(has,T2,{exp,sin,r});

$$T3 := \frac{A\, e^{(-a^2 k^2 t)} \sin(k\,r)}{r}$$

Taking the radius of the cannon ball to be R, the surface of the ball is held at 0 degrees, so $\sin(k\,R) = 0$, and therefore $k = n\,\pi/R$, with $n = 1,\ 2,\ ....$ Entering this result, the nth normal mode of the temperature is displayed in $T4$.

>   k:=n*Pi/R: T4:=T3;

$$T4 := \frac{A\, e^{\left(-\frac{a^2 n^2 \pi^2 t}{R^2}\right)} \sin\left(\dfrac{n\,\pi\,r}{R}\right)}{r}$$

The initial temperature is $100°$ throughout the cannon ball. Noting that the weight function ([AS72]) for the spherical Bessel functions is r^2, orthogonality leads to the following expression for the coefficients:

$$A = \int_0^R r^2\, 100\, (\sin(k\,r)/r)\, dr \Big/ \int_0^R r^2\, (\sin(k\,r)/r)^2\, dr.$$

A is now calculated, assuming that n is an integer.

>   A:=int(r^2*100*sin(k*r)/r,r=0..R)/int(r^2*(sin(k*r)/r)^2,
     r=0..R) assuming n::integer;

$$A := \frac{200\,(-1)^{(1+n)}\, R}{n\,\pi}$$

The formal series representation of the temperature distribution T inside the cannon ball is now completely determined. Retaining 200 terms in the series, T has the following structure.

>   T:=Sum(T4,n=1..200);

$$T := \sum_{n=1}^{200} \left(\frac{200\,(-1)^{(1+n)}\, R\, e^{\left(-\frac{a^2 n^2 \pi^2 t}{R^2}\right)} \sin\left(\dfrac{n\,\pi\,r}{R}\right)}{n\,\pi\,r} \right)$$

---

[3]The spherical Bessel functions j_n of the first kind are related to the "ordinary" Bessel functions by the relation $j_n(x) \equiv \sqrt{(\pi/2x)}\, J_{n+1/2}(x)$.

Taking $R = 20$ and $a = \sqrt{0.185}$, T is now evaluated, but not displayed.

```
> T:=eval(value(T),{R=20,a=sqrt(0.185)}):
```

An arrow operator `gr` is formed to plot T at 60 s (1 min) intervals.

```
> gr:=i->plot(eval(T,t=i*60),r=0..20,numpoints=1000):
```

Using `gr` and the sequence command, the temperature T is displayed as a function of radius r in Figure 4.13 at $t = 0$ (top curve), $t = 1$ min (next lowest curve), etc, to $t = 15$ minutes (bottom curve).

```
> display(seq(gr(i),i=0..15),labels=["r","T"]);
```

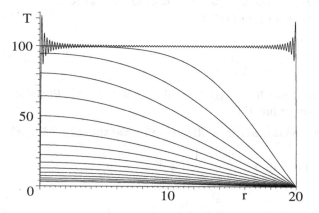

Figure 4.13: Time evolution of the temperature T inside the cannon ball.

The "ringing" in the initial temperature distribution is, of course, the Gibb's phenomenon due to the initial step function temperature profile. Taking the limit of T as $r \rightarrow 0$, the temperature at the center of the sphere at 15 minutes, or 900 seconds, is calculated and found to be about 3.3°C.

```
> Tcenter:=eval(limit(T,r=0),t=900);
```

$$Tcenter := 3.287377363$$

4.3.5   Variation on a Split-sphere Potential

Would you convey my compliments to the purist who reads your proofs and tell him or her ... that when I split an infinitive, God damn it, I split it so it will stay split.
Raymond Chandler, American writer of detective fiction, (1888–1959)

The following recipe is based on a simple variation of a standard problem in electrostatics. The objective is to completely determine the potential V inside and outside a hollow sphere of unit radius with a specified piecewise potential on the spherical surface. With θ measured from the positive z-axis (pointing vertically upwards), the spherical surface between $\theta = 0$ and $45°$ has the constant potential $V0$, the intermediate section between $45°$ and $135°$ has a variable potential given by $\sqrt{2}\cos(\theta)\,V0$, and the lower portion between $135°$ and $180°$ held at $-V0$. Keeping the first 6 non-zero terms in V and taking $V0 = 1$, plot the equipotentials corresponding to $V = 0.8, 0.6, ..., -0.6, -0.8$.

After loading the necessary library packages, Laplace's equation is entered in spherical coordinates, the origin taken at the center of the sphere. By symmetry, V must be independent of the azimuthal[4] angle ϕ, i.e., $V = V(r, \theta)$.

```
> restart: with(plots): with(VectorCalculus):
```

```
> pde:=expand(Laplacian(V(r,theta),'spherical'
          [r,theta,phi]))=0;
```

$$pde := \frac{2\left(\frac{\partial}{\partial r}V(r,\theta)\right)}{r} + \left(\frac{\partial^2}{\partial r^2}V(r,\theta)\right) + \frac{\cos(\theta)\left(\frac{\partial}{\partial \theta}V(r,\theta)\right)}{r^2 \sin(\theta)} + \frac{\frac{\partial^2}{\partial \theta^2}V(r,\theta)}{r^2} = 0$$

Then *pde* is analytically solved, assuming that $V(r,\theta) = R(r)\,\Theta(\theta)$, the result involving Legendre functions.

```
> V:=rhs(pdsolve(pde,HINT=R(r)*Theta(theta),INTEGRATE,build));
```

$$V := \frac{\_C3\,\text{LegendreP}(-\frac{1}{2} + \frac{1}{2}\sqrt{1 + 4\_c_1},\cos(\theta))\_C1\,r^{(1/2\sqrt{1+4\_c_1})}}{\sqrt{r}} + \cdots$$

The separation constant $\_c_1$ is replaced in V with $-1/4 + (n + 1/2)^2$ and the result simplified with the **symbolic** option and then expanded.

```
> V:=expand(simplify(subs(_c[1]=-1/4+(n+1/2)^2,V),symbolic));
```

$$V := \_C3\,\text{LegendreP}(n,\cos(\theta))\_C1\,r^n + \frac{\_C3\,\text{LegendreP}(n,\cos(\theta))\_C2}{r\,r^n}$$
$$+ \_C4\,\text{LegendreQ}(n,\cos(\theta))\_C1\,r^n + \frac{\_C4\,\text{LegendreQ}(n,\cos(\theta))\_C2}{r\,r^n}$$

V is expressed in terms of the Legendre functions of the first $(P_n(\cos\theta))$ and second $(Q_n(\cos\theta))$ kinds. The Q_n diverge at the end points of the θ range and must be rejected. The redundant constant $\_C3$ in the P_n terms is set equal to 1 and $\_C1$ and $\_C2$ are replaced with the symbols A and B.

```
> V:=subs({LegendreQ(n,cos(theta))=0,_C3=1,_C1=A,_C2=B},V);
```

$$V := \text{LegendreP}(n,\cos(\theta))\,A\,r^n + \frac{\text{LegendreP}(n,\cos(\theta))\,B}{r\,r^n}$$

For the inside $(r < 1)$ solution Vin, we set $B = 0$ in V so that Vin doesn't diverge at the origin. For $r > 1$, we take $A = 0$ so that $Vout \to 0$ as $r \to \infty$.

```
> Vin:=subs(B=0,V); Vout:=subs(A=0,V);
```

$$Vin := \text{LegendreP}(n,\cos(\theta))\,A\,r^n$$

---

[4]The angle between the projection of the radius vector into the x-y (horizontal) plane and the x-axis.

$$Vout := \frac{\mathrm{LegendreP}(n,\cos(\theta))\,B}{r\,r^n}$$

Setting $u = \cos\theta$, the angular distributions in each region are entered.

```
>  f1:=-V0: f2:=sqrt(2)*u*V0: f3:=V0:
```

Making use of orthogonality of the $P_n(u)$, the coefficients in the series representation of the solution are evaluated using $A_n = ((2n+1)/2)\int_{-1}^{1} f\,P_n(u)\,du$. An operator AA is introduced to evaluate the coefficients for a given n value.

```
>  AA:=n->((2*n+1)/2)*(int(f1*LegendreP(n,u),u=-1..-1/sqrt(2))
          +int(f2*LegendreP(n,u),u=-1/sqrt(2)..1/sqrt(2))
          +int(f3*LegendreP(n,u),u=1/sqrt(2)..1)):
```

Then, employing AA(n), the inside and outside solutions are determined, the series being terminated at $n = 12$.

```
>  VIN:=sum(eval(Vin,A=AA(n)),n=0..12);
```

$$VIN := \frac{5}{4}\cos(\theta)V0\,r - \frac{7}{32}\mathrm{LegendreP}(3,\cos(\theta))V0\,r^3$$

$$-\frac{11}{128}\mathrm{LegendreP}(5,\cos(\theta))V0\,r^5 + \frac{85}{2048}\mathrm{LegendreP}(7,\cos(\theta))V0\,r^7$$

$$+\frac{323}{8192}\mathrm{LegendreP}(9,\cos(\theta))V0\,r^9 - \frac{1219}{65536}\mathrm{LegendreP}(11,\cos(\theta))V0\,r^{11}$$

```
>  VOUT:=sum(eval(Vout,B=AA(n)),n=0..12);
```

$$VOUT := \frac{5}{4}\frac{\cos(\theta)V0}{r^2} - \frac{7}{32}\frac{\mathrm{LegendreP}(3,\cos(\theta))V0}{r^4} + \cdots$$

The equipotentials will now be plotted in the x-z plane, by setting $r = \sqrt{x^2 + z^2}$ and $\cos\theta = z/r$. We also set $V0 = 1$.

```
>  r:=sqrt(x^2+z^2): cos(theta):=z/r: V0:=1.0:
```

A piecewise potential function VPW is formed with $V = VIN$ for $r < 1$ and $VOUT$ for $r > 1$.

```
>  VPW:=piecewise(r<1,VIN,r>1,VOUT):
```

Loading the plottools package, a blue circle of radius 1 centered on the origin is produced in c to represent the spherical surface (a circle in 2 dimensions).

```
>  with(plottools): c:=circle([0,0],1,color=blue,thickness=2):
```

The contourplot command is used in cp to plot the requested equipotentials.

```
>  cp:=contourplot(VPW,x=-3..3,z=-4..4,contours=
      [seq(0.8-0.2*i,i=0..8)],grid=[60,60],thickness=2,color=red):
```

The graphs c and cp are now displayed together with constrained scaling,

```
>  display({c,cp},scaling=constrained);
```

the resulting picture being shown in Figure 4.14.

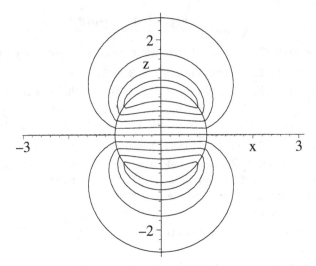

Figure 4.14: Equipotentials for the split-sphere in the x-z plane.

The 3-dimensional equipotential surfaces are obtained by mentally rotating the picture about the z-axis. The recipe is easily adjusted to handle other variations on the angular distribution and can even be modified to handle cylindrical geometry.

4.3.6   Another Poisson Recipe

Every man is a potential genius until he does something.
Sir Herbert Beerbohm Tree, English actor-manager, (1853–1917)

In magnetostatics, the vector potential $\vec{A}(\vec{R})$ at a point $\vec{R}$, associated with a current density $\vec{J}$ in free space, satisfies the *vector Poisson equation*, [Gri99]

$$\nabla^2 \vec{A} = -\mu_0 \vec{J}, \tag{4.9}$$

where μ_0 is the permeability of free space. Equation (4.9) is three scalar Poisson equations, one for each Cartesian component, e.g., $\nabla^2 A_x = -\mu_0 J_x$. In any other curvilinear coordinate system, the unit vectors are functions of position. Thus, e.g., in spherical polar coordinates it is not true that $\nabla^2 A_r = -\mu_0 J_r$. Assuming that $\vec{J} \to 0$ at infinity, the solution of Eq. (4.9) is given by,

$$\vec{A}(\vec{R}) = \frac{\mu_0}{4\pi} \int \frac{\vec{J}(\vec{R}_1)}{|\vec{R} - \vec{R}_1|} \, dv_1, \tag{4.10}$$

again representing three 3-dimensional integrals, e.g.,

$$A_x(x, y, z) = \frac{\mu_0}{4\pi} \int\!\!\int\!\!\int \frac{J_x(x_1, y_1, z_1) \, dx_1 \, dy_1 \, dz_1}{\sqrt{(x - x_1)^2 + (y - y_1)^2 + (z - z_1)^2}}.$$

If you want to calculate the integrals in Equation (4.10) in other curvilinear coordinates, you must first express $\vec{J}$ in terms of its Cartesian components. Once $\vec{A}$ is determined, the magnetic field is then given by $\vec{B} = \nabla \times \vec{A}$. As a representative example, consider the following magnetostatic problem.

A uniformly charged solid sphere of radius a carries a total charge Q and is spinning with angular velocity ω about the vertical z-axis. Determine the magnetic vector potential $\vec{A}$ inside and outside the sphere and then calculate the magnetic field $\vec{B}$ in both regions. Plot $\vec{B}/(\mu_0 Q\omega)$ for $a=1$.

Taking the origin at the sphere's center, we let $\vec{R}$ (r, θ, ϕ) be the location of the "observation" point P inside or outside the sphere and $\vec{R}1$ $(r1, \theta1, \phi1)$ be the location of a current density "source" point P_1 inside the sphere. The polar angles θ and $\theta1$ are measured from the z-axis and the azimuthal angles ϕ and $\phi1$ from the projection of the radius vector into the (horizontal) x-y plane with the x-axis. To ensure a later simplification, it is assumed that $r > 0$.

```
>   restart: with(plots): with(VectorCalculus): assume(r>0):
```

Since the charge is uniformly distributed in the sphere, the charge density ρ is equal to the total charge Q divided by the volume $4\pi a^3/3$ of the sphere. At a radial distance $r1$ and angle $\theta1$ with the z-axis, the source point P_1 has a linear velocity $r1 \sin(\theta1)\omega$. So the current density magnitude J is equal to ρ times the linear velocity, the entered form of ρ being automatically substituted. The corresponding vector $\vec{J}$ points in the $\hat{\phi}$ direction for every P_1.

```
>   rho:=Q/((4*Pi*a^3)/3); J:=rho*r1*sin(theta1)*omega;
```

$$\rho := \frac{3Q}{4\pi a^3} \qquad J := \frac{3}{4}\frac{Q\, r1 \sin(\theta1)\,\omega}{\pi a^3}$$

Without loss of generality, let's take the observation point P to be in the x-z plane, so that $\phi=0$. The position vector $\vec{R}$ of P is now entered, being expressed in terms of the Cartesian unit vectors along the x and z axes.

```
>   R:=<r*sin(theta),0,r*cos(theta)>;
```

$$R := r\sin(\theta)\,e_x + r\cos(\theta)\,e_z$$

The position vector $\vec{R}1$ of the source point P_1 is also entered.

```
>   R1:=<r1*sin(theta1)*cos(phi1),r1*sin(theta1)*sin(phi1),
        r1*cos(theta1)>;
```

$$R1 := r1 \sin(\theta1) \cos(\phi1)\,e_x + r1 \sin(\theta1) \sin(\phi1)\,e_y + r1 \cos(\theta1)\,e_z$$

To calculate $\vec{A}$, let's first evaluate $1/(|\vec{R}-\vec{R}1|)=1/\sqrt{(\vec{R} - \vec{R}1) \cdot (\vec{R} - \vec{R}1)}$. This is done in f using the DotProduct command, the result then being simplified with the symbolic option.

```
>   f:=simplify(1/sqrt(DotProduct(R-R1,R-R1)),symbolic);
```

$$f := \frac{1}{\sqrt{r^2 - 2r\sin(\theta)\,r1 \sin(\theta1) \cos(\phi1) + r1^2 - 2r\cos(\theta)\,r1 \cos(\theta1)}}$$

We could attempt to evaluate the integral in $\vec{A}$ directly, but it is more instructive to Taylor expand f about a specified value of $r1$, out to some given order, and see what the various orders contribute to the overall answer. This approach is equivalent to the *multipole expansion* discussed in standard electromagnetic texts such as Griffiths [Gri99]. So, a functional operator T is formed to Taylor expand f to order n about a specified point $r1 = d$. The convert(,polynom) command is included to remove the order of term which would otherwise appear.

```
>   T:=(n,d)->convert(taylor(f,r1=d,n),polynom):
```

The current density vector can be resolved into the Cartesian components $J_x = J \sin(\phi1)$, $J_y = J \cos(\phi1)$, and $J_z = 0$. At the observation point (chosen in the x-z plane), the J_x contribution to $\vec{A}$ will add up to zero, but the J_y contribution will not. But since, our choice of observation point in the x-z plane was arbitrary and there is complete rotational symmetry about the z-axis, the resultant component J_y will yield the ϕ component of $\vec{A}$. A functional operator A is created to perform the volume integration in (4.10) using J_y and the Taylor expansion of f and taking spherical polar coordinates. The volume element is $r1^2 \sin(\theta1)\, d\theta1\, d\phi1\, dr1$. The angular coordinate $\theta1$ ranges from 0 to π, while $\phi1$ varies from 0 to 2π. The order n, the radial distance d about which Taylor expansion is taking place, and the lower and upper limits, $d1$ and $d2$, of the $r1$ integration must be specified. Again the result is simplified.

```
>   A:=(n,d,d1,d2)->simplify((mu[0]/(4*Pi))*int(int(int(T(n,d)
        *J*cos(phi1)*r1^2*sin(theta1),theta1=0..Pi),r1=d1..d2),
        phi1=0..2*Pi),symbolic):
```

First, let's take the observation point P to be outside the sphere, i.e., $r > a$. Since $r1 \leq a$, then $r1/r < 1$ and we can Taylor expand f about $r1 = d = 0$. The limits of the $r1$ integration are $d1 = 0$ and $d2 = a$. Making uses of the operator A with the above arguments, the vector potential is evaluated in *Out* to order $n = 1$, 2, and 3 and the result assigned.

```
>   Out:=seq(A||n=A(n,0,0,a),n=1..3); assign(Out):
```

$$Out := A1 = 0, \quad A2 = \frac{1}{20}\, \frac{\mu_0 \sin(\theta)\, a^2\, Q\, \omega}{\pi\, r^2}, \quad A3 = \frac{1}{20}\, \frac{\mu_0 \sin(\theta)\, a^2\, Q\, \omega}{\pi\, r^2}$$

For $n = 1$, the so-called *monopole* contribution $A1$ to the vector potential is 0, a well-known general result. For $n = 2$, there is a non-zero *dipole* contribution $A2$. For $n = 3$ (and higher), there is no additional contribution to the vector potential, indicating that outside the sphere the vector potential (and hence, the magnetic field) is that of a "pure" magnetic dipole. The vector potential *Aout* outside the sphere is now expressed in spherical polar coordinates, having only a ϕ component.

```
>   Aout:=VectorField(<0,0,A2>,'spherical'[r,theta,phi]);
```

$$Aout := \frac{1}{20}\, \frac{\mu_0 \sin(\theta)\, a^2\, Q\, \omega}{\pi\, r^2}\, \bar{e}_\phi$$

Now, consider P to be inside the sphere, i.e., $r < a$. For $r1 < r$, we can again Taylor expand f about $r1 = 0$, the integration being from $r1 = 0$ to r. But for $r1 > r$, f is Taylor expanded about $r1 = \infty$, the $r1$ integration being from r to a. Adding the two contributions, $\vec{A}$ inside the sphere is calculated for $n = 1$, 2, and 3. No further contribution occurs for higher n values.

> In:=seq(AA||n=A(n,0,0,r)+A(n,infinity,r,a),n=1..3);
    assign(In):

$$In := AA1 = 0, \quad AA2 = \frac{1}{20} \frac{\mu_0 \sin(\theta) Q \omega r^3}{\pi a^3},$$

$$AA3 = \frac{1}{20} \frac{\mu_0 \sin(\theta) Q \omega r^3}{\pi a^3} - \frac{1}{8} \frac{\mu_0 Q \omega r \sin(\theta) (-a^2 + r^2)}{\pi a^3}$$

In terms of spherical polar coordinates, the vector potential Ain inside the sphere takes the following form.

> Ain:=VectorField(<0,0,AA||3>,'spherical'[r,theta,phi]);

$$Ain := \left(\frac{1}{20} \frac{\mu_0 \sin(\theta) Q \omega r^3}{\pi a^3} - \frac{1}{8} \frac{\mu_0 Q \omega r \sin(\theta) (-a^2 + r^2)}{\pi a^3} \right) \bar{e}_\phi$$

Using Curl, the magnetic field is calculated outside and inside the sphere.

> Bout:=Curl(Aout); Bin:=simplify(Curl(Ain));

$$Bout := \frac{1}{10} \frac{\mu_0 a^2 Q \omega \cos(\theta)}{r^3 \pi} \bar{e}_r + \frac{1}{20} \frac{\sin(\theta) \mu_0 a^2 Q \omega}{r^3 \pi} \bar{e}_\theta$$

$$Bin := -\frac{1}{20} \frac{\cos(\theta) \mu_0 Q \omega (3 r^2 - 5 a^2)}{\pi a^3} \bar{e}_r + \frac{1}{20} \frac{\sin(\theta) \mu_0 Q \omega (6 r^2 - 5 a^2)}{\pi a^3} \bar{e}_\theta$$

To plot $\vec{B}/(\mu_0 Q \omega)$, the MapToBasis command is used to convert the normalized magnetic field to Cartesian coordinates.

> F:=u->MapToBasis(u/(mu[0]*Q*omega),'cartesian'[x,y,z]):

Taking $a = 1$, we will plot the magnetic field in the x-z plane. To accomplish this, let's set $r = \sqrt{x^2 + z^2}$ and form an operator G to evaluate the magnetic field for $y = 0$.

> a:=1: r:=sqrt(x^2+z^2): G:=B->simplify(eval(F(B),y=0)):

The magnetic field inside and outside the sphere then takes the following forms, expressed in Cartesian coordinates.

> Bin2:=G(Bin); Bout2:=G(Bout);

$$Bin2 := \frac{3 x z}{20 \pi} e_x - \frac{6 x^2 + 3 z^2 - 5}{20 \pi} e_z$$

$$Bout2 := \frac{3 z x}{20 (x^2 + z^2)^{(5/2)} \pi} e_x - \frac{-2 z^2 + x^2}{20 (x^2 + z^2)^{(5/2)} \pi} e_z$$

The complete magnetic field $\vec{B}$ is formed with the piecewise operator PW, taking the nth component of $Bin2$ and $Bout2$ for $r < a$ and $r > a$, respectively.

> PW:=n->piecewise(r<a,Bin2[n],r>a,Bout2[n]):

The x and z components of $\vec{B}$ are obtained by taking $n = 1$ and 3 in PW.

> B[1]:=PW(1): B[3]:=PW(3):

In c, a blue circle of radius a is plotted to represent the spherical surface.

> c:=plot(a,theta=0..2*Pi,coords=polar,color=blue,thickness=2):

The fieldplot command is used in fp to plot the magnetic field as thick red arrows, the grid density being taken to be 10×10.

```
>  fp:=fieldplot([B[1],B[3]],x=-1.1..1.1,z=-1.1..1.1,
         arrows=THICK,grid=[10,10],color=red):
```
The two graphs are superimposed with the `display` command to produce Figure 4.15. The picture should be mentally rotated around the z-axis to obtain the 3-dimensional magnetic field of the rotating uniformly charged sphere.

```
>  display({c,fp});
```

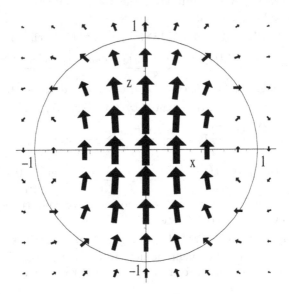

Figure 4.15: Magnetic field of rotating charged sphere in x-z plane.

Along the spin axis, the magnetic field points vertically upwards, having maximum strength at the center of the sphere. The magnetic field rapidly drops in strength outside the sphere. The arrows form curved paths which leave the top of the sphere and loop back on the outside to re-enter the sphere at the bottom. The magnetic field behavior is characteristic of a dipole field.

4.4 Supplementary Recipes

04-S01: General Solutions
Consider a second order linear PDE of the general form (a, b, c are constants and the subscripts denote derivatives),

$$a\,\psi_{xx} + b\,\psi_{xy} + c\,\psi_{yy} = 0.$$

(a) By assuming a solution of the form $\psi = f(x+r\,y)$ where f is arbitrary, show that the general solution of the PDE is $\psi = f(x+r_1\,y) + g(x+r_2\,y)$ where

r_1 and r_2 are distinct roots of the quadratic equation $cr^2 + br + a = 0$. Determine the two roots. If $r_2 = r_1$, show that the general solution is $\psi = f(x + r_1 y) + y g(x + r_1 y)$, where f and g are arbitrary functions.

(b) Making use of (a), find the general solutions of the following PDEs. In each case, confirm the solution by solving the PDE directly with pdsolve.

(i) $\psi_{xy} - \psi_{yy} = 0$; (ii) $\psi_{xx} + \psi_{xy} - 2\psi_{yy} = 0$; (iii) $\psi_{xx} - 2\psi_{xy} + \psi_{yy} = 0$.

04-S02: Balalaika Blues

While trying to create music with a small balalaika, Justine plucks one of the strings which is initially at rest. If the string is fixed at $x = 0$ and L and has an initial transverse profile $\psi(x, 0) = h x^3 \sin(3\pi x/2L)(L - x)^2/L^5$, determine the subsequent displacement $\psi(x, t)$ of the string using a Fourier series approach. Taking $L = 20$ cm, $h = 30$ cm, and wave speed $c = 1$ cm/s, animate the motion of the string. Leave the scaling unconstrained for better viewing of the vibrations.

04-S03: Damped Oscillations

If damping (damping coefficient R) is included, the transverse vibrations of a light, homogeneous, string under tension are governed by

$$\frac{\partial^2 \psi}{\partial x^2} = \frac{1}{c^2}\frac{\partial^2 \psi}{\partial t^2} + \frac{2R}{c^2}\frac{\partial \psi}{\partial t}.$$

Consider a horizontal string of length L fixed at $x = 0$ and L. If it is initially at rest and is given the initial shape $f(x) = h x (x - L)$, use the separation of variables method to determine $\psi(x, t)$ for $t > 0$. Taking $L = 2$ m, $h = 0.1$ m^{-1}, speed $c = 5$ m/s, and $R = 1$ s^{-1}, animate the motion of the string over the time interval $t = 0$ to 10 s. You should observe *underdamped* oscillations. Show that *overdamping* occurs for $R = 10$ s^{-1}.

04-S04: Kids Will Be Kids

Consider a very long uniform clothesline under sufficient tension T to keep it horizontal, with Justine's recently washed soccer jersey (mass m) attached with a single clothes peg to the middle of the line at $x = 0$. In an unsuccessful attempt to dump her sister's jersey onto the ground, Gabrielle shakes the clothesline in such a way that a plane wave with amplitude A and phase velocity $c = \omega/k$ is incident from $x < 0$. Show that reflection and transmission occur at $x = 0$ and that the energy reflection coefficient $R = \sin^2(\theta)$ and transmission coefficient $Tr = \cos^2(\theta)$, where $\theta = \arctan(m\omega^2/2kT)$. Determine the phase angle changes for the reflected and transmitted waves in terms of θ.

04-S05: Energy of a Vibrating String

For a finite horizontal string, with linear density $\epsilon(x)$ and under tension T, stretched between $x = x_1$ and x_2, the kinetic energy for small transverse vibrations is KE $= (1/2)\int_{x_1}^{x_2} \epsilon(x)(\partial\psi/\partial t)^2 dx$. If the ends of the string are either fixed ($\psi = 0$) or free ($\psi' = 0$), the potential energy is PE $= (1/2)\int_{x_1}^{x_2} T (\partial\psi/\partial x)^2 dx$.

Consider a horizontal string of constant density ϵ and under tension T fixed at $x=0$ and L. If at time $t=0$, the string is given the initial triangular profile $f(x) = 2\,h\,x/L$ for $0 \leq x \leq L$, $f(x) = 2\,h\,(L-x)/L$ for $L/2 \leq x \leq L$, and is released from rest, calculate explicit analytic expressions for the kinetic energy, potential energy, and total energy. What fraction of the total energy is in the fundamental (lowest) frequency? first harmonic (next highest frequency)? second harmonic? Plot the kinetic and potential energies divided by the total energy as a function of time for $L=1$ m and wave velocity $c=\sqrt{T/\epsilon}=1$ m/s.

04-S06: Vibrations of a Tapered String

A horizontal tapered string fixed at $x=0$ and L has a linear density $\epsilon = a(1+b\,x)$. Determine the normal modes for transverse oscillations of the string. Identify the functions which occur in the analytic solution. If $L=1$ m, $a=1/1000$ kg/m, $b=2$ m^{-1}, and the tension $T=1$ N, determine the three lowest eigenfrequencies. Animate the fundamental mode over one period, taking 50 frames.

04-S07: Green Function for Forced Vibrations

A stretched (tension T) string fixed at $x = 0$ and L is subjected to forced vibrations by an external force per unit length $f(x,t) = F\,\delta(x - \zeta)\,e^{-i\omega t}$ with the force amplitude $F = T$ and $0 < \zeta < L$. Solve this Green function problem and show that the Green function (spatial part) is

$$G = \frac{\sin[k(L-\zeta)]\,\sin(k\,x)}{k\,\sin(k\,L)}, \quad x \leq \zeta, \qquad G = \frac{\sin[k(L-x)]\,\sin(k\,\zeta)}{k\,\sin(k\,L)}, \quad x \geq \zeta,$$

where $k = \omega/v$, v being the wave velocity. G has singularities at certain values of k. Determine these values. Explain, in physical terms, the origin of these singularities and how in real life, they would be "removed".

04-S08: Plane-wave Propagation in a 5-Piece String

Consider an infinitely long string which has a linear density $\epsilon_1 = 1$ in region 1 ($x < 0$), density $\epsilon_2 = 4$ in region 2 ($0 < x < L$), density $\epsilon_3 = 1$ in region 3 ($L < x < 2L$), density $\epsilon_4 = 4$ in region 4 ($2L < x < 3L$), and density $\epsilon_5 = 1$ in region 5 ($x > 3L$). For a plane wave of incident amplitude 1, frequency ω, and wave number K, coming from $x = -\infty$:

(a) Derive the energy transmission coefficient T into region 5 and simplify the result as much as possible.

(b) Show that $T + R = 1$, where R is the energy reflection coefficient.

(c) Plot R and T in the same graph as a function of $K L$ up to $K L = 6$. At what values of $K L$ is there 100% transmission? Discuss your answer.

04-S09: Transverse Vibrations of a Whirling String

Derive the wave equation for transverse vibrations of a light string of length L pivoted at one end and whirling in a horizontal plane about that pivot with an angular velocity ν. Determine the normal modes of oscillation of the whirling string and the five lowest allowed frequencies. Animate the mode with the second lowest frequency, taking $L=1$ m and $\nu=1$ s^{-1}.

04-S10: Newton Would Think That This Recipe Is Cool

Newton's law of cooling says that the heat flux (amount of heat crossing a unit area per unit time) across a surface is proportional to the temperature difference $T_S - T_0$ between the surface (temperature T_S) and the surrounding medium (temperature T_0). Explicitly this gives the boundary condition $\hat{n} \cdot \nabla T|_S = -h(T_S - T_0)$, where $\hat{n}$ is the outward unit normal to the surface and h is the heat exchange coefficient. Consider an infinitely long uniform bar of rectangular cross-section with axis along the z direction. The two opposite faces at $y=0$ and $y=b$ are held at the temperatures $T=0$ and $T=A$, while the other two faces at $x=\pm a$ radiate heat according to Newton's cooling law into the surrounding medium which is held at $T_0=0$. Derive the solution which takes the form

$$T(x,y) = \sum_{n=1}^{\infty} C_n \sinh(g_n\, y/a)\, \cos(g_n\, x/a),$$

where the coefficients C_n remain to be identified and g_n are the positive roots of the transcendental equation $g \tan(g) = a\, h$. Taking $a=1$, $b=0.5$, $h=1$, $A=100$, solve the transcendental equation and plot the isotherms inside the bar.

04-S11: Locomotive on a Bridge

As a model of the motion of a locomotive across a railway bridge, consider a periodically vibrating point load $A \sin(\omega t)$ moving with constant velocity V along a horizontal, uniform, rectangular steel beam of density ρ, cross-sectional area S, and length L. The beam is fixed at $x=0$ and L and it is assumed that at time $t=0$ the beam is at rest and the locomotive is at $x=0$. The equation of motion for transverse vibrations (amplitude ψ) of the beam is given by

$$a^4 \frac{\partial^4 \psi}{\partial t^4} + \frac{\partial^2 \psi}{\partial t^2} = \frac{q(x,t)}{\epsilon},$$

with $q(x,t) = A \sin(\omega t)\, \delta(x - V t)$. Here $\epsilon = \rho\, S$ is the linear density of the beam and $a = (\kappa^2\, Y/\rho)^{1/4}$ with κ the radius of gyration of the beam and Y its Young's modulus. Assuming a Fourier series of the form $\psi(x,t) = \sum_{n=1}^{\infty} b_n(t) \sin(n\pi x/L)$, determine the transverse vibrations of the beam. Taking $\rho=7800$ kg/m^3, $Y=2.1 \times 10^{11}$ N/m^2, $S=1$ m^2, $\kappa=0.5$ m, $A=5 \times 10^4$ N, $\omega=1$ s^{-1}, $L=1000$ m, and $V=1$ m/s, animate the motion of the beam with the motion of the locomotive along the beam superimposed. Keep 20 terms in the series solution and use 100 frames in the animation. To see the oscillations, unconstrained scaling should be used.

04-S12: The Temperature Switch

The temperature at the ends $x = 0$ and $x = 100$ of a rod (insulated on its sides) 100 cm long is held at $0°$ and $100°$, respectively, until steady-state is achieved. Then, at the instant $t = 0$, the temperature of the two ends is interchanged. Determine the resultant temperature distribution $T(x,t)$. If $a = \sqrt{K/\rho C} = 10/\pi$ cm/s$^{1/2}$, where K is the thermal conductivity of the rod, ρ is its density, and C the specific heat, animate the temperature distribution. What is the temperature at $x = 20$ cm after 25 seconds?

04-S13: Telegraph Equation Revisited

For an audio-frequency submarine cable [SR66], the telegraph equation applies with the leakage constant $G=0$ and the self inductance (per unit length) $L=0$. As shown in recipe **04-2-1**, in this case the telegraph equation reduces to a 1-dimensional diffusion equation for the potential V with a diffusion constant $d = 1/(RC)$. Consider a submarine cable $\ell = 1000$ km in length, and let the voltage at the source (at $x=0$), under steady-state conditions be 1200 volts and at the receiving end (at $x=\ell$) be 1100 volts. At time $t=0$, the receiving end is grounded, so that its voltage is reduced to zero, but the potential at the source is maintained at its constant value of 1200 volts. If $R=2$ ohms/km and $C=3 \times 10^{-7}$ farad/km, determine the current and voltage in the line after the grounding of the receiving end and animate the results.

04-S14: Another "Trampoline" Example

A "trampoline" consists of a uniform, light, horizontal, stretched, rectangular membrane with edges of length a and $2a$. The edges at $x=0$, a are fixed, while those at $y=0$, $2a$ are free. Using the separation of variables method, determine the trampoline's subsequent motion if it has the initial shape $f = 2\,x\,h/a$ for $0 \le x \le a/2$ and $f = 2\,h\,(a - x)/a$ for $a/2 \le x \le a$ and is released from rest. Taking $h=1/5$ m, $a=2$ m, and $c=2$ m/s, animate the motion of the membrane over the time interval $t=0$ to 5 seconds taking 100 frames.

04-S15: An Electrostatic Poisson Problem

The faces of a rectangular box, $0 \le x \le a$, $0 \le y \le b$, $0 \le z \le c$ are held at the potential $\phi=0$ and the interior is filled with charge with a charge density $\rho= A \sin(\pi\,x/a) \sin(\pi\,z/c)\,[y\,(y - b)]$ Coulombs/meter^3. Using the electrostatic Poisson equation $\nabla^2 \phi=-\rho/\epsilon_0$, where ϵ_0 is the permittivity of free space, determine the potential at an arbitrary point inside the box. Taking $A=10^{-9}$ C/m^5, $a = b = c = 1$ m, and $\epsilon_0 = 8.85 \times 10^{-12}$ Farads/meter, determine the potential Φ (in volts) at the center of the box. Plot the equipotentials $\Phi/10$, $2\,\Phi/10$,..., $9\,\Phi/10$ in the mid-plane $x=a/2$.

04-S16: SHE Does Not Want to Separate

Demonstrate that the scalar Helmholtz equation (SHE) does not separate in bispherical coordinates even with the modified assumption that was successful for Laplace's equation in recipe **04-3-1**.

04-S17: WE Can Separate

A 2-dimensional curvilinear coordinate system (u, v) can be defined through the equations $x=\sqrt{u\,v}$, $y=(u - v)/2$, where the range of u, v is from 0 to ∞.

(a) Plot the contours in the x-y plane corresponding to holding u and v constant. Show that this curvilinear coordinate system is orthogonal.

(b) Calculate the scale factors and the wave equation (WE) for this system.

(c) Show that WE is separable. Identify the separated ODEs and solutions.

04-S18: The Stark Effect

A hydrogenic atom consists of an electron of charge $-e$ and mass m moving in the attractive Coulomb field of a nucleus (atomic number Z) of charge $Z\,e$ and mass M. If $Z=1$, one has the hydrogen atom, $Z=2$ corresponds to the He^+ ion, $Z=3$ to the Li^{++} ion, and so on. The time-independent Schrödinger equation for the wave function ψ then takes the form $\nabla^2\psi + (2\,m/\hbar^2)[E + Z\,e^2/r]\,\psi = 0$, where $\hbar$ is Planck's constant divided by $2\,\pi$, E is the total energy, and r the radial distance of the electron from the nucleus. When a hydrogenic atom is placed in an external electric field, the energy levels are found to shift. This phenomenon is referred to as the *Stark effect*. If the electric field (magnitude E_0) is oriented in the positive z direction, a potential energy term $-e\,E_0\,z$ must be added to the hydrogenic atom problem. Show that the time-independent Schrödinger equation is still separable in *parabolic* (or *paraboloidal*) coordinates $(\zeta,\,\eta,\,\phi)$ which are related to Cartesian coordinates by the relations $x = \zeta\,\eta\,\cos\phi$, $y = \zeta\,\eta\,\sin\phi$, and $z = (\eta^2 - \zeta^2)/2$, with $0 \le \zeta < \infty$, $0 \le \eta < \infty$, $0 \le \phi \le 2\,\pi$.

04-S19: Annular Temperature Distribution

An annular region, of inner radius $r = 10$ cm and outer radius 20 cm, has its inner boundary maintained at the temperature (in degrees Celsius) $T = 20\cos\theta$ and the outer boundary held at $T = 30\sin\theta$. Determine the steady-state temperature distribution in the annular region and plot the isotherms corresponding to -30, -20, -15,..., 0, 5, 10,..., 30 degrees.

04-S20: Split-boundary Temperature Problem

A thin circular plate of radius 1 m, whose two faces are insulated, has half of its circular boundary kept at the constant temperature $T1$ and the other half at the constant temperature $T2$. Find the steady-state temperature distribution in the plate. Taking $T1 = 300$ degrees Celsius and $T2 = 200$ degrees Celsius, plot the isotherms in 5 degree increments.

04-S21: Fluid Flow Around a Sphere

A solid sphere of radius a is placed in a fluid which was flowing uniformly with speed V_0 in the z direction. The velocity potential U for the fluid in the region outside the sphere satisfies Laplace's equation in spherical polar coordinates and the velocity field is given by $\vec{v} = -\nabla U$. If the sphere is assumed to be rigid, the normal component of $\vec{v}$ must vanish at the surface of the sphere. Determine the velocity field for the fluid outside the sphere and plot the velocity vectors. Take $a = 1$ m and $V_0 = 1$ m/s.

04-S22: Sound of Music?

Some musically inclined people like to sing in the shower stall when taking their shower. In this problem, the shower stall is empty without the water running and consists of a completely enclosed hollow vertical metal cylinder of radius a and height h with (approximately) rigid walls. The speed of sound for the air inside the cylinder is c. By solving the scalar Helmholtz equation for the spatial part of the velocity potential, determine the allowed normal modes inside the cylinder. For rigid walls, the normal component of the fluid velocity (or the normal derivative of the potential) must vanish at each wall. Taking $a = 1.83$ m, $h = 3.04$ m, and $c = 344$ m/s, determine the three lowest eigenfrequencies. By either consulting a musically inclined friend or a music reference book, or going to the Internet, find the closest musical notes on the equal-tempered scale to these eigenfrequencies.

Chapter 5

Complex Variables

5.1 Introduction

Let $z = x + iy$ be a *complex variable*, with $i \equiv \sqrt{-1}$ and x and y real. For given values of x and y, z is represented by a point in the complex z-*plane* (the *Argand plane*) with x and y its coordinates on the real and imaginary axes, respectively. In polar form, z can be written as $z = r\,e^{i\theta} = r\,\cos\theta + i\,r\,\sin\theta$, with $r = \sqrt{x^2 + y^2}$ the radial distance (called the *modulus*) of z from the origin and $\theta = \arctan(y/x)$ the angle (called the *argument*) with the real axis . The argument is not unique. If one writes it as $\theta + n(2\pi)$, where $0 \le \theta \le 2\pi$ and $n = 0, \pm 1, ...$, then θ is called the *principal argument*.

A *complex function* of z is of the general form $w = f(z) = u(x,y) + i\,v(x,y)$, with f specified and u and v real. If, e.g., $f(z) = z^2$, then $w = (x + iy)^2 = x^2 - y^2 + 2\,i\,x\,y$, so $u = x^2 - y^2$ and $v = 2\,x\,y$. $f(z)$ is *single-valued* in a region R if for each z in R there is only one value of w. Otherwise, $f(z)$ is *multi-valued*. An example of the latter is $f(z) = z^{1/2} = r^{1/2}e^{i\theta/2}$. For a single-valued $f(z)$, increasing θ from 0 to 2π yields the same value of the function. However, for our example, $f(z) = r^{1/2}$ for $\theta = 0$, but $f(z) = r^{1/2}e^{i\pi} = -r^{1/2}$ for $\theta = 2\pi$. One must increase θ by 4π (2 revolutions around the origin) to regain $+r^{1/2}$. $f(z) = z^{1/2}$ is said to have two *branches* and the origin is called a *branch point*.

$f(z)$ is *analytic* at a point z if a unique derivative exists there, independent of how z is approached. A *necessary* and *sufficient* condition for $w = f(z)$ to be *regular* (single-valued and analytic) in R is that u and v satisfy the *Cauchy–Riemann* (C-R) *conditions*,

$$\frac{\partial u}{\partial x} = \frac{\partial v}{\partial y}, \quad \frac{\partial v}{\partial x} = -\frac{\partial u}{\partial y}, \tag{5.1}$$

provided that the partial derivatives exist and are continuous in R. Assuming the second derivatives exist, differentiating (5.1) with respect to x and y yields

$$\nabla^2 u = \frac{\partial^2 u}{\partial x^2} + \frac{\partial^2 u}{\partial y^2} = 0, \quad \nabla^2 v = \frac{\partial^2 v}{\partial x^2} + \frac{\partial^2 v}{\partial y^2} = 0, \tag{5.2}$$

so u and v satisfy Laplace's equation and thus represent possible real potentials.

5.1.1   Jennifer Tests Basics

There's a basic rule which runs through all kinds of music, kind of an unwritten rule. I don't know what it is. But I've got it.
Ron Wood, British rock musician, *Independent (London, 10 Sept. 1992)*

As a test of the basic ideas presented in the introduction, Jennifer has prepared the following quiz for her complex variables class:

Consider the single-valued, analytic, function

$$\omega = u + i\,v = f(z) = z^2 e^{z^2} - 5\cos^3 z, \text{ where } z = x + i\,y.$$

(a) Determine u and v. Confirm that the Cauchy–Riemann conditions are satisfied and that u and v satisfy Laplace's equation.

(b) Form a new function $F = f(z)\,z^* = U + iV$, where $z^* = x - i\,y$. Show that U and V do not satisfy the C-R conditions, so F is non-analytic.

Ms. Curious, who is currently in Jennifer's class, has kindly supplied us with her solution to this quiz. I will let her explain it to you in her own words.

"I will begin by assuming that x and y are real, to simplify later results, and then enter the complex variable $z = x + i\,y$. On inputting the given complex function f, z is automatically substituted.

```
>   restart: assume(x::real,y::real):
>   z:=x+I*y: f:=z^2*exp(z^2)-5*cos(z)^3;
```

$$f := (x + y\,I)^2\,e^{((x+y\,I)^2)} - 5\cos(x + y\,I)^3$$

Although I could grind out the forms of u and v by hand, it's simpler to apply the complex evaluation command to f, which splits it into real and imaginary parts, and then extract u and v by taking the real and imaginary parts.

```
>   f:=evalc(f); u:=Re(f); v:=Im(f);
```

$$u := (x^2 - y^2)\,e^{(x^2-y^2)}\cos(2\,x\,y) - 2\,x\,y\,e^{(x^2-y^2)}\sin(2\,x\,y)$$
$$- 5\cos(x)^3\cosh(y)^3 + 15\cos(x)\cosh(y)\sin(x)^2\sinh(y)^2$$
$$v := 2\,x\,y\,e^{(x^2-y^2)}\cos(2\,x\,y) + (x^2 - y^2)\,e^{(x^2-y^2)}\sin(2\,x\,y)$$
$$+ 15\cos(x)^2\cosh(y)^2\sin(x)\sinh(y) - 5\sin(x)^3\sinh(y)^3$$

I will check the C-R conditions using two different approaches. For the first, $\partial u/\partial x$ is calculated in ux and $\partial v/\partial y$ in vy (output suppressed). Forming $ux - vy$ in $CR1$, and simplifying, yields 0, confirming the first C-R condition.

```
>   ux:=diff(u,x); vy:=diff(v,y): CR1:=simplify(ux-vy);
```

$$ux := 2\,x\,e^{(x^2-y^2)}\cos(2\,x\,y) + 2\,(x^2 - y^2)\,x\,e^{(x^2-y^2)}\cos(2\,x\,y)$$
$$- 2\,(x^2 - y^2)\,e^{(x^2-y^2)}\sin(2\,x\,y)\,y - 2\,y\,e^{(x^2-y^2)}\sin(2\,x\,y)$$
$$- 4\,x^2\,y\,e^{(x^2-y^2)}\sin(2\,x\,y) - 4\,x\,y^2\,e^{(x^2-y^2)}\cos(2\,x\,y)$$
$$+ 15\cos(x)^2\cosh(y)^3\sin(x) - 15\sin(x)^3\cosh(y)\sinh(y)^2$$
$$+ 30\cos(x)^2\cosh(y)\sin(x)\sinh(y)^2$$

$$CR1 := 0$$

For the second C-R condition, the "is" command is used to ask whether $\partial v/\partial x = -\partial u/\partial y$, or not. The possible answers are either true or fail. As can be seen in $CR2$, the answer is true, thus confirming the second C-R condition.

```
> CR2:=is(diff(v,x)=-diff(u,y));
```

$$CR2 := true$$

To answer the last part of (a), I form an operator L which will calculate the two-dimension Laplacian of an input function w and simplify the result.

```
> L:=w->simplify(diff(w,x,x)+diff(w,y,y));
```

$$L := w \rightarrow simplify((\frac{\partial^2}{\partial x^2} w) + (\frac{\partial^2}{\partial y^2} w))$$

Applying L to u and v confirms that $\nabla^2 u = 0$ and $\nabla^2 v = 0$.

```
> Lu:=L(u); Lv:=L(v);
```

$$Lu := 0 \qquad Lv := 0$$

For part (b), the function $F = f\,z^*$ is formed and the real and imaginary parts of F extracted in U and V, respectively.

```
> F:=f*conjugate(z): U:=Re(F); V:=Im(F);
```

$$U := e^{(x^2-y^2)} \cos(2\,x\,y)\,x^3 + x\,y^2\,e^{(x^2-y^2)} \cos(2\,x\,y) - x^2\,y\,e^{(x^2-y^2)} \sin(2\,x\,y)$$
$$- e^{(x^2-y^2)} \sin(2\,x\,y)\,y^3 - 5\,x\cos(x)^3\cosh(y)^3 + 15\,x\cos(x)\cosh(y)\sin(x)^2\sinh(y)^2$$
$$+ 15\,y\cos(x)^2\cosh(y)^2\sin(x)\sinh(y) - 5\,y\sin(x)^3\sinh(y)^3$$

$$V := e^{(x^2-y^2)} \cos(2\,x\,y)\,y^3 + x^2\,y\,e^{(x^2-y^2)} \cos(2\,x\,y) + x\,y^2\,e^{(x^2-y^2)} \sin(2\,x\,y)$$
$$+ e^{(x^2-y^2)} \sin(2\,x\,y)\,x^3 + 5\,y\cos(x)^3\cosh(y)^3 - 15\,y\cos(x)\cosh(y)\sin(x)^2\sinh(y)^2$$
$$+ 15\,x\cos(x)^2\cosh(y)^2\sin(x)\sinh(y) - 5\,x\sin(x)^3\sinh(y)^3$$

The combination $\partial U/\partial x - \partial V/\partial y$ is calculated in $CR1b$ and simplified. The answer is non-zero for arbitrary (you can check it for specific values, if you desire) values of x and y, so the first C-R condition isn't satisfied.

```
> CR1b:=simplify(diff(U,x)-diff(V,y));
```

$$CR1b := 30\cos(x)\cosh(y)^3 - 30\cos(x)\cosh(y) - 40\cos(x)^3\cosh(y)^3$$
$$+ 30\cos(x)^3\cosh(y) - 4\,x\,y\,e^{((x-y)\,(x+y))} \sin(2\,x\,y)$$
$$+ 2\,x^2\,e^{((x-y)\,(x+y))} \cos(2\,x\,y) - 2\,y^2\,e^{((x-y)\,(x+y))} \cos(2\,x\,y)$$

The second C-R condition, calculated in $CR2b$, fails when the "is " command is applied.

```
> CR2b:=is(diff(V,x)=-diff(U,y));
```

$$CR2b := FAIL$$

The above failures imply that the single-valued function F is non-analytic."

5.1.2   The Stream Function

Human kindness is like a defective tap, the first gush may be
impressive but the stream soon dries up.
P. D. James, British mystery writer, *Devices and Desires, (1989)*

In aerodynamics and fluid mechanics, the functions ϕ and ψ in the analytic function $f(z = x + i\,y) = \phi(x,y) + i\,\psi(x,y)$ are called the *velocity potential* and *stream function*, respectively. The velocity potential was introduced in Recipes **04-S21** and **04-S22**. The curves $\psi(x,y) = $ constant represent the tracks of the fluid particles and are called *streamlines*. Consider $\phi = x^2 + 4x - y^2 + 2y$.

(a) Confirm that ϕ satisfies Laplace's equation so can represent the velocity potential for steady-state fluid flow.

(b) Using the Cauchy–Riemann conditions, determine $\psi(x,y)$.

(c) Make a contour plot, showing curves of constant ϕ and ψ. Use constrained scaling to show that the two families of curves appear to be orthogonal. Suggest a fluid flow problem where these contours might apply.

(d) Analytically show that the contours in **(c)** are orthogonal.

(e) Express f completely in terms of z.

The plots and VectorCalculus packages are loaded, the former needed for the contourplot command, the latter for the Laplacian.

```
>   restart: with(plots): with(VectorCalculus):
```
The given function $\phi(x,y)$ is entered.

```
>   phi(x,y):=x^2+4*x-y^2+2*y;
```
$$\phi(x,\,y) := x^2 + 4x - y^2 + 2y$$

Applying the Laplacian operator to ϕ in Cartesian coordinates yields 0, so ϕ satisfies Laplace's equation. Thus, ϕ is indeed a velocity potential for fluid flow.

```
>   LE:=Laplacian(phi(x,y),'cartesian'[x,y]);
```
$$LE := 0$$

The first Cauchy–Riemann condition, $\partial\psi/\partial y = \partial\phi/\partial x$, is calculated in *CR1*.

```
>   CR1:=diff(psi(x,y),y)=diff(phi(x,y),x);
```
$$CR1 := \frac{\partial}{\partial y}\,\psi(x,\,y) = 2x + 4$$

The form of $\psi(x,y)$ is easily obtained by applying pdsolve to *CR1*.

```
>   sol1:=pdsolve(CR1,psi(x,y));
```
$$sol1 := \psi(x,\,y) = 2yx + 4y + \_F1(x)$$

The second C-R condition, $\partial\psi/\partial x = -\partial\phi/\partial y$, is calculated in *CR2*.

```
>   CR2:=diff(psi(x,y),x)=-diff(phi(x,y),y);
```
$$CR2 := \frac{\partial}{\partial x}\,\psi(x,\,y) = 2y - 2$$

An alternate form of ψ follows on applying pdsolve to $CR2$.

```
> sol2:=pdsolve(CR2,psi(x,y));
```

$$sol2 := \psi(x, y) = 2\,y\,x - 2\,x + \_F1(y)$$

For the rhs of $sol1$ and $sol2$ to be the same, one must have $\_F1(x) = -2\,x + C$ and $\_F1(y) = 4\,y + C$, where C is an arbitrary constant. These forms are substituted into $sol1$ and $sol2$.

```
> sol1b:=subs(_F1(x)=-2*x+C,sol1);
  sol2b:=subs(_F1(y)=4*y+C,sol2);
```

$$sol1b := \psi(x, y) = 2\,y\,x + 4\,y - 2\,x + C$$

$$sol2b := \psi(x, y) = 2\,y\,x + 4\,y - 2\,x + C$$

Taking the arbitrary constant $C = 0$ for plotting purposes, then ψ is given by the right-hand side of $sol1b$ (or $sol2b$).

```
> C:=0: psi(x,y):=rhs(sol1b);
```

$$\psi(x, y) := 2\,y\,x + 4\,y - 2\,x$$

The contourplot command, with 19 contours and a 60×60 grid, is used to plot $\phi(x,y)$ and $\psi(x,y)$. The plot is colored by including the option filled=true and boxed axes are chosen. The resulting picture is shown in Figure 5.1.

```
> contourplot({phi(x,y),psi(x,y)},x=-8..4,y=-5..7,
  scaling=constrained,contours=19,grid=[60,60],
  filled=true,axes=box,tickmarks=[3,3]);
```

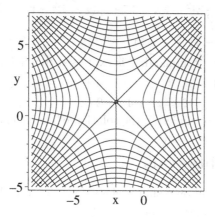

Figure 5.1: Streamlines and equipotentials.

The rectangular hyperboli in each quadrant are the streamlines and the curves intersecting them at apparently $90°$ are the equipotentials. Considering, say, the upper-right quarter of the figure, the streamlines could represent fluid flow directions near the corner of two planar surfaces intersecting at right angles.

If, say $\phi(x,y) = \text{constant}$, then $d\phi = (\partial\phi/\partial x)\,dx + (\partial\phi/\partial y)\,dy = 0$, so the slope of these curves is given by $dy/dx = -(\partial\phi/\partial x)/(\partial\phi/\partial y)$. The slope of the constant ϕ curves is calculated at an arbitrary point (x, y).

```
>    slope[phi]:=-diff(phi(x,y),x)/diff(phi(x,y),y);
```

$$slope_\phi := -\frac{2x+4}{-2y+2}$$

The slope of the constant ψ curves is similarly determined.

```
>    slope[Psi]:=-diff(psi(x,y),x)/diff(psi(x,y),y);
```

$$slope_\Psi := -\frac{2y-2}{2x+4}$$

The product of the two slopes is calculated and found to be equal to -1, so the two sets of curves are orthogonal.

```
>    slope_product:=simplify(slope[phi]*slope[Psi]);
```

$$slope\_product := -1$$

The complex function $f = \phi(x, y) + i\,\psi(x, y)$ is entered, the forms of ϕ and ψ being automatically entered. It is desired to express f entirely in terms of z.

```
>    f:=phi(x,y)+I*psi(x,y);
```

$$f := x^2 + 4x - y^2 + 2y + (2yx + 4y - 2x)\,I$$

Now, $x = (z + z^*)/2$ and $y = (z - z^*)/(2i)$, where z^* is the complex conjugate of z. These forms are substituted into f. In the output, the symbol $\overline{z}$ represents the conjugate of z.

```
>    f:=subs({x=(z+conjugate(z))/2,y=(z-conjugate(z))/(2*I)},f);
```

$$f := (\frac{z}{2} + \frac{1}{2}\overline{z})^2 + 2z + 2\overline{z} + \frac{1}{4}(z-\overline{z})^2 - (z-\overline{z})\,I$$

$$+ (-(z-\overline{z})(\frac{z}{2} + \frac{1}{2}\overline{z})\,I - 2I(z-\overline{z}) - z - \overline{z})\,I$$

On applying the `simplify` command, f is expressed completely in terms of z.

```
>    f:=simplify(f);
```

$$f := z^2 + 4z - 2Iz$$

So, $f(z) = z^2 + 4z - 2iz$, completing the solution of the problem.

5.2 Contour Integrals

If $f(z)$ is regular within and on a simple closed curve C, the Cauchy–Riemann conditions lead to *Cauchy's theorem*,

$$\oint_C f(z)\,dz = 0. \tag{5.3}$$

Equivalently, $\int_{z_1}^{z_2} f(z)\,dz$ has a value independent of the path joining two points z_1 and z_2. If z_0 is any point inside C, Cauchy's theorem may be used in turn to derive *Cauchy's integral formula*,

$$\oint_C \frac{f(z)}{(z - z_0)}\,dz = 2\pi i\,f(z_0), \tag{5.4}$$

where C is traversed in a counter-clockwise direction. Note that the integrand is not analytic at $z = z_0$. This point is called a *first order pole* or a *simple pole* of the integrand. It is an example of an *isolated singularity*.

Cauchy's formula may be differentiated $n - 1$ times to yield

$$\oint_C \frac{f(z)}{(z - z_0)^n} \, dz = \frac{2 \pi i}{(n - 1)!} \left(\frac{d^{n-1} f(z)}{dz^{n-1}} \right)_{z = z_0}, \tag{5.5}$$

or, on setting $g(z) \equiv f(z)/(z - z_0)^n$,

$$\oint_C g(z) \, dz = \frac{2 \pi i}{(n - 1)!} \left(\frac{d^{n-1}[(z - z_0)^n g(z)]}{dz^{n-1}} \right)_{z = z_0}. \tag{5.6}$$

The isolated singularity in $g(z)$ is called an nth order pole, while the coefficient of $2 \pi i$ on the rhs of (5.6) is referred to as the *residue* of $g(z)$ at $z = z_0$.

The above formulas may be used to prove *Cauchy's residue theorem*: If $g(z)$ is regular within and on the closed contour C, except for a finite number of poles, then $\oint g(z) \, dz = 2\pi i \times$(Sum of the residues of $g(z)$ at its poles within C).

5.2.1   Jennifer Tests Cauchy's Theorem

The test of a real comedian is whether you laugh at him before he opens his mouth.
George Jean Nathan, American critic, (1882–1958)

As a follow up to her earlier quiz, Jennifer has asked her complex variables class to confirm that Cauchy's theorem is satisfied for the regular complex function $f(z = x + i y = r\, e^{i\theta}) = z^2 e^{z^2} - 5 \cos^3 z$ for the following two closed contours C:

(a) C_1: (i) along the x-axis from the origin to $x = 1$, (ii) vertically upwards along $x = 1$ to $y = 1$, (iii) along $y = 1$ from $x = 1$ to $x = 0$, (iv) vertically downwards along $x = 0$ from $y = 1$ back to the origin;

(b) C_2: (i) radially outwards along the x-axis from the origin to radius $r = R$, (ii) along a circular arc of radius R from the x-axis to the y-axis, (iii) radially inwards along the y-axis from R to the origin.

Again, Ms. Curious has been asked to present and discuss the recipe which she has created to solve this problem.

"In **(a)**, C_1 is such that rectangular coordinates should be used. Letting $f = u + i v$, the general line integral is $\int f(z) \, dz = \int (u + i v) (dx + i \, dy) = \int (u \, dx - v \, dy) + i \int (v \, dx + u \, dy)$. Assuming that x and y are real, f is entered along with $z = x + i y$."

```
>   restart: assume(x::real,y::real):
>   f:=z^2*exp(z^2)-5*cos(z)^3; z:=x+I*y:
```

$$f := z^2 \, e^{(z^2)} - 5 \cos(z)^3$$

"The real and imaginary parts of f are determined (output suppressed here)."

>   `u:=Re(f); v:=Im(f);`

"For the first leg, the integral is $I1 = \int_0^1 (u+iv)\, dx$, with the integrand evaluated at $y=0$. The answer is expressed in terms of the error function (erf)."

>   `I1:=int(eval(u+I*v,y=0),x=0..1);`

$$I1 := \frac{1}{2}e + \frac{1}{4}I\sqrt{\pi}\,\mathrm{erf}(I) - \frac{5}{3}\cos(1)^2\sin(1) - \frac{10}{3}\sin(1)$$

"For the second leg, the integral is $I2 = \int_0^1 (-v + iu)\, dy$, with $x=1$."

>   `I2:= int(eval(-v+I*u,x=1),y=0..1);`

$$I2 := -\frac{1}{24}(-90\sin(1)\,e^{(3I)} - 10\sin(3)\,e^{(3I)} + 12\,e^{(1+3I)}$$
$$+ 6I\sqrt{\pi}\,\mathrm{erf}(I)\,e^{(3I)} - 12I\,e^{(5I)} - 12\,e^{(5I)} + 6I\sqrt{\pi}\,\mathrm{erf}(1-I)\,e^{(3I)}$$
$$+ 5I\,e^3 + 45I\,e^{(1+2I)} - 5I\,e^{(-3+6I)} - 45I\,e^{(-1+4I)})e^{(-3I)}$$

"For the third leg, the integral is $I3 = \int_1^0 (u + iv)\, dx$, with $y=1$."

>   `I3:= int(eval(u+I*v,y=1),x=1..0);`

$$I3 := \frac{1}{24}(5I\,e^{(-3+3I)} - 45I\,e^{(1+3I)} + 57I\,e^{(-1+3I)} - 6I\sqrt{\pi}\,\mathrm{erf}(1)\,e^{(3I)}$$
$$- 5I\,e^{(3+3I)} - 12I\,e^{(5I)} - 12\,e^{(5I)} + 6I\sqrt{\pi}\,\mathrm{erf}(1-I)\,e^{(3I)}$$
$$+ 5I\,e^3 + 45I\,e^{(1+2I)} - 5I\,e^{(-3+6I)} - 45I\,e^{(-1+4I)})e^{(-3I)}$$

"For the fourth leg, the integral is $I4 = \int_1^0 (-v + iu)\, dy$, with $x=0$."

>   `I4:=int(eval(-v+I*u,x=0),y=1..0);`

$$I4 := -\frac{5}{24}I\,e^{(-3)} + \frac{15}{8}I\,e - \frac{19}{8}I\,e^{(-1)} + \frac{1}{4}I\sqrt{\pi}\,\mathrm{erf}(1) + \frac{5}{24}I\,e^3$$

"Adding the four integrals and simplifying the complete contour integral CI"

>   `CI:=simplify(I1+I2+I3+I4);`

$$CI := 0$$

"yields 0, thus confirming Cauchy's theorem. To deal with part **(b)**, I will unassign the Cartesian form of z, by enclosing z in right quotes, and enter its polar form, $z=re^{i\theta}$. The complex function f then is as follows."

>   `z:='z': z:=r*exp(I*theta): f:=f;`

$$f := r^2\,(e^{(\theta I)})^2\,e^{(r^2\,(e^{(\theta I)})^2)} - 5\cos(r\,e^{(\theta I)})^3$$

"For the first leg of the new contour C_2, f is evaluated at $\theta=0$ and integrated from $r=0$ to R."

>   `I1b:=int(eval(f,theta=0),r=0..R);`

$$I1b := \frac{1}{2}\,e^{(R^2)}\,R + \frac{1}{4}I\sqrt{\pi}\,\mathrm{erf}(R\,I) - \frac{5}{3}\cos(R)^2\sin(R) - \frac{10}{3}\sin(R)$$

"For the second leg, the integral is of the form $\int f(z)\, dz = \int_0^{\pi/2} f(Re^{i\theta})\, i\,Re^{i\theta}\, d\theta$. This integration is carried out in $I2b$."

```
> I2b:=int(eval(f*I*z,r=R),theta=0..Pi/2);
```

$$I2b := \frac{1}{12}(-6\,e^{(2\,R^2)}\,R - 3\,I\,\sqrt{\pi}\,\mathrm{erf}(R\,I)\,e^{(R^2)} + 20\cos(R)^2\sin(R)\,e^{(R^2)}$$
$$+ 40\sin(R)\,e^{(R^2)} + 6\,I\,R - 3\,I\,\sqrt{\pi}\,\mathrm{erf}(R)\,e^{(R^2)}$$
$$- 20\,I\cosh(R)^2\sinh(R)\,e^{(R^2)} - 40\,I\sinh(R)\,e^{(R^2)})e^{(-R^2)}$$

"The third leg is radially inwards along the y axis, for which $\theta = \pi/2$. The relevant integral is performed in *I3b*."

```
> I3b:=int(eval(f*z/r,theta=Pi/2),r=R..0);
```

$$I3b := \frac{-1}{12}\,I(6\,R - 3\,\sqrt{\pi}\,\mathrm{erf}(R)\,e^{(R^2)} - 20\cosh(R)^2\sinh(R)\,e^{(R^2)}$$
$$- 40\sinh(R)\,e^{(R^2)})e^{(-R^2)}$$

"Adding the three integrals and simplifying, the complete contour integral *CIb*"

```
> CIb:=simplify(I1b+I2b+I3b);
```

$$CIb := 0$$

"is zero, once again confirming Cauchy's theorem."

5.2.2 Cauchy's Residue Theorem

To know yet to think that one does not know is best;
Not to know yet to think that one knows will lead to difficulty.
Lao–Tzu, Chinese philosopher, 6th century BC

As a follow-up quiz to that posed in the last recipe, here's one provided by Jennifer for which the integrands of the contour integrals contain isolated singularities, resulting in non-zero values for the integrals. The contours are to be traversed in a counterclockwise sense

(a) Evaluate $\oint_C(5\,z^4 - 3\,z^2 + 2)/(z - 1)^n$, with $n = 2, 3, 4, 5$, where C is any simple closed curve enclosing $z=1$. Identify the types of singularities.

(b) Evaluate $\oint_C(2\,z^3 + z)/((z^2 - 1/4)(z^2 + 2\,z + 2))$, where the contour C is given by (i) $|z|=3/2$, (ii) $|z|=5/8$. Identify the singular points and make a plot showing their locations in the z-plane and the two contours.

The plottools library package is needed to plot the contours in part (b).

```
> restart: with(plots): with(plottools):
```

Now, let's tackle part (a). An operator g is introduced for generating the given integrand $(5\,z^4 - 3\,z^2 + 2)/(z - 1)^n$ for different input values of n. Clearly, since the numerator remains finite there, the integrand has second, third, fourth and fifth order poles at $z=1$ for $n=2, 3, 4, 5$, respectively.

```
> g:=n->(5*z^4-3*z^2+2)/(z-1)^n;
```

$$g := n \to \frac{5\,z^4 - 3\,z^2 + 2}{(z - 1)^n}$$

A second operator F is formed for evaluating the integral for the nth order pole using the Cauchy integral result in Equation (5.5).

```
>  F:=n->2*Pi*I*eval(diff(numer(g(n)),z$(n-1)),z=1)/(n-1)!;
```

$$F := n \to \frac{2\,I\,\pi\,(\dfrac{d^{n-1}}{dz^{n-1}}\,\text{numer}(g(n)))\Big|_{z\,=\,1}}{(n-1)!}$$

Making use of F, the contour integral is evaluated for $n = 2, ..., 5$, the results being displayed in II(2), ..., II(5).

```
>  seq(II(n)=F(n),n=2..5);
```

$$\text{II}(2) = 28\,I\,\pi, \quad \text{II}(3) = 54\,I\,\pi, \quad \text{II}(4) = 40\,I\,\pi, \quad \text{II}(5) = 10\,I\,\pi$$

An even easier approach to evaluating the contour integrals is to use the `residue` command. An operator G is created to evaluate the residue of $g(n)$ at $z = 1$ and multiply the result by $2\,\pi\,i$.

```
>  G:=n->2*Pi*I*residue(g(n),z=1);
```

$$G := n \to 2\,I\,\pi\,\text{residue}(g(n),\,z = 1)$$

As a check, $G(n)$ is calculated for $n = 2$ to 5, the answers given for II(2) to II(5) being the same as obtained above.

```
>  check:=seq(II(n)=G(n),n=2..5);
```

$$check := \text{II}(2) = 28\,I\,\pi, \quad \text{II}(3) = 54\,I\,\pi, \quad \text{II}(4) = 40\,I\,\pi, \quad \text{II}(5) = 10\,I\,\pi$$

The integrand for part **(b)** is now entered in $g2$.

```
>  g2:=(2*z^3+z)/((z^2-1/4)*(z^2+2*z+2));
```

$$g2 := \frac{2\,z^3 + z}{\left(z^2 - \dfrac{1}{4}\right)\left(z^2 + 2\,z + 2\right)}$$

One approach to evaluating the contour integral is to first convert $g2$ to a *partial fraction* form in terms of z. The *complex* option allows a complete decomposition (in floating point form) in terms of the complex roots.

```
>  g2b:=convert(g2,parfrac,z,complex);
```

$$g2b := \frac{0.5846153848 - 1.323076923\,I}{z + 1. + 1.000000000\,I} + \frac{0.6000000002}{z + 0.5000000000}$$
$$+ \frac{0.5846153848 + 1.323076923\,I}{z + 1. - 1.000000000\,I} + \frac{0.2307692307}{z - 0.5000000000}$$

$g2b$ can be converted to a somewhat more compact rational form.

```
>  g2c:=convert(g2b,rational);
```

$$g2c := \frac{\dfrac{38}{65} - \dfrac{86}{65}\,I}{z + 1 + I} + \frac{3}{5\left(z + \dfrac{1}{2}\right)} + \frac{\dfrac{38}{65} + \dfrac{86}{65}\,I}{z + 1 - I} + \frac{3}{13\left(z - \dfrac{1}{2}\right)}$$

Examining $g2c$, the integrand $g2$ has four first order (simple) poles located at $z = 1/2, -1/2, -1 + i$ and $-1 - i$. Using the **pointplot** command, the four simple poles are plotted as size 16 blue circles in the complex z-plane in **pp**.

```
> pp:=pointplot([[1/2,0],[-1/2,0],[-1,1],[-1,-1]],
      symbol=circle,symbolsize=16,color=blue):
```

The specified contours $|z| = 5/8$ and $|z| = 3/2$ are circles centered on the origin of radii 5/8 and 3/2, respectively. These circles are created in c1 and c2, the smaller circle being colored red, the larger one colored green.

```
> c1:=circle([0,0],5/8,color=red,thickness=2):
```

```
> c2:=circle([0,0],3/2,color=green,thickness=2):
```

The three graphs, pp, c1, and c2, are superimposed with the display command, the resulting picture being shown in Figure 5.2. The larger circular contour encloses all four poles, the smaller circle enclosing only the poles at $z = \pm 1/2$.

```
> display({pp,c1,c2},labels=["x","y"]);
```

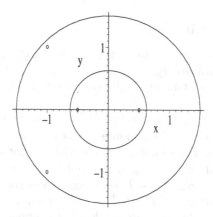

Figure 5.2: Four simple poles of $g2$ and two circular contours.

Cauchy's residue theorem states that the contour integral is equal to $2\pi i$ times the sum of the residues of the poles enclosed by the contour. Let's first consider the contour $|z| = 3/2$, which encloses all 4 poles. In $g2c$, this means that one simply has to add the numerators of all 4 terms. The numerator of the ith term in $g2c$ can be extracted with the operand command op[i,1]. The 4 numerators are then added and multiplied by $2\pi i$, yielding the answer $4\pi i$ for the integral.

```
> I2:=2*Pi*I*add(op([i,1],g2c),i=1..4);
```

$$I2 := 4 I \pi$$

As shown in *check2*, the same answer follows on applying the residue command directly to $g2$ at each pole, adding the four residues, and multiplying by $2\pi i$.

```
> check2:=2*Pi*I*(residue(g2,z=1/2)+residue(g2,z=-1/2)
      +residue(g2,z=-1+I)+residue(g2,z=-1-I));
```

$$check2 := 4 I \pi$$

Using the residue command, the value of the contour integral is now obtained for the contour $|z| = 5/8$ by only keeping the residues of the poles at $z = \pm 1/2$.

```
>   I2b:=2*Pi*I*(residue(g2,z=1/2)+residue(g2,z=-1/2));
```

$$I2b := \frac{108}{65} I \pi$$

In this latter case, the contour integral has the value $(108/65)\pi i$.

5.3   Definite Integrals

By performing a closed contour integration in the complex z-plane with a suitably chosen path and using Cauchy's residue theorem, it is possible to easily evaluate some real definite integrals. The choice of path depends on the form of the integral. A few representative examples will now be considered.

5.3.1   Infinite Limits

God does not care about our mathematical difficulties.
He integrates empirically.
Albert Einstein, Nobel laureate in physics, (1879–1955)

To evaluate an integral of the form $I = \int_{-\infty}^{\infty} f(x)\, dx$ using contour integration, a "standard" approach is to evaluate $J = \oint_C f(z)\, dz$ with C a closed semi-circle of radius R in the complex z-plane with its flat portion along the real axis. Along the real axis, $z = x$ and the contribution to J is $J_1 = \int_{-R}^{R} f(x)\, dx$. In the limit as $R \to \infty$, $J_1 \to I$, the original integral whose value we seek. Along the semi-circular arc, $z = Re^{i\theta}$ and the line integral contribution to J is $J_2 = \int_0^{\pi} f(Re^{i\theta})\, iRe^{i\theta}\, d\theta$, if the arc is taken in the upper-half z-plane. If $J_2 \to 0$ as $R \to \infty$, evaluating J will be equivalent to evaluating I. The value of J is then determined by calculating the sum of the residues of its poles inside C and multiplying by $2\pi i$. As an illustrative example, let's evaluate the integral

$$I = \int_{-\infty}^{\infty} dx/(1 + x^4).$$

As suggested, we consider the contour integral $J = \oint_C dz/(1 + z^4)$ with C a closed semi-circle of radius R in the upper-half z-plane and its flat portion along the real x-axis. The plottools library package is loaded so that the semi-circular arc can be plotted, and the integrand of J is then entered.

```
>   restart: with(plots): with(plottools):
```

```
>   integrand:=1/(1+z^4);
```

$$integrand := \frac{1}{1 + z^4}$$

Along the semi-circle, the line integral is $J_2 = \int_0^{\pi} iRe^{i\theta}\, d\theta/(1 + (Re^{i\theta})^4)$. For large R, the integrand of J_2 has a $1/R^3$ dependence, so $J_2 \to 0$ as $R \to \infty$. The real axis contribution becomes the original integral. To find its value, we must determine the sum of the residues of any poles inside C. To this end, the denominator of the integrand is set equal to zero and solved for the z roots.

```
> Z:=solve(denom(integrand)=0,z);
```

$$Z := \frac{\sqrt{2}}{2} + \frac{1}{2}I\sqrt{2}, \quad \frac{1}{2}I\sqrt{2} - \frac{\sqrt{2}}{2}, \quad -\frac{\sqrt{2}}{2} - \frac{1}{2}I\sqrt{2}, \quad -\frac{1}{2}I\sqrt{2} + \frac{\sqrt{2}}{2}$$

There are four complex roots which locate four simple poles in the complex z-plane. To see which poles contribute to the contour integral, a semi-circular path of radius $R=2$ will be drawn in the upper-half z-plane.

```
> R:=2:
```

The `arc` command is used to draw a red semi-circle of radius R centered on the origin $(0,0)$. In the command, the angle is allowed to vary from 0 to π radians.

```
> a:=arc([0,0],R,0..Pi,color=red,thickness=2):
```

The flat portion of the contour between $(-R,0)$ and $(R,0)$ is plotted by entering the end points as a list of lists and choosing a line style.

```
> b:=plot([[-R,0],[R,0]],style=line,color=red,thickness=2):
```

Taking the real and imaginary parts of each root and putting them into a list of lists, the locations of the four poles are plotted as size 16 blue circles.

```
> c:=pointplot([seq([Re(Z[i]),Im(Z[i])],i=1..4)],
        symbol=circle,symbolsize=16,color=blue):
```

The three graphs are superimposed with constrained scaling, the resulting picture being shown in Figure 5.3.

```
> display({a,b,c},scaling=constrained,view=[-R..R,-R..R]);
```

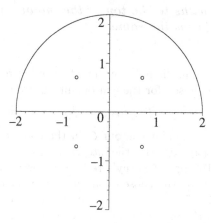

Figure 5.3: Locations of the four poles and the semi-circular contour.

Two of the poles lie inside the contour, the remaining two outside. Only the former poles contribute to the closed contour integral. The sum r of their residues is now calculated.

```
> r:=residue(integrand,z=Z[1])+residue(integrand,z=Z[2]);
```

$$r := \frac{1}{2I\sqrt{2} - 2\sqrt{2}} + \frac{1}{2I\sqrt{2} + 2\sqrt{2}}$$

The value of the integral follows on multiplying r by $2\pi i$ and applying the complex evaluation command to simplify the answer to a real number.

>   V:=evalc(2*Pi*I*r);

$$V := \frac{\pi\sqrt{2}}{2}$$

As a check, we can directly evaluate the integral using Maple's int command. This is done in the next two command lines, the resulting number being the same as that obtained above using contour integration.

>   I1:=Int(integrand,z=-infinity..infinity);

$$I1 := \int_{-\infty}^{\infty} \frac{1}{1+z^4}\, dz$$

>   check:=value(I1);

$$check := \frac{\pi\sqrt{2}}{2}$$

If the semi-circular contour had been chosen to be in the lower-half z-plane, exactly the same answer would be generated using the poles below the real axis, provided that the result is multiplied by -1 to take into account that the contour then is taken in a clockwise sense, rather than counterclockwise.

5.3.2 Poles on the Contour

There are many paths to the top of the mountain,
but the view is always the same.
Chinese Proverb

In the previous example, the poles of the integrand were either definitely inside or outside the path chosen for the contour integration. Suppose that we want to evaluate the integral $I = \int_{-\infty}^{\infty} \cos(x)\, dx/(a^2 - x^2)$ with a real. The infinite limits suggest a semi-circular contour C in the complex z-plane with the flat portion on the real (x) axis. But the integrand has two simple poles lying right on C at $x = \pm a$. To apply Cauchy's residue theorem, the poles must be definitely inside or outside the chosen contour. So what do we do? A standard approach is to "indent" the contour in the vicinity of each pole, taking a small semi-circular "detour" around the pole which either puts the pole inside or outside the contour. Then one performs the contour integration using the residue theorem, and afterwards lets the radii of the small semi-circles go to zero. As an illustration of this method, let's evaluate the above integral I.

The plottools library package is again loaded so the various semi-circles can be drawn for the indented contour.

>   restart: with(plots): with(plottools):

Instead of tackling I directly, let's consider the integral $\oint_C e^{iz} dz/(a^2 - z^2)$, with z the complex variable. At the end of the calculation the real integral I can be obtained by taking the real and imaginary parts. The integrand of the complex closed contour integral is now entered.

```
>  integrand:=exp(I*z)/(a^2-z^2);
```

$$integrand := \frac{e^{(z\,I)}}{a^2 - z^2}$$

The integrand's poles can be extracted by applying the discontinuity command.

```
>  pole:=discont(integrand,z);
```

$$pole := \{a, -a\}$$

To draw the indented semi-circular contour, let's take the poles to be at ± 1 by setting $a = 1$. The small semi-circle around each pole is given a radius $r = 0.2$, and the larger semi-circle which completes the path is given a radius $R = 2$.

```
>  R:=2: r:=0.2: a:=1:
```

An operator A is formed to draw a thick red semi-circle of arbitrary radius r around a specified point $(x, 0)$ on the real axis.

```
>  A:=(x,r)->arc([x,0],r,0..Pi,color=red,thickness=2):
```

Using A, a large semi-circle of radius R centered on the origin is produced in A1, and a small semi-circle of radius r around each pole in A2 and A3.

```
>  A1:=A(0,R): A2:=A(pole[1],r): A3:=A(pole[2],r):
```

In B, a thick red line is drawn along the real axis between, $x = -R$ and $-a - r$, between $-a + r$ and $a - r$, and between $a + r$ and R.

```
>  B:=plot([[[-R,0],[-a-r,0]],[[-a+r,0],[a-r,0]],
         [[a+r,0],[R,0]]],style=line,color=red,thickness=2):
```

In C, the two poles are plotted as size 16 blue circles.

```
>  C:=pointplot([[pole[1],0],[pole[2],0]],symbol=circle,
         symbolsize=16,color=blue):
```

The graphs are superimposed, generating the full contour shown in Figure 5.4.

```
>  display({A1,A2,A3,B,C},scaling=constrained,
     view=[-R..R,0..R],tickmarks=[2,2],labels=["x","y"]);
```

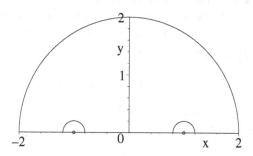

Figure 5.4: Semi-circular contour indented around the two poles on the x-axis.

Note that I have chosen to indent the small semi-circles around each pole in such a way that both poles lie outside the contour C. So the line integral around C will yield zero, since no poles lie inside C. I could just as well have chosen to indent the small semi-circles in such a way that both poles were inside the contour, or even so that one pole was inside and the other outside. As you may verify, the final answer is the same, irrespective of the choice of indentation.

The quantities a and r are now unassigned. On each small semi-circle, the complex variable will be written as $Z = re^{i\theta}$.

```
>   a:='a': r:='r': Z:=r*exp(I*theta):
```

For our choice of C, starting at the point $(-R, 0)$ the complete counterclockwise contour integral will be as follows:

$$\oint_C \frac{e^{iz} dz}{(a^2 - z^2)} = 0 = \int_{-R}^{-a-r} \frac{e^{ix} dx}{(a^2 - x^2)} + \int_{\pi}^{0} \frac{e^{i(-a+Z)} iZ \, d\theta}{(a^2 - (-a+Z)^2)} + \int_{-a+r}^{a-r} \frac{e^{ix} dx}{(a^2 - x^2)}$$

$$+ \int_{\pi}^{0} \frac{e^{i(a+Z)} iZ \, d\theta}{(a^2 - (a+Z)^2)} + \int_{a+r}^{R} \frac{e^{ix} dx}{(a^2 - x^2)} + \int_{0}^{\pi} \frac{e^{iRe^{i\theta}} iRe^{i\theta} \, d\theta}{(a^2 - (Re^{i\theta})^2)}.$$

In the limit that $R \to \infty$ and $r \to 0$, the sum of the first, third, and fifth integrals on the right-hand side of 0 will yield $\int_{-\infty}^{\infty} e^{ix} dx/(a^2 - x^2)$. Taking the real part will produce the original integral $\int_{-\infty}^{\infty} \cos(x) \, dx/(a^2 - x^2)$, whose value we are trying to determine.

Let's look at the last integral on the right-hand side, which is written as

$$\int_{0}^{\pi} \frac{e^{iR\cos\theta - R\sin\theta} iRe^{i\theta} \, d\theta}{(a^2 - R^2 e^{2i\theta})}.$$

Since the large semi-circle is in the upper-half z-plane, $\sin\theta > 0$. Therefore, as $R \to \infty$, the integrand $\to 0$ and the integral contribution is zero. Finally, we have to evaluate the second and fourth integrals on the right-hand side. A function F is formed for substituting $z = \pm a + Z$ into the integrand.

```
>   F:=n->subs(z=pole[n]+Z,integrand):
```

The first pole extracted was $+a$ (the ordering varies from one run to the next), so F(1) is used to form the fourth integral and simplify it.

```
>   I1:=simplify(Int(F(1)*I*Z,theta=Pi..0));
```

$$I1 := -I \int_{\pi}^{0} \frac{e^{((a + r\, e^{(\theta\, I)}) I)}}{2\, a + r\, e^{(\theta\, I)}} \, d\theta$$

Taking the limit of $I1$ at $r = 0$ completes the evaluation of the integral in $I1b$.

```
>   I1b:=limit(I1,r=0);
```

$$I1b := \frac{\frac{1}{2} I\, e^{(a\, I)}\, \pi}{a}$$

Using F(2), the second integral is similarly formed in $I2$ and evaluated in $I2b$.

> `I2:=simplify(Int(F(2)*I*Z,theta=Pi..0));`

$$I2 := -I \int_{\pi}^{0} \frac{e^{((-a+r\,e^{(\theta\,I)})\,I)}}{-2\,a + r\,e^{(\theta\,I)}}\,d\theta$$

> `I2b:=limit(I2,r=0);`

$$I2b := \frac{\frac{-1}{2}\,I\,e^{(-I\,a)}\,\pi}{a}$$

The two integral terms are added and simplified assuming that a is real.

> `terms:=simplify(I1b+I2b) assuming a::real;`

$$terms := -\frac{\pi\,\sin(a)}{a}$$

The answer to the original real integral is just minus the above output.

> `answer:=Int(cos(x)/(a^2-x^2),x=-infinity..infinity)=-terms;`

$$answer := \int_{-\infty}^{\infty} \frac{\cos(x)}{a^2 - x^2}\,dx = \frac{\pi\,\sin(a)}{a}$$

5.3.3   An Angular Integral

Science is an integral part of culture. ... It's one of the glories of the human intellectual tradition.
Stephen Jay Gould, American scientist, (b. 1941)

Integrals whose integrands are rational functions of $\cos\theta$ and $\sin\theta$ in the range of integration can be evaluated by making the transformation $z = e^{i\theta}$. If θ varies from 0 to 2π, the integration in the complex z-plane will be counter-clockwise around a circle of unit radius. As an example, consider

$$J = \int_{0}^{2\pi} \frac{\sin^2\theta\,d\theta}{(5 + 4\,\cos\theta)}.$$

For variety, let's load the Student[Calculus1] package, which contains two useful commands, `Integrand` and `Roots`, which haven't been used yet.

> `restart: with(Student[Calculus1]):`

The integral J is now entered and, as a check on the contour integration, evaluated directly with the `value` command.

> `J:=Int(sin(theta)^2/(5+4*cos(theta)),theta=0..2*Pi);`

$$J := \int_{0}^{2\pi} \frac{\sin(\theta)^2}{5 + 4\,\cos(\theta)}\,d\theta$$

> `check:=value(J);`

$$check := \frac{\pi}{4}$$

A contour integration should yield the value $\pi/4$ for J. To carry out this integration, note that $\sin\theta = (e^{i\theta} - e^{-i\theta})/(2i) = (z - 1/z)/(2i)$ and $\cos\theta = (e^{i\theta} + e^{-i\theta})/2 = (z + 1/z)/2$. These two transformations are entered.

> `sin(theta):=(z-1/z)/(2*I); cos(theta):=(z+1/z)/2;`

$$\sin(\theta) := \frac{-1}{2} I \left(z - \frac{1}{z}\right) \qquad \cos(\theta) := \frac{z}{2} + \frac{1}{2z}$$

Using the `Integrand` command to extract the integrand of J and using the fact that $d\theta = dz/(iz)$, the integrand of the contour integral is as follows.

> `integrand:=simplify(Integrand(J)/(I*z));`

$$integrand := \frac{\frac{1}{4} I \left(z^2 - 1\right)^2}{z^2 \left(5z + 2z^2 + 2\right)}$$

The `Roots` command is used to determine the zeros of the reciprocal of the integrand, and thus the poles of the integrand.

> `R:=Roots(1/integrand);`

$$R := [-2, \frac{-1}{2}, 0]$$

There are three poles, viz., two simple poles at $z = -2$ and $z = -1/2$ and a second order pole at $z = 0$. Since the contour C is a unit circle in the z-plane, the pole at $z = -2$ will not contribute to the answer since it's outside C. The residue of the integrand at the two poles inside C is calculated in $r1$ and $r2$.

> `r1:=residue(integrand,z=R[2]); r2:=residue(integrand,z=R[3]);`

$$r1 := \frac{3}{16} I \qquad r2 := \frac{-5}{16} I$$

By Cauchy's residue theorem, the value of the integral J is just $2\pi i$ times the sum of the two residues. The answer is $\pi/4$ as expected.

> `answer:=2*Pi*I*(r1+r2);`

$$answer := \frac{\pi}{4}$$

5.3.4   A Branch Cut

Trust everybody, but cut the cards.
Finley Peter Dunne, American journalist, (1867–1936)

Cauchy's residue theorem is built on the assumption that the integrand $f(z)$ in the contour integral $\oint_C f(z)\,dz$ is a single-valued function. What if $f(z)$ is multi-valued with more than one mathematical branch? The answer is to choose a contour which keeps $f(z)$ on a single branch. Typically, a line is drawn in the complex z-plane, called a *branch cut*, which the contour is not permitted to cross in order to keep $f(z)$ single-valued. Section 5.5 will elaborate on this.

To see how the procedure is implemented for a specific integral, consider $J = \int_0^\infty \frac{\sqrt{x}\,dx}{1+x^2}$. The corresponding complex integral is $\oint_C (\sqrt{z}/(1+z^2))\,dz$, which has a double-valued integrand because of the $\sqrt{z}$ factor. The branch cut will be taken from $z = 0$ (the branch point) to $-\infty$ along the negative x-axis. (The branch cut could be taken in any direction from $z = 0$ to ∞.) The contour

C is chosen to consist of an inner circle of radius ϵ and an outer circle of radius R (both centered on the origin) joined by horizontal lines a distance η above and below the branch cut. Ultimately, we let both ϵ and $\eta \to 0$ and $R \to \infty$.

In order to draw the circular arcs in C, the plottools library package is first loaded. The complex integrand is then entered.

```
> restart: with(plots): with(plottools):
```

```
> integrand:=sqrt(z)/(1+z^2);
```

$$integrand := \frac{\sqrt{z}}{1 + z^2}$$

Before drawing C, let's use the **singular** command to extract the singularities of the integrand, needed for the application of Cauchy's residue theorem.

```
> pole:=singular(integrand);
```

$$pole := \{z = I\}, \{z = -I\}$$

There are two singular points at $z = i$ and $z = -i$, corresponding to the locations of two simple poles. The operand command, op, is now used to separately obtain the locations, $z1$ and $z2$, of the two poles.

```
> z1:=op([1,2],pole[1]); z2:=op([1,2],pole[2]);
```

$$z1 := I \qquad z2 := -I$$

The pole locations may be converted into polar form.

```
> z1_polar:=convert(z1,polar); z2_polar:=convert(z2,polar);
```

$$z1\_polar := \mathrm{polar}(1, \frac{\pi}{2}) \qquad z2\_polar := \mathrm{polar}(1, -\frac{\pi}{2})$$

The poles lie on a circle of radius 1 at the polar angles $\theta = +\pi/2$ and $-\pi/2$. To schematically draw the contour, let's set $R = 2$, $\epsilon = 0.2$, and $\eta = 0.05$.

```
> R:=2: epsilon:=0.2: eta:=0.05:
```

Thick red circular arcs of radius ϵ and R, respectively, are drawn in A1 and A2.

```
> A1:=arc([0,0],epsilon,Pi-0.2..-Pi+0.2,color=red,thickness=2):
```

```
> A2:=arc([0,0],R,Pi-0.02..-Pi+0.02,color=red,thickness=2):
```

A thick blue line is drawn along the real axis between $x = -R - 0.5$ and the origin to represent the branch cut.

```
> B:=plot([[-R-0.5,0],[0,0]],style=line,color=blue,thickness=3):
```

Red lines are drawn between $(-R, \eta)$, $(-\epsilon, \eta)$ and between $(-R, -\eta)$, $(-\epsilon, -\eta)$.

```
> B2:=plot([[[-R,eta],[-epsilon,eta]],[[-R,-eta],
          [-epsilon,-eta]]],style=line,color=red,thickness=2):
```

The two poles are represented by size 16 blue circles,

```
> C:=plot({[0,z1/I],[0,z2/I]},style=point,symbol=circle,
          symbolsize=16,color=blue):
```

and the textplot command is used to add labels to the figure.

```
> tp:=textplot([[-1.9,-.15,"a"],[-1.9,.15,"b"],[-.25,.15,"c"],
          [-.25,-.15,"d"],[-2.3,.1,"cut"]]):
```

The six plots are superimposed, the resulting picture being shown in Figure 5.5.

```
>  display({A1,A2,B,B2,C,tp},scaling=constrained,
   view=[-R-0.5..R,-R..R],labels=["x","y"]);
```

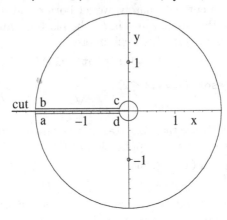

Figure 5.5: Closed contour around branch cut (stretching from $z=0$ to $-\infty$).

The complete contour C consists of four legs, viz., $a \to b$, $b \to c$, $c \to d$, and $d \to a$. Traversing C counter-clockwise, in the limit that $\eta \to 0$, the increase of 2π in θ in going from a to b is canceled out by the -2π contribution in going from c to d. Since the net angular change is zero, $f(z)$ remains single-valued. Multi-valuedness of $f(z) = \sqrt{z}/(1+z^2)$ only occurs if the net change in θ exceeds 2π. Since $f(z)$ is single-valued, Cauchy's residue theorem can be applied. The poles lie inside C, so the line integral around C will yield a non-zero result. Let's calculate the residues of the integrand at the two poles.

```
>  r1:=residue(integrand,z=z1); r2:=residue(integrand,z=z2);
```

$$r1 := \frac{-1}{2} I \left(\frac{\sqrt{2}}{2} + \frac{1}{2} I \sqrt{2}\right) \qquad r2 := \frac{1}{2} I \left(\frac{\sqrt{2}}{2} - \frac{1}{2} I \sqrt{2}\right)$$

Now let's look at each contribution to the line integral in the limit that $\eta \to 0$.

(a) The integral $I_{a \to b} = \int_{-\pi}^{\pi} (i R^{3/2} e^{3i\theta/2}/(1 + R^2 e^{2i\theta})) \, d\theta \to 0$ as $R \to \infty$.

(b) $I_{b \to c} = \int_{R}^{\epsilon} (r^{1/2} e^{3\pi i/2}/(1 + r^2 e^{2\pi i})) \, dr \to i J$ as $\epsilon \to 0$ and $R \to \infty$.

(c) $I_{c \to d} = \int_{\pi}^{-\pi} (i \epsilon^{3/2} e^{3i\theta/2}/(1 + \epsilon^2 e^{2i\theta})) \, d\theta \to 0$ as $\epsilon \to 0$.

(d) $I_{d \to a} = \int_{\epsilon}^{R} (r^{1/2} e^{-3\pi i/2}/(1 + r^2 e^{-2\pi i})) \, dr \to i J$ as $\epsilon \to 0$ and $R \to \infty$.

Since these four contributions add up to $2 i J$, the original integral must equal $2\pi i$ times the sum of the two residues divided by $2 i$.

```
>  answer:=Int(sqrt(x)/(1+x^2),x=0..infinity)
           =evalc(2*Pi*I*(r1+r2)/(2*I));
```

$$answer := \int_{0}^{\infty} \frac{\sqrt{x}}{1 + x^2} \, dx = \frac{\pi \sqrt{2}}{2}$$

As a check, we apply the **value** command to the left-hand side of *answer*,

```
>   check:=value(lhs(answer));
```

$$check := \frac{\pi \sqrt{2}}{2}$$

obtaining exactly the same result as in the contour integration.

5.4 Laurent Expansion

Using Cauchy's integral formula, one can prove *Laurent's theorem*: If $g(z)$ is regular in an annular region R between two concentric circles with center z_0, then $g(z)$ may be represented in R by a *Laurent expansion*

$$g(z) = \sum_{n=-\infty}^{\infty} a_n (z - z_0)^n, \text{ with } a_n = \frac{1}{2\pi i} \oint_C \frac{g(z)\, dz}{(z - z_0)^{n+1}}, \qquad (5.7)$$

C being any simple closed path encircling z_0 counter-clockwise within R.

If $g(z)$ has a finite number of terms with negative exponents, it's behavior is dominated as z approaches z_0 by the term with the largest negative exponent. If, e.g., it has the form

$$g(z) = \frac{a_{-n}}{(z - z_0)^n} + \frac{a_{-(n-1)}}{(z - z_0)^{n-1}} + \cdots + \frac{a_{-1}}{(z - z_0)} + a_0 + a_1(z - z_0) + \cdots, \quad (5.8)$$

then it has a pole of order n at $z = z_0$. Note that if $g(z)$ in (5.8) is multiplied by $(z - z_0)^n$, the result then differentiated $n - 1$ times and z set equal to z_0, then the coefficient a_{-1} is just the residue of the nth order pole, i.e.,

$$a_{-1} = \frac{1}{(n-1)!} \left(\frac{d^{n-1}[(z - z_0)^n g(z)]}{dz^{n-1}} \right)_{z = z_0}. \qquad (5.9)$$

If $g(z)$ has an infinite number of terms with negative exponents, it is said to have an *essential singularity* at $z = z_0$. For example, $g(z) = e^{\frac{1}{z}} = 1 + \frac{1}{z} + \frac{1}{z^2} + \cdots$ has an essential singularity at $z = 0$.

Instead of doing the contour integrations in (5.7), the coefficients a_n can often be more easily obtained by considering the Taylor series of $(z - z_0)^n g(z)$.

5.4.1 Ms. Curious Meets Mr. Laurent

Isn't life a series of images that change as they repeat themselves?
Andy Warhol, American pop artist, (1928–87)

Jennifer has posed the following question for her complex variables class:

Keeping 10 terms, determine the Laurent expansion of $f = \cos(z)\, e^{z - 2z^2}/z^2$ about $z = 0$. Identify the singularity and give the region of convergence of the series. Determine the residue of f at the singularity from the series. Confirm the value of the residue by alternate means.

Here is the solution presented by Ms. Curious, expressed in her own words.

"The number N of terms to be retained in the series is set equal to 10.

```
>  restart: N:=10:
```

Since the numerator remains finite there, clearly f has a second-order pole at $z = 0$. To obtain the Laurent series about this singularity, I will enter the coefficient of $1/z^2$ in G.

```
>  G:=cos(z)*exp(z-2*z^2);
```

$$G := \cos(z)\, e^{(z-2\,z^2)}$$

Then G is Taylor expanded in powers of z to order N, and the order of term removed with the convert(g,polynom) command.

```
>  g:=taylor(G,z,N);
```

```
>  g:=convert(g,polynom);
```

$$g := 1+z - 2\,z^2 - \frac{7}{3}\,z^3+\frac{11}{6}\,z^4+\frac{79}{30}\,z^5 - z^6 - \frac{1217}{630}\,z^7+\frac{841}{2520}\,z^8+\frac{23617}{22680}\,z^9$$

The Laurent expansion follows on dividing g by z^2 and expanding the result.

```
>  LS:=expand(g/z^2);
```

$$LS := \frac{1}{z^2}+\frac{1}{z} - 2 - \frac{7\,z}{3}+\frac{11\,z^2}{6}+\frac{79\,z^3}{30} - z^4 - \frac{1217\,z^5}{630}+\frac{841\,z^6}{2520}+\frac{23617\,z^7}{22680}$$

The Laurent expansion LS converges for all values of $z \neq 0$. By inspecting the series and recalling that the residue is the coefficient of the $1/z$ term, the residue of f at $z=0$ is 1. It should be noted that the Laurent series can also be obtained by loading the numapprox library package, applying the laurent command to G/z^2 as in *check1*, and removing the order of term as in *LSb*.

```
>  with(numapprox):
```

```
>  check1:=laurent(G/z^2,z=0,N);
```

```
>  LSb:=convert(check1,polynom);
```

$$LSb := \frac{1}{z^2}+\frac{1}{z} - 2 - \frac{7\,z}{3}+\frac{11\,z^2}{6}+\frac{79\,z^3}{30} - z^4 - \frac{1217\,z^5}{630}+\frac{841\,z^6}{2520}+\frac{23617\,z^7}{22680}$$

The residue of $f(z)$ at $z=0$ can also be obtained by differentiating $z^2 f$ or G once (since the singularity is a pole of order 2) with respect to z, evaluating the result at $z=0$, and dividing by 1!

```
>  r1:=eval(diff(G,z),z=0)/1!;
```

$$r1 := 1$$

An even simpler method of determining the residue of $f = G/z^2$ at $z=0$ is to use the residue command."

```
>  r1b:=residue(G/z^2,z=0);
```

$$r1b := 1$$

5.4.2 Converge or Diverge?

I shall be telling this with a sigh, Somewhere ages and ages hence:
Two roads diverged in a wood, and I –
I took the one less traveled by, And that has made all the difference.
Robert Frost, American poet, *The Road Not Taken* (1874-1963)

Jennifer has given a second quiz question on the Laurent expansion.

Find the Laurent series for $f = z^2/((z-1)^2(z+3))$ about $z=1$. Determine the residue and the region of convergence. Confirm the latter by creating plots for different z values which show convergence and divergence.

Again, Ms. Curious will present her recipe for solving this problem.

"The numapprox package is loaded so that the `laurent` command can be used. The given function f is then entered. It has a second order pole at $z=1$ and a first order pole at $z=-3$.

```
>   restart: with(plots): with(numapprox):
>   f:=z^2/((z-1)^2*(z+3));
```

$$f := \frac{z^2}{(z-1)^2(z+3)}$$

An arrow operator L is formed for calculating the Laurent expansion of f about $z=1$ to terms of order $(z-1)^N$. For example, for $N=5$, the Laurent series has the explicit form given in LS.

```
>   L:=N->laurent(f,z=1,N): LS:=L(5);
```

$$LS := \frac{1}{4}(z-1)^{-2} + \frac{7}{16}(z-1)^{-1} + \frac{9}{64} - \frac{9}{256}(z-1) + \frac{9}{1024}(z-1)^2$$
$$- \frac{9}{4096}(z-1)^3 + \frac{9}{16384}(z-1)^4 + O((z-1)^5)$$

By examining the $(z-1)^{-1}$ term, I can see that the residue is 7/16. This may be confirmed by applying the `residue` command to f at $z=1$.

```
>   r:=residue(f,z=1);
```

$$r := \frac{7}{16}$$

The region of convergence for the Laurent series is $0 < |z-1| < 4$. To test this condition for some specified value Z of z, I first create a functional operator `gr1` which plots the Laurent expansion evaluated at $z=Z$ over the range $N=1$ to 10, connecting consecutive plotting points with a thick blue straight line.

```
>   gr1:=Z->plot([seq([N,eval(convert(L(N),polynom),z=Z)],
                N=1..10)],color=blue,thickness=2):
```

A second operator `gr2` evaluates f at $z=Z$ and plots a horizontal line at the calculated value over the same range of N.

```
>   gr2:=Z->plot(eval(f,z=Z),N=1..10,thickness=2):
```

A third operator P superimposes `gr1(Z)` and `gr2(Z)` for a specified Z value.

```
>   P:=Z->display({gr1(Z),gr2(Z)},labels=["N","f"]):
```

According to the convergence criterion given above, the Laurent expansion should converge for $z = Z = 3$ and diverge for $z = 7$. Entering P(3) yields the plot on the left of Figure 5.6, while P(7) generates the plot on the right.

> P(3); P(7);

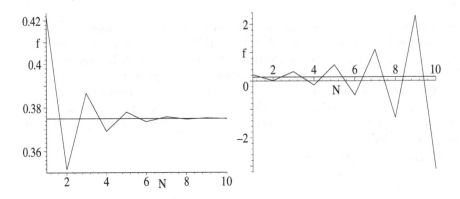

Figure 5.6: Left plot: Convergence for $z = Z = 3$; Right: Divergence for $z = 7$.

The Laurent expansion clearly converges to the exact f value for $z = 3$ and diverges for $z = 7$."

5.5   Conformal Mapping

The relation $\omega = f(z)$ describes a *mapping* of points in the complex z-plane into corresponding points in the ω-plane. If $f(z)$ is a single-valued function of $z = x + iy$, each point in the z-plane maps into a single point in the ω-plane. If $f(z)$ is a multi-valued function with n branches, each point in the z-plane maps into n points in the ω-plane. For example, $\omega = z^{1/2}$ has two branches with each point in the z-plane producing two points in the ω-plane. The mapping from the z-plane can be made one to one by introducing a branch cut in the z-plane and restricting our attention to a single branch of the function, usually referred to as the *principal branch*. For $\omega = f(z) = z^{1/2}$, the principal branch is traditionally obtained by introducing a branch cut along the negative x-axis and restricting the polar angle θ to the range $-\pi < \theta < \pi$. The principal branch of $z^{1/2}$ maps only into the $u > 0$ half of the ω-plane, i.e., the polar angle in this plane is restricted to the range $-\pi/2 < \theta < \pi/2$.

If $f(z)$ is analytic (and $df/dz \neq 0$) at a point z_0 in the z-plane, the angle between two curves intersecting at z_0 is not changed on transforming to the corresponding point ω_0 in the ω-plane, even though the shapes of the two curves will generally change. Transformations which have this angle-preserving property are called *conformal transformations* or *conformal mapping*. The orthogonality of the equipotential and field lines in potential problems is preserved under a conformal mapping. It is not surprising therefore that a solution of Laplace's equation in one plane remains a solution in the other.

Conformal mapping may be used for solving 2-dimensional potential problems. Suppose that we are given a potential problem with a somewhat complicated boundary condition. We then try to transform the problem to a new plane where the boundary configuration is simpler. On solving this simpler situation, we can transform the results back to the original plane, thus determining the field and potential configuration for the original problem.

Many important conformal mappings, such as in the following recipe, can be discovered by experimenting with different mathematical forms.

5.5.1 Field Around a Semi-infinite Plate

If all the ways I have been along were marked on a map and joined up with a line, it might represent a minotaur.
Pablo Picasso, Spanish artist, (1881–1973)

In this recipe, I will use the mapping $w = \sqrt{z}$ to determine the equipotentials and electric field (represented by field lines and by vectors) in the vicinity of a thin semi-infinite grounded (held at zero potential) conducting plate. The problem can be treated as 2-dimensional with the cross-section of the plate as schematically shown in Figure 5.7.

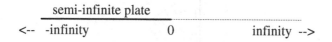

semi-infinite plate

<-- -infinity   0   infinity -->

Figure 5.7: Cross-section of the thin semi-infinite plate.

The VectorCalculus package is loaded, because the `Gradient` command will be used to calculate the electric field vector $\vec{E} = -\nabla u$, with u the potential.

```
> restart: with(plots): with(VectorCalculus):
```
The given mapping function is entered.

```
> w:=sqrt(z);
```
$$w := \sqrt{z}$$

Before solving the problem, it is instructive to see the effect of the transformation $w = u + iv = \sqrt{z} = \sqrt{x + iy}$ on the grid lines $x = c_1$, $y = c_2$, where c_1 and c_2 are real constants. Forming the inverse transformation $z = w^2$ and separating into real and imaginary parts, we have $x = u^2 - v^2$ and $y = 2uv$. A grid line $x = c_1$ is mapped into that portion of the hyperbola $u^2 - v^2 = c_1$ for which $u > 0$, while $y = c_2$ is mapped into a branch of the rectangular hyperbola $2uv = c_2$, the branch depending on whether c_2 is positive or negative.

A graphical confirmation of the transformation of grid lines from the z-plane to the w-plane can be generated by applying the following `conformal` command to the grid lines in a particular region of the z-plane , e.g., the region $-2 \leq x \leq 2$, $-2 \leq y \leq 2$.

```
>   conformal(w,z=-2-2*I..2+2*I,grid=[14,14],numxy=[100,100],
    color=[red,blue],scaling=constrained,labels=["u","v"]);
```

The `grid` option is used to specify the number of grid lines in both the x and y directions, the default being 11×11. The option `numxy=[m,n]`, with $m = 100$ and $n = 100$ here, is employed to specify the number of points to be plotted in each grid line, with m points in the x direction and n points in the y direction. The default is 15 points in each direction. The transformed curves corresponding to constant x will be colored red, while those corresponding to constant y are colored blue. The scaling is constrained and labels u, v, added.

The rectangular region $-2 \leq x \leq 2$, $-2 \leq y \leq 2$ in the z-plane maps into the curved grid region in the w-plane shown in Figure 5.8.

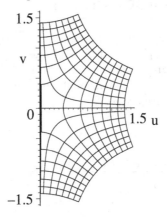

Figure 5.8: Rectangular grid in z-plane maps into curved grid in w-plane.

As expected, the curves $y = c_2$ for $c_2 > 0$ map into the rectangular hyperboli in the first quadrant ($u > 0$, $v > 0$) of the w-plane, while those for $c_2 < 0$ map into rectangular hyperboli in the fourth quadrant ($u > 0$, $v < 0$). The $x = c_1$ lines map into the hyperbolic curves in the figure which intersect the rectangular hyperboli. The $90°$ angle between grid lines is preserved by the conformal transformation.

Now, let's tackle the problem of the semi-infinite grounded conducting plate. Setting $z = x + i\,y$, w takes the following form,

```
>   z:=x+I*y:   w:=w;
```

$$w := \sqrt{x + y\,I}$$

which can be split into real and imaginary parts by applying the complex evaluation (`evalc`) command.

```
> w2a:=evalc(w) assuming y>0; w2b:=evalc(w) assuming y<0;
```

$$w2a := \frac{\sqrt{2\sqrt{x^2 + y^2} + 2x}}{2} + \frac{1}{2}I\sqrt{2\sqrt{x^2 + y^2} - 2x}$$

$$w2b := \frac{\sqrt{2\sqrt{x^2 + y^2} + 2x}}{2} - \frac{1}{2}I\sqrt{2\sqrt{x^2 + y^2} - 2x}$$

w2a shows the form of ω for $y > 0$, while *w2b* applies when $y < 0$. The potential u relevant to our problem is obtained by removing the imaginary part from *w2a*.

```
> u:=remove(has,w2a,I);
```

$$u := \frac{\sqrt{2\sqrt{x^2 + y^2} + 2x}}{2}$$

The electric field lines for $y > 0$ and $y < 0$ are obtained in *v2a* and *v2b* by selecting the imaginary parts of *w2a* and *w2b*, respectively, and dividing by i.

```
> v2a:=select(has,w2a,I)/I: v2b:=select(has,w2b,I)/I:
```

The electric field vector $\vec{E} = -\nabla u$ is calculated in Cartesian coordinates.

```
> E:=-Gradient(u,'cartesian'[x,y]);
```

$$E := -\frac{\frac{2x}{\sqrt{x^2 + y^2}} + 2}{4\sqrt{2\sqrt{x^2 + y^2} + 2x}}\bar{e}_x - \frac{y}{2\sqrt{2\sqrt{x^2 + y^2} + 2x}\sqrt{x^2 + y^2}}\bar{e}_y$$

Now, let's plot the results. A thick blue line is plotted between $(-2, 0)$ and $(0, 0)$ to represent a portion of the semi-infinite plate.

```
> gr1:=plot([[-2,0],[0,0]],style=line,color=blue,thickness=4):
```

The fieldplot command is used to plot the electric field vector $\vec{E}$, the vectors being represented by thick red arrows.

```
> gr2:=fieldplot([E[1],E[2]],x=-2..2,y=-2..2,arrows=THICK,
         grid=[10,10],color=red):
```

A functional operator G is formed to produce a contour plot of any input function V, the color C of the curves to be specified. The range is taken to be $x = -2$ to 2 and $y = -2$ to 2. The contours are chosen to be $V = 0, 0.2, 0.4, ..., 2$.

```
> G:=(V,C)->contourplot(V,x=-2..2,y=-2..2,contours=
        [seq(0.2*n,n=0..10)],grid=[60,60],color=C):
```

Using G, the equipotentials (u =const.) and electric field lines are plotted together, along with the plate, using the display command.

```
> display({gr1,G(u,green),G(v2a,red),G(v2b,red)});
```

On the computer screen, the equipotentials are colored green and the electric field lines red. The corresponding black and white version is shown on the left of Figure 5.9.

The semi-infinite plate is represented by the thick horizontal line between $x = -2$ and 0. The equipotentials are the family of parabolas opening to the left, the closest one to the plate being for $\phi = 0.2$, the next furthest one for $\phi = 0.4$, and so on. The electric field lines are the family of parabolas opening to the right. Naturally, they intersect the equipotential at right angles. This includes the plate which is the zero equipotential. The electric field lines terminate on the surface charge on the plate.

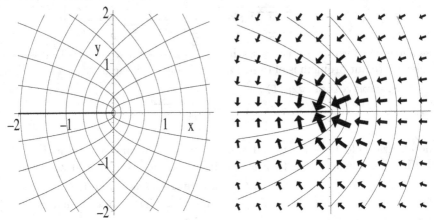

Figure 5.9: Equipotentials and (a) field lines (left plot), (b) field vectors (right).

The field lines convey the sense of the electric field but do not indicate how the field strength varies with distance from the plate. To this end, the following `display` command is used to superimpose the equipotentials and the electric field ($\vec{E}$) vectors.

```
>  display({gr1,gr2,G(u,green)});
```

It can be seen from the resulting picture shown on the right of Figure 5.9 that the electric field is strongest near the edge of the plate, i.e., near the tip of the thick line in the figure.

5.5.2 A Clever Transformation

We are obliged to regard many of our original minds as crazy at least until we have become as clever as they are.
G. C. Lichtenberg, German physicist, philosopher, (1742–99)

An infinitely long conducting cylinder of unit radius, with its axis horizontal, has its upper-half held at the potential $\Phi = +V$ and its lower-half held at $\Phi = -V$, the two halves being separated by an infinitesimally thin layer of insulation. What are the equipotentials and electric field inside the cylinder?

The fact that this problem is 2-dimensional in nature, and has a certain symmetry to it, allows us to solve it rather easily by using a conformal transformation approach. Loading some needed library packages,

```
>  restart: with(plots): with(VectorCalculus):
```

the complex variable $z = x + iy$ is entered. As will be seen, the mapping $w = \ln((1 + z)/(1 - z))$, which is inputted, will transform the original circular boundary value problem into a planar boundary value problem which can be solved by inspection.

```
>  z:=x+I*y: w:=ln((1+z)/(1-z));
```

$$w := \ln(\frac{x + yI + 1}{-x - yI + 1})$$

The complex function $w = u + iv$ is split into real and imaginary parts with the complex evaluation command, and u and v separately extracted with the `remove` and `select` commands, respectively.

```
>  w:=simplify(evalc(w));
```

$$w := \frac{1}{2} \ln(\frac{x^2 + 2x + 1 + y^2}{x^2 - 2x + 1 + y^2}) + \arctan(\frac{2y}{x^2 - 2x + 1 + y^2}, -\frac{x^2 - 1 + y^2}{x^2 - 2x + 1 + y^2}) I$$

```
>  u:=remove(has,w,I); v:=select(has,w,I)/I;
```

$$u := \frac{1}{2} \ln(\frac{x^2 + 2x + 1 + y^2}{x^2 - 2x + 1 + y^2})$$

$$v := \arctan(\frac{2y}{x^2 - 2x + 1 + y^2}, -\frac{x^2 - 1 + y^2}{x^2 - 2x + 1 + y^2})$$

Recognizing that the first argument of the arctangent in v is the numerator and the second argument the denominator, v is expressed in a more familiar mathematical form by using the operand (op) command.

```
>  v:=arctan(op(1,v)/op(2,v));
```

$$v := -\arctan(\frac{2y}{x^2 - 1 + y^2})$$

Since v satisfies Laplace's equation, it can be taken as the potential. To see how the potential on the circular boundary transforms under the action of w, let's make the algebraic substitution $y^2 = 1 - x^2$ in w.

```
>  w2:=algsubs(y^2=1-x^2,w);
```

$$w2 := \frac{1}{2} \ln(\frac{2x + 2}{-2x + 2}) + \arctan(\frac{2y}{-2x + 2}, 0) I$$

The result $w2$ is simplified in $w3a$, assuming that $y > 0$ and $x < 1$, and in $w3b$, assuming that $y < 0$ and $x < 1$.

```
>  w3a:=simplify(w2) assuming y>0,x<1;
```

$$w3a := -\frac{1}{2} \ln(-x + 1) + \frac{1}{2} \ln(x + 1) + \frac{1}{2} I \pi$$

```
>  w3b:=simplify(w2) assuming y<0,x<1;
```

$$w3b := -\frac{1}{2} \ln(-x + 1) + \frac{1}{2} \ln(x + 1) - \frac{1}{2} I \pi$$

The `select` command is used to extract the imaginary parts of $w3a$ and $w3b$ which are then divided by i in $v2a$ and $v2b$, respectively.

```
>  v2a:=select(has,w3a,I)/I; v2b:=select(has,w3b,I)/I;
```

$$v2a := \frac{\pi}{2} \qquad v2b := -\frac{\pi}{2}$$

The upper (lower) surface of the split cylinder, which is at the potential $+V$ ($-V$), maps into the infinite plane $v2a = +\pi/2$ ($v2b = -\pi/2$). The region inside the split cylinder maps into the region between these infinite planes. The potential problem in the w plane is easily solved. By inspection, the equipotentials must be parallel planes lying between $v = +\pi/2$ and $-\pi/2$, $v=0$ (the u-axis) corresponding to the zero potential. Analytically, the potential must be given by $\Phi = (2v/\pi)V$, since $v = \pi/2$ generates the equipotential $\Phi = V$, and so on. A picture of the equipotentials in the w plane are generated for $V = 1$ with the `contourplot` command.

```
>  contourplot(2*Y/Pi,X=-2..2,Y=v2b..v2a);
```

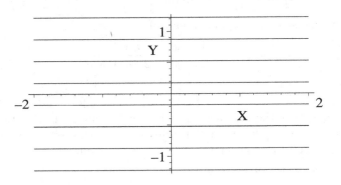

Figure 5.10: Planar equipotentials in the w-plane.

Entering $\Phi = 2Vv/\pi$ generates the potential in terms of x and y, i.e., in the z-plane. It can be converted to polar form if so desired, as in $\Phi2$.

```
>  Phi:=2*V*v/Pi;
```

$$\Phi := -\frac{2V \arctan(\dfrac{2y}{x^2 - 1 + y^2})}{\pi}$$

```
>  Phi2:=simplify(subs({x=r*cos(theta),y=r*sin(theta)},Phi));
```

$$\Phi2 := -\frac{2V \arctan(\dfrac{2r\sin(\theta)}{-1 + r^2})}{\pi}$$

The electric field expression follows on calculating $\vec{E} = -\nabla\Phi$.

```
>  E:=-Gradient(Phi,'cartesian'[x,y]);
```

$$E := -\frac{8Vyx}{(x^2-1+y^2)^2(1+\frac{4y^2}{(x^2-1+y^2)^2})\pi}\,\bar{e}_x + \frac{2V(\frac{2}{x^2-1+y^2}-\frac{4y^2}{(x^2-1+y^2)^2})}{(1+\frac{4y^2}{(x^2-1+y^2)^2})\pi}\,\bar{e}_y$$

A normalized piecewise potential function is formed, made up of Φ/V inside the unit circle and zero outside.

>   `pot:=piecewise(x^2+y^2<1,Phi/V,x^2+y^2>1,0);`

The x and y components of the normalized electric field are also put into a similar piecewise form in *E1* and *E2*.

>   `E1:=piecewise(x^2+y^2<1,E[1]/V,x^2+y^2>1,0):`

>   `E2:=piecewise(x^2+y^2<1,E[2]/V,x^2+y^2>1,0):`

The normalized equipotentials are plotted with the `contourplot` command.

>   `cp:=contourplot(pot,x=-1..1,y=-1..1,grid=[80,80],`
>   `contours=15):`

The normalized electric field vectors are plotted using the `fieldplot` command.

>   `fp:=fieldplot([E1,E2],x=-1..1,y=-1..1,grid=[10,10],`
>   `arrows=THICK,color=red):`

The two graphs, cp and fp, are superimposed to produce Figure 5.11.

>   `display({cp,fp},scaling=constrained);`

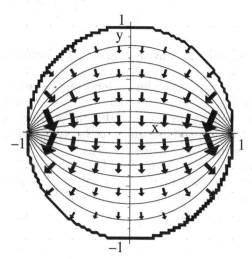

Figure 5.11: Equipotentials and electric field vectors for the split cylinder.

Although this split cylinder problem was easily solved using a conformal transformation, it should be noted that this approach is limited in its usefulness, being applicable to two-dimensional potential problems of relatively simple geometry or symmetry.

5.5.3   Schwarz–Christoffel Transformation

Physical concepts are free creations of the human mind, and are not, however it may seem, uniquely determined by the external world.
Albert Einstein, German-American physicist, *Evolution of Physics, 1938*

To this point, the conformal transformations have been drawn out of "thin air". A systematic approach for obtaining many specific transformations is to apply the general *Schwarz–Christoffel transformation* which can be used to transform the inside[1] of a closed polygon in the $z = x + iy$ plane into the upper half of the $w = u + iv$ plane. Referring to Figure 5.12, the vertices of the polygon are

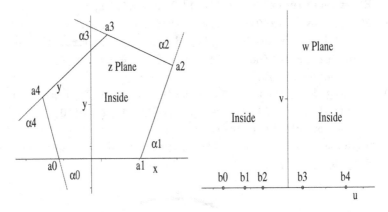

Figure 5.12: Schwarz–Christoffel transformation.

labeled a0, a1,..., the corresponding exterior angles $\alpha 0$, $\alpha 1$,..., and the transforms of the vertices to the w plane b0, b1,.... Traversing the exterior of the polygon in a counterclockwise sense, so that the interior of the polygon is to the left, angles which correspond to turning further to the left are taken as positive, while turns to the right are regarded as negative.

The Schwarz–Christoffel (S-C) transformation which maps the interior of the polygon in the z plane onto the upper-half of the w plane and the boundary of the polygon onto the real (u) axis is given by

$$z = A \int (w - b0)^{-\alpha 0/\pi} (w - b1)^{-\alpha 1/\pi} \cdots (w - bn)^{-\alpha n/\pi} \, dw + B, \qquad (5.10)$$

where A and B are arbitrary complex constants. Note the following facts:

- Any three of the points b0, b1,...,bn may be chosen at will. We can place b0 at infinity which removes the first factor in (5.10).

- A and B are adjusted to fix the polygon size, orientation, and position.

- Infinite open polygons are regarded as limiting cases of closed polygons.

---

[1]The method can be extended to map the outside of the polygon, see, e.g., [MF53].

As a simple example, let's determine the S-C transformation which maps the interior of the semi-infinite rectangular strip shown on the left of Figure 5.13 into the upper-half of the w plane. Then, this transformation will be used to plot the equipotentials and lines of force inside a straight slit, with the same geometry, cut in an infinite conducting sheet.

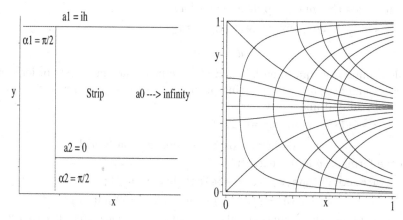

Figure 5.13: Geometry (left) and equipotentials and field lines (right) for strip.

The infinite strip can be regarded as the limit of a triangle with vertices $a0 \to \infty$, $a1 = ih$, and $a2 = 0$. The exterior angles then are $\alpha0 = \pi$, $\alpha1 = \pi/2$, and $\alpha2 = \pi/2$. Choosing $b0 = \infty$, $b1 = -1$, and $b2 = 1$, the S-C transformation is given by $z = A \int (w+1)^{-1/2}(w-1)^{-1/2} \, dw + B = \int (1/\sqrt{w^2-1}) \, dw + B$.

After loading the plots package, needed for the contour plot,

```
> restart: with(plots):
```

the integral in the S-C transformation is explicitly carried out.

```
> Z:=A*int(1/sqrt(w^2-1^2),w)+B;
```

$$Z := A \ln(w + \sqrt{w^2 - 1}) + B$$

Now, when $z = a2 = 0$, then $w = b2 = 1$. Similarly, when $z = a1 = ih$, then $w = b1 = -1$. These two boundary conditions are entered,

```
> bc1:=eval(Z,w=1)=0; bc2:=eval(Z,w=-1)=I*h;
```

$$bc1 := B = 0 \qquad bc2 := A\pi I + B = hI$$

and solved for the constants A and B.

```
> sol:=solve({bc1,bc2},{A,B});
```

$$sol := \{A = \frac{h}{\pi}, B = 0\}$$

On assigning the solution, the S-C transformation is given by Z.

```
> assign(sol): Z:=Z;
```

$$Z := \frac{h \ln(w + \sqrt{w^2 - 1})}{\pi}$$

$w(z)$ may be obtained by setting $Z=z$, solving for w, and simplifying.

> `w:=simplify(solve(Z=z,w),symbolic);`

$$w := \frac{1}{2}\left(1 + e^{\left(\frac{2z\pi}{h}\right)}\right)e^{\left(-\frac{z\pi}{h}\right)}$$

Converting w to trig form and applying the `combine` command with the trig option, yields the final simplified form of w.

> `w:=combine(convert(w,trig),trig);`

$$w := \cosh(\frac{z\pi}{h})$$

To determine the equipotentials and electric field lines in the strip, let's enter $z=x+iy$ and set $h=1$ for plotting purposes. Then, $w=u+iv$ is expressed in terms of real and imaginary parts

> `z:=x+I*y: h:=1: w:=evalc(w);`

$$w := \cosh(\pi x)\cos(\pi y) + \sinh(\pi x)\sin(\pi y)\,I$$

and the forms of u and v extracted,

> `u:=remove(has,w,I); v:=select(has,w,I)/I;`

$$u := \cosh(\pi x)\cos(\pi y) \quad v := \sinh(\pi x)\sin(\pi y)$$

In the w plane, the equipotentials $v = $ const. are clearly parallel to the real axis and the field lines $u = $ const. perpendicular to this axis. In the z plane, the equipotential and field lines can be obtained by using the `contourplot` command, and selecting some fixed values for the contours. A function operator `cp` is created to perform this task, where V will be taken to be u and v and the color C chosen to be blue and red, respectively.

> `cp:=(V,C)->contourplot(V,x=0..1,y=0..1,`
>   `[contours=seq(n,n=-4..4),-1/4,-1/2,1/4,-1/2],`
>   `grid=[100,100],color=C,thickness=2):`

Using the operator `cp` in the following `display` command produces the picture shown on the right of Figure 5.13. The conducting sheet is an equipotential so the electric field lines in the strip intersect the edges of the sheet perpendicularly as expected, as well as all other equipotential curves.

> `display({cp(u,blue),cp(v,red)},axes=frame);`

5.6 Supplementary Recipes

05-S01: Roots
Determine all the roots of $z^{1/9}$, where $z=(1+\sqrt{3}\,i)$. Plot the roots and z in the same figure, superimposing them on circles centered on the origin of radii $|z|$ and $|z|^{1/9}$, respectively. What is the value in radians of the principal argument?

05-S02: Fluid Flow Around a Cylinder
Consider the complex function $w = u + iv = V_0\,(z + a^2/z)$, with $z=x+iy$.

(a) Determine u and v and demonstrate that they satisfy both Cauchy–Riemann conditions and Laplace's equation.

(b) Graphically show that v may be used to represent the streamlines outside and perpendicular to an infinitely long rigid cylinder of radius a. Take $V_0 = a = 1$.

(c) By calculating $\vec{V} = -\nabla u$, show that the fluid velocity components are given by $V_x = V_0(-r^2 + a^2\cos(2\,\theta))/r^2$ and $V_y = V_0\,a^2\sin(2\,\theta)/r^2$, where r is the radial distance from the center of the cylinder and θ is the angle with respect to the x-axis.

(d) Calculate the fluid speed V and show that $V = 0$ at $r = a$ and $\theta = 0,\ \pi$. These two points are called *stagnation points*.

05-S03: Constructing f(z)

If $5\,x^4\,y - 10\,x^2\,y^3 + y^5 + 2\,a^2\,x\,y/((x^2 - y^2)^2 + 4\,x^2\,y^2)$ is the imaginary part of an analytic function $f(z = x + i\,y)$, determine $f(z)$.

05-S04: Analytic or Non-analytic?

Consider the function $f(z) = z^2\,e^z\,\cos(z^*)$. Determine whether f is analytic or non-analytic by (a) applying the Cauchy–Riemann conditions, (b) performing the contour integral of $f(z)$ along the two paths $y_1 = x$ and $y_2 = x^2$ between $z = 0$ and $z = 1 + i$.

05-S05: A Contour Integral

Evaluate the integral $J = \displaystyle\int_{-\infty}^{\infty} ((x^2 - x + 2)/(x^4 + 10\,x^2 + 9))\,dx$ using contour integration. Confirm your answer by evaluating J with the `int` command.

05-S06: A Higher-order Pole

Evaluate the integral $J = \displaystyle\int_{-\infty}^{\infty} (x^2/(x^2 + a^2)^3)\,dx$, with $a > 0$, using contour integration. Confirm your answer by evaluating J with the `int` command.

05-S07: Another Angular Integral

Evaluate the integral $J = \displaystyle\int_{0}^{\pi} (\cos^2(3\,\theta)/(5 - 4\,\cos(\theta))\,d\theta$ using contour integration. Confirm your answer by evaluating J with the `int` command.

05-S08: A Removable Singularity

Evaluate the integral $J = \displaystyle\int_{0}^{\infty} (\sin x/x)\,dx$ using contour integration and confirm your answer by evaluating J with the `int` command. The corresponding complex integrand $\sin z/z$ has a singularity at $z = 0$. Because the integrand is finite as $z \to 0$, the singularity is called a *removable singularity*.

05-S09: Another Contour Integral

Evaluate the integral $J = \displaystyle\int_{0}^{\infty} ((x^4 + 3\,x^2 - 2)/(x^6 + 1))\,dx$ using contour integration and check the value by using the `int` command.

05-S10: Fluid Flow & Electric Field Around a Plate

Consider the function $w = u + i\,v = \sqrt{z^2 - 1}$, with $z = x + i\,y$.

(a) By creating a suitable figure in the x-y plane, show that the constant u curves represent the streamlines for fluid flow around an infinitely long plate of finite width, lying between $x=-1$ and $+1$, inserted perpendicular to a previously uniform fluid flow in the y direction. Include the plate and the equipotentials in your figure.

(b) Taking u to represent the equipotentials, create a plot showing the electric field vectors and equipotentials around a thin infinite conducting plate of finite width, lying between $x=-1$ and $+1$.

05-S11: Another Branch Cut

By taking a branch cut along the negative real axis from the branch point $x=0$ to $x=-\infty$, evaluate the integral $J=\int_0^\infty (x^{1/3}/(1+x^4))\,dx$ using contour integration. Check your answer with the int command.

05-S12: Laurent Expansion

Determine the singularities of the function $f=\cos(\pi z)\,e^{-z^3}/(z^2\,(z-1)^5)$. Determine the Laurent series about each singularity, keeping eight terms in each expansion. Calculate the residue of f at each singular point.

05-S13: Capacitor Edge Effects

In elementary treatments of the parallel-plate capacitor, it is assumed that the two plates are infinitely large so that "edge effects" can be ignored. Using the conformal command, show that the conformal transformation $w=z+e^z$ when applied to the region $-7 < x < 2$, $-\pi < y < \pi$ produces the equipotentials and electric field lines near one edge of a finite parallel-plate capacitor. Suggest another physical problem to which this transformation applies.

Chapter 6

Integral Transforms

Applications of the Fourier, Laplace, and Hankel transforms and their inverses are illustrated in this chapter. The integral transform (`inttrans`) library package plays the key role in enabling us to perform these transforms. In some cases, assumptions must be provided about the nature of the parameters (e.g., whether they are positive) in order for the transforms to be explicitly done.

6.1   Fourier Transforms

Using the Euler identity $e^{i\theta} = \cos\theta + i\sin\theta$, where $i = \sqrt{-1}$, the Fourier series representation (2.8) of the function $f(x)$, defined over the interval $-L \le x \le L$, may be expressed in the *complex Fourier series* form

$$f(x) = \sum_{n=-\infty}^{\infty} c_n\, e^{in\pi x/L}, \quad \text{where} \quad c_n = \frac{1}{2L} \int_{-L}^{L} f(x)\, e^{-in\pi x/L}\, dx. \qquad (6.1)$$

Setting $k_n = n\pi/L$ and $\Delta k_n \equiv k_{n+1} - k_n = \pi/L$, Eq. (6.1) may be written as

$$f(x) = \frac{1}{2\pi} \sum_{k_n=-\infty}^{\infty} \Delta k_n \left[\int_{-L}^{L} f(y)\, e^{-ik_n y}\, dy \right] e^{ik_n x}. \qquad (6.2)$$

In the limit that $L \to \infty$, i.e., the interval becomes infinite, Eq. (6.2) yields

$$f(x) = \frac{1}{2\pi} \int_{-\infty}^{\infty} \int_{-\infty}^{\infty} f(y)\, e^{ik\,(x-y)}\, dy\, dk, \qquad (6.3)$$

which is known as *Fourier's integral theorem*. From (6.3), it follows that if

$$F(k) = \frac{1}{\sqrt{2\pi}} \int_{-\infty}^{\infty} f(x)\, e^{-ikx}\, dx, \quad \text{then} \quad f(x) = \frac{1}{\sqrt{2\pi}} \int_{-\infty}^{\infty} F(k)\, e^{ikx}\, dk. \quad (6.4)$$

The function $F(k)$ is called the *Fourier transform* of $f(x)$, while $f(x)$ is the *inverse Fourier transform* of $F(k)$. It should be noted that the factor $1/(2\pi)$ in (6.3) has been split symmetrically here. This fairly standard convention differs from that used in Maple's `fourier` and `invfourier` commands, where the factor is split asymmetrically, with 1 in the Fourier transform and $1/(2\pi)$ in the inverse. Since, in solving most problems of physical interest, usually both the Fourier transform and its inverse are performed, this numerical factor splitting issue normally need not concern us.

If $f(x)$ is an even function, Equation (6.4) yields

$$F_c(k) = \sqrt{\frac{2}{\pi}} \int_0^\infty f(x)\,\cos(kx)\,dx, \quad f(x) = \sqrt{\frac{2}{\pi}} \int_0^\infty F_c(k)\,\cos(kx)\,dk. \quad (6.5)$$

$F_c(k)$ and $f(x)$ are the *Fourier cosine transforms* of each other. On the other hand, if $f(x)$ is an odd function one obtains the *Fourier sine transforms*,

$$F_s(k) = \sqrt{\frac{2}{\pi}} \int_0^\infty f(x)\,\sin(kx)\,dx, \quad f(x) = \sqrt{\frac{2}{\pi}} \int_0^\infty F_s(k)\,\sin(kx)\,dk. \quad (6.6)$$

The Fourier sine and cosine transforms are useful for solving problems involving semi-infinite domains, while the "full" Fourier transforms (6.4) can be applied to situations involving infinite domains.

Some of the more important properties[1] of Fourier transforms are as follows:

- If the Fourier transform of $f(x)$ (denoted by $\mathcal{F}[f(x)]$) is $F(k)$, then

$$\mathcal{F}\left[\frac{d^n f(x)}{dx^n}\right] = (ik)^n\, F(k),$$

  provided that $f(x)$ and its first $(n-1)$ derivatives vanish at $x = \pm\infty$.

- The function

$$C(x) = \frac{1}{\sqrt{2\pi}} \int_{-\infty}^\infty f(x-y)\, g(y)\, dy$$

  is called the *convolution* of the functions f and g (often denoted $f \star g$) over the interval $-\infty$ to ∞. If $F(k)$ and $G(k)$ are the Fourier transforms of $f(x)$ and $g(x)$, respectively, the *convolution theorem* is

$$C(x) = \frac{1}{\sqrt{2\pi}} \int_{-\infty}^\infty F(k)\, G(k)\, e^{ikx}\, dk.$$

  Alternately, the theorem may be stated as follows: $\mathcal{F}(f \star g) = \mathcal{F}(f)\,\mathcal{F}(g)$.

- If one sets $x = 0$ in the convolution theorem, the *Parseval relation*

$$\int_{-\infty}^\infty f(-y)\, g(y)\, dy = \int_{-\infty}^\infty F(k)\, G(k)\, dk$$

  results. Another Parseval relation is

$$\int_{-\infty}^\infty f(y)\, g^\star(y)\, dy = \int_{-\infty}^\infty F(k)\, G^\star(k)\, dk.$$

  If $g = f$, this last result yields *Parseval's theorem*,

$$\int_{-\infty}^\infty |f(y)|^2\, dy = \int_{-\infty}^\infty |F(k)|^2\, dk.$$

---

[1] Similar properties hold for the sine and cosine transforms.

6.1.1 Some Fourier Transform Shapes

Spoon feeding in the long run teaches us nothing
but the shape of the spoon.
E. M. Forster, British novelist, (1879–1970)

In this recipe, the Fourier transforms of some common $f(x)$ are examined.

Calculate the Fourier transform $F(k)$ of the following functions, using the symmetric factor convention:

(a) $f1 = A e^{-a^2 x^2}$; (b) $f2 = A e^{-a|x|}$; (c) $f3 = A$ for $|x| < a$, 0 for $|x| > a$.

Plot the $f(x)$ and their transforms for $a = A = 1$. Confirm Parseval's theorem.

The `inttrans` library package contains the Fourier transform command, `fourier`, so is loaded. To accomplish the transforms, the assumption that both a and A are positive is entered.

```
>   restart: with(inttrans): assume(a>0,A>0):
```

A functional operator `F` is formed to take the Fourier transform of an arbitrary function f and simplify the result. The result is divided by $\sqrt{2\pi}$ to agree with the symmetric factor convention.

```
>   F:=f->simplify(fourier(f,x,k)/sqrt(2*Pi)):
```

The Gaussian function $f1$ is entered, and its Fourier transform $F1$ calculated using the operator `F`. The transform is a Gaussian profile centered on $k = 0$.

```
>   f1:=A*exp(-a^2*x^2); F1:=F(f1);
```

$$f1 := A e^{(-a^2 x^2)} \qquad F1 := \frac{1}{2} \frac{A e^{\left(-\frac{k^2}{4a^2}\right)} \sqrt{2}}{a}$$

The symmetric exponentially decaying function $f2$ is entered and its transform calculated. The resulting profile $F2$ is a *Lorentzian* line shape centered on $k = 0$.

```
>   f2:=A*exp(-a*abs(x)); F2:=F(f2);
```

$$f2 := A e^{(-a|x|)} \qquad F2 := \frac{A a \sqrt{2}}{(a^2 + k^2) \sqrt{\pi}}$$

Making use of the `Heaviside` command, the rectangular profile $f3$ is entered, and its transform $F3$ determined.

```
>   f3:=A*Heaviside(x+a)-A*Heaviside(x-a); F3:=F(f3);
```

$$f3 := A \operatorname{Heaviside}(x + a) - A \operatorname{Heaviside}(x - a)$$

$$F3 := \frac{A \sin(a k) \sqrt{2}}{k \sqrt{\pi}}$$

To plot each function and its transform, an arrow operator `G` is introduced to evaluate an arbitrary function FF at $a = 1$, $A = 1$.

```
>   G:=FF->eval(FF,{a=1,A=1}):
```

Making use of `G`, the three given functions $f1$, $f2$, and $f3$ are plotted together, the curves being given different colors and line styles. The resulting picture is shown on the left of Figure 6.1.

```
> plot([G(f1),G(f2),G(f3)],x=-4..4,color=[red,blue,green],
  linestyle=[1,2,3],thickness=2,labels=["x","f"]);
```

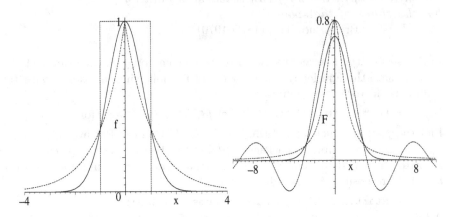

Figure 6.1: Left: the three $f(x)$. Right: Corresponding Fourier transforms.

The three Fourier transforms *F1*, *F2*, and *F3* are plotted with matching colors and line styles and shown on the right of the figure.

```
> plot([G(F1),G(F2),G(F3)],k=-10..10,color=[red,blue,green],
  linestyle=[1,2,3],thickness=2,labels=["x","F"]);
```

The oscillatory curve is the transform of the rectangle *f3*, the shorter smooth curve the transform of the Gaussian *f1*, and the taller smooth curve is the transform *F2*. The correspondence is more evident on the computer screen.

Using the complex conjugate (`conjugate`) command, an operator L is formed to calculate the left-hand side of Parseval's theorem for a given function f.

```
> L:=f->int(f*conjugate(f),x=-infinity..infinity);
```

$$L := f \rightarrow \int_{-\infty}^{\infty} f\,\overline{(f)}\,dx$$

Similarly, an operator R is created to calculate the right-hand side of Parseval's theorem for a specified Fourier transform FT.

```
> R:=FT->int(FT*conjugate(FT),k=-infinity..infinity);
```

$$R := FT \rightarrow \int_{-\infty}^{\infty} FT\,\overline{(FT)}\,dk$$

Using L and R, Parseval's theorem is now confirmed for the function *f1* and its transform *F1*, the results *L1* and *R1* being identical.

```
> L1:=L(f1); R1:=R(F1);
```

$$L1 := \frac{A^2\sqrt{\pi}\sqrt{2}}{2\,a} \qquad R1 := \frac{A^2\sqrt{\pi}\sqrt{2}}{2\,a}$$

It is left as an exercise for you to confirm that Parseval's theorem holds for the other two cases.

6.1.2 A Northern Weenie Roast

Wedding: the point at which a man stops toasting a woman and begins roasting her.
Helen Rowland, American journalist, *A Guide to Men, "Syncopations" (1922)*

While visiting his inlaws in Northern Ontario over the Christmas holidays, Russell went snowmobiling with his family. After a few hours, feeling slightly cold and hungry, they stopped at one of the warming huts along the track and soon had a roaring fire going in the fireplace. Using some long thin steel rods which had been conveniently leaning against the outer wall of the hut, they satisfied their hunger by roasting weenies in the fire.

Idealizing the cooking situation, let's calculate and animate the temperature distribution inside a thin insulated semi-infinite ($0 \leq x \leq \infty$) steel rod which has the $x=0$ end held at a fixed temperature $T0$ and whose interior is initially at 0 degrees Celsius. For steel, the thermal conductivity $K=46$ W/(m K°), the density $\rho = 8 \times 10^3$ kg/m^3, and the specific heat $C = 500$ J/(kg K°).

The plots and integral transform packages are loaded, the former needed for the animation, the latter for the Fourier sine (`fouriersin`) command.

```
>  restart: with(plots): with(inttrans):
```
The one-dimensional heat diffusion equation is entered in *de*, $T(x,t)$ being the temperature at a distance x from the hot end of the rod at time t. The parameter $a = \sqrt{K/(\rho C)}$.

```
>  de:=diff(T(x,t),t)=a^2*diff(T(x,t),x,x);
```

$$de := \frac{\partial}{\partial t} T(x,\,t) = a^2 \left(\frac{\partial^2}{\partial x^2} T(x,\,t) \right)$$

Since the temperature is specified at $x=0$, i.e., $T(0,t)=T0$, we will solve the diffusion equation by first taking the Fourier sine transform of *de*, yielding the time-dependent ODE shown in the output of *eq*.

```
>  eq:=fouriersin(de,x,k);
```

$$eq := \frac{\partial}{\partial t} \text{fouriersin}(T(x,\,t),\,x,\,k) =$$
$$\frac{a^2 k \left(\sqrt{2}\,T(0,\,t) - k\,\text{fouriersin}(T(x,\,t),\,x,\,k)\,\sqrt{\pi} \right)}{\sqrt{\pi}}$$

The notation is simplified by replacing fouriersin($T(x,\,t),\,x,\,k$) with $F(k,\,t)$.

```
>  eq2:=subs(fouriersin(T(x,t),x,k)=F(k,t),eq);
```

$$eq2 := \frac{\partial}{\partial t} F(k,\,t) = \frac{a^2 k \left(\sqrt{2}\,T(0,\,t) - k\,F(k,\,t)\,\sqrt{\pi} \right)}{\sqrt{\pi}}$$

The boundary condition $T(0,t) = T0$ is substituted into *eq2*.

```
>  eq3:=subs(T(0,t)=T0,eq2);
```

$$eq3 := \frac{\partial}{\partial t} F(k,\,t) = \frac{a^2 k \left(\sqrt{2}\,T0 - k\,F(k,\,t)\,\sqrt{\pi} \right)}{\sqrt{\pi}}$$

Then, the ODE *eq3* is analytically solved for $F(k, t)$.

>   eq4:=dsolve(eq3,F(k,t));

$$eq4 := F(k, t) = \frac{\sqrt{2}\, T0}{k\,\sqrt{\pi}} + e^{(-a^2\, k^2\, t)}\, \_F1(k)$$

Since $T(x,0)=0$ for $x > 0$, then $F(k, 0)=0$. So, the arbitrary function $\_F1(k)$ is determined by evaluating the right-hand side of *eq4* at time $t=0$, setting the result equal to 0, and solving for $\_F1(k)$.

>   _F1(k):=solve(eval(rhs(eq4),t=0),_F1(k));

$$\_F1(k) := -\frac{\sqrt{2}\, T0}{k\,\sqrt{\pi}}$$

$\_F1(k)$ is automatically substituted into *eq4*, the result being labeled *eq5*.

>   eq5:=eq4;

$$eq5 := F(k, t) = \frac{\sqrt{2}\, T0}{k\,\sqrt{\pi}} - \frac{e^{(-a^2\, k^2\, t)}\,\sqrt{2}\, T0}{k\,\sqrt{\pi}}$$

The temperature distribution inside the rod at arbitrary time $t \geq 0$ follows on applying the following Fourier sine command to the right-hand side of *eq5*, and simplifying with the symbolic option.

>   T:=simplify(fouriersin(rhs(eq5),k,x),symbolic);

$$T := -T0\left(-1 + \operatorname{erf}(\frac{x}{2\,a\,\sqrt{t}})\right)$$

The temperature T is expressed in terms of the error function. As a partial check on our calculation, note that for $x > 0$ and $t=0$, $T(x, 0) = T0(1 - \operatorname{erf}(\infty)) = T0(1 - 1) = 0$ as it should. The thermal conductivity (K), density (ρ), and specific heat (C) values for steel are now entered and the diffusion coefficient $a = \sqrt{K/(\rho C)}$ calculated. In the MKS system, a has the units m/s$^{1/2}$. For convenience, a is converted to CGS units by multiplying by 100, since 1 m = 100 cm. The units of a are then cm/s$^{1/2}$, the distance x being expressed in cm.

>   K:=46: rho:=8*10^3: C:=500: a:=100*sqrt(K/(rho*C));

$$a := \frac{\sqrt{46}}{20}$$

The normalized temperature distribution $T/T0$ is given in TT,

>   TT:=T/T0;

$$TT := 1 - \operatorname{erf}(\frac{5\,x\,\sqrt{46}}{23\,\sqrt{t}})$$

which is plotted at 60 second (1 minute) time intervals up to 5 minutes.

>   plot([seq(eval(TT,t=60*i),i=1..5)],x=0..25,thickness=2,
      labels=["x","T/T0"]);

The resulting time evolution of the temperature inside the rod is shown in Figure 6.2, the curve furthest to the left corresponding to 1 minute, the curve furthest to the right to 5 minutes. Even at 5 minutes the temperature at $x=25$ cm is a tiny fraction of the temperature at the hot end.

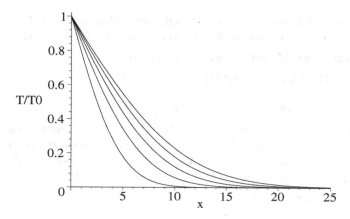

Figure 6.2: Normalized temperature distribution inside rod at different times.

Clearly, the semi-infinite approximation to the long rod is not too bad. The normalized temperature distribution is now animated over the same time interval, 100 frames being used.

```
>   animate(TT,x=0..25,t=0..300,frames=100,thickness=2,
    labels=["x","T/T0"]);
```
You will have to execute the recipe on your computer to see the animation.

6.1.3 Turn Off the Boob Tube and Concentrate

Television's perfect. You turn a few knobs, a few of those mechanical adjustments at which the higher apes are so proficient, and lean back and drain your mind of all thought. And there you are watching the bubbles in the primeval ooze. You don't have to concentrate. You don't have to react. You don't have to remember. You don't miss your brain because you don't need it.
Raymond Chandler, American novelist, (1888–1959)

Since you are reading this book, you have turned your back (at least temporarily) on watching TV. Having your undivided attention, I would like you to concentrate on the following example involving the Fourier cosine transform.

Consider a thin semi-infinite ($0 \leq x \leq \infty$) hollow tube, closed at the $x=0$ end, containing water. A drop of blue ink is injected at the point $x = x0 > 0$ at time $t = 0$. Suppose that the drop has an initial concentration profile given approximately by $C(x,0) = f(x) = A\,\delta(x - x0)$. If the diffusion constant of ink in water is d, determine the concentration of ink for arbitrary time $t \geq 0$. At what time T is the concentration of ink at $x = 0$ a maximum? Taking the nominal values $A = 1$, $d = 1$, and $x0 = 10$, evaluate T and animate the concentration profile $C(x, t)$ and plot $C(0, t)$ over a time interval up to $2T$.

The plots and integral transform packages are loaded and the 1-dimensional diffusion equation entered in *pde* for the concentration $C(x,t)$ of ink.

```
>  restart: with(plots): with(inttrans):
>  pde:=diff(C(x,t),t)=d*diff(C(x,t),x,x);
```

$$pde := \frac{\partial}{\partial t} C(x,\,t) = d\,(\frac{\partial^2}{\partial x^2} C(x,\,t))$$

Since there can be no flow of ink through the $x = 0$ end, the concentration gradient $\partial C/\partial x$ must be zero there. This boundary condition implies that we should use a Fourier cosine transform approach on the diffusion equation. The Fourier cosine transform (`fouriercos`) command is applied to *pde*.

```
>  eq:=fouriercos(pde,x,k);
```

$$eq := \frac{\partial}{\partial t} \text{fouriercos}(C(x,\,t),\,x,\,k) =$$

$$-\frac{d\,(\sqrt{2}\,D_1(C)(0,\,t) + k^2\,\text{fouriercos}(C(x,\,t),\,x,\,k)\,\sqrt{\pi})}{\sqrt{\pi}}$$

The result in *eq* is expressed in terms of the concentration gradient at $x = 0$, viz., $D_1(C)(0,\,t)$. This term is set equal to zero in *eq2*. The notation is also simplified by replacing fouriercos($C(x,\,t),\,x,\,k$) with $F(k,\,t)$.

```
>  eq2:=subs({fouriercos(C(x,t),x,k)=F(k,t),D[1](C)(0,t)=0},eq);
```

$$eq2 := \frac{\partial}{\partial t} F(k,\,t) = -d\,k^2\,F(k,\,t)$$

The initial concentration profile $f = A\,\delta(x - x0)$ of the ink is entered,

```
>  f:=A*Dirac(x-x0);
```

$$f := A\,\text{Dirac}(-x + x0)$$

and its Fourier cosine transform taken and simplified assuming that $x0 > 0$.

```
>  F1:=simplify(fouriercos(f,x,k)) assuming x0>0;
```

$$F1 := \frac{A\,\sqrt{2}\,\cos(k\,x0)}{\sqrt{\pi}}$$

The ODE *eq2* is analytically solved for $F(k,\,t)$.

```
>  eq3:=dsolve(eq2,F(k,t));
```

$$eq3 := F(k,\,t) = \_F1(k)\,e^{(-d\,k^2\,t)}$$

At $t = 0$, $F(k,\,0) = \_F1(k)$, so $\_F1(k) = F1$ is substituted into *eq3*.

```
>  eq4:=subs(_F1(k)=F1,eq3);
```

$$eq4 := F(k,\,t) = \frac{A\,\sqrt{2}\,\cos(k\,x0)\,e^{(-d\,k^2\,t)}}{\sqrt{\pi}}$$

The general desired expression for the concentration follows on applying the Fourier cosine transform to the right-hand side of *eq4*, taking us from k-space back into x-space.

```
>  C:=fouriercos(rhs(eq4),k,x);
```

$$C := \dfrac{A \sqrt{\dfrac{\pi}{t\,d}}\, e^{\left(-\frac{x0^2+x^2}{4\,t\,d}\right)} \cosh\left(\dfrac{x0\,x}{2\,t\,d}\right)}{\pi}$$

The time T at which the ink concentration is an extremum at $x=0$ is obtained by evaluating C at this point, differentiating with respect to t, setting the result equal to zero, and solving for the time t. The plot of the concentration at $x=0$ will confirm that the time T corresponds to the maximum ink concentration at that end of the tube.

```
>  T:=solve(diff(eval(C,x=0),t)=0,t);
```

$$T := \frac{x0^2}{2\,d}$$

The parameter values $A=1$, $d=1$, and $x0=10$ are entered and the concentration profile $C(x,t)$ and time T then determined.

```
>  A:=1: d:=1: x0:=10: C:=C; T:=T;
```

$$C := \dfrac{\sqrt{\dfrac{\pi}{t}}\, e^{\left(-\frac{100+x^2}{4\,t}\right)} \cosh\left(\dfrac{5\,x}{t}\right)}{\pi}$$

$$T := 50$$

The concentration reaches its maximum value at the $x=0$ end 50 time units after the ink drop is injected.

The concentration profile is animated over the time interval $t=1$ to $2T$, 100 frames being taken. The animation does not start at $t=0$ because the initial profile is a Dirac delta function. The initial frame is shown in Figure 6.3. Run the animation and see how the profile evolves with time.

```
>  animate(C,x=0..35,t=1..2*T,frames=100,thickness=2,
   numpoints=200,labels=["x","C"]);
```

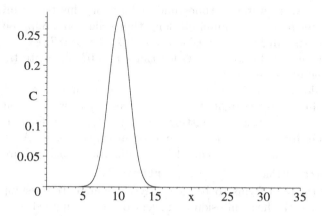

Figure 6.3: Initial frame in the animation of the ink concentration.

The concentration of ink at $x = 0$ is now plotted in Figure 6.4 over the time interval $t=0$ to $2T$. The maximum concentration occurs at 50 time units.

```
>  plot(eval(C,x=0),t=0..2*T,thickness=2,
      labels=["t","C(0,t)"]);
```

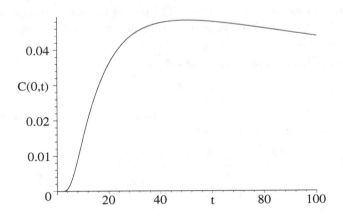

Figure 6.4: Time evolution of the ink concentration at $x=0$.

6.1.4   Diffusive Heat Flow

In my own time there have been inventions of this sort,...tubes for diffusing warmth equally through all parts of a building...
Seneca, Roman writer, philosopher, and statesman, (5BC–65AD)

In contrast to the radiant heating referred to by Seneca, involving the convective flow of warm water in tubes under the floor, this recipe involves the diffusive flow of heat in an infinitely long thin solid rod. Suppose that the initial temperature distribution inside the rod is $T(x, 0) = T_0\, e^{-b^2 x^2}$. What is the temperature distribution $T(x, t)$ for times $t \geq 0$? Animate the suitably normalized temperature distribution.

Since the domain is infinite, the full Fourier transform will be used. Because the inverse will also be calculated, it is not necessary to worry about the fact that the Maple convention for splitting the factor $1/(2\pi)$ in Fourier's integral theorem differs from the convention adopted in this chapter. As before, the integral transform and plots (needed for the animation) packages are loaded.

```
>  restart: with(plots): with(inttrans):
```

In order for the transforms to be explicitly calculated, the assumption that $b > 0$, $t > 0$, and the heat diffusion coefficient $a > 0$, is supplied.

```
>  assume(a>0,b>0,t>0):
```

The one-dimensional heat diffusion equation is entered.

```
>  de:=diff(T(x,t),t)=a^2*diff(T(x,t),x,x);
```

$$de := \frac{\partial}{\partial t} T(x, t) = a^2 \left(\frac{\partial^2}{\partial x^2} T(x, t)\right)$$

The Fourier transform of the PDE *de* is taken in *eq* and notationally simplified in *eq2* by substituting $F(k, t)$ for fourier($T(x, t), x, k$).

```
>  eq:=fourier(de,x,k);
```

$$eq := \frac{\partial}{\partial t} \text{fourier}(T(x, t), x, k) = -a^2 k^2 \text{fourier}(T(x, t), x, k)$$

```
>  eq2:=subs(fourier(T(x,t),x,k)=F(k,t),eq);
```

$$eq2 := \frac{\partial}{\partial t} F(k, t) = -a^2 k^2 F(k, t)$$

Using the **dsolve** command, the ODE *eq2* is analytically solved for $F(k, t)$.

```
>  sol:=dsolve(eq2,F(k,t));
```

$$sol := F(k, t) = \_F1(k) e^{(-a^2 k^2 t)}$$

The arbitrary function $\_F1(k)$ is determined by Fourier transforming the initial temperature profile, $T_0 e^{-b^2 x^2}$. Its functional form is automatically substituted into *sol*, the result being labeled *sol2*.

```
>  _F1(k):=fourier(T0*exp(-b^2*x^2),x,k); sol2:=sol;
```

$$\_F1(k) := T0 \, e^{\left(-\frac{k^2}{4b^2}\right)} \sqrt{\frac{\pi}{b^2}}$$

$$sol2 := F(k, t) = T0 \, e^{\left(-\frac{k^2}{4b^2}\right)} \sqrt{\frac{\pi}{b^2}} \, e^{(-a^2 k^2 t)}$$

The temperature profile inside the rod for $t > 0$ follows on taking the inverse Fourier transform (**invfourier**) of the right-hand side of *sol2*.

```
>  T:=invfourier(rhs(sol2),k,x);
```

$$T := \frac{T0 \, e^{\left(-\frac{b^2 x^2}{4 b^2 a^2 t + 1}\right)}}{\sqrt{4 b^2 a^2 t + 1}}$$

The temperature is normalized (made dimensionless) by dividing T by $T0$, and normalized spatial (ζ) and time (τ) coordinates are introduced by substituting $x = \zeta/b$ and $t = \tau/(b^2 a^2)$.

```
>  Tnorm:=subs({x=zeta/b,t=tau/(b^2*a^2)},T/T0);
```

$$Tnorm := \frac{e^{\left(-\frac{\zeta^2}{4\tau + 1}\right)}}{\sqrt{4\tau + 1}}$$

The normalized temperature, *Tnorm*, is animated over the time interval $\tau = 0$ to 10, the spatial range being $\zeta = -10$ to 10.

```
>  animate(Tnorm,zeta=-10..10,tau=0..10,frames=100,
      thickness=2,numpoints=200,labels=["zeta","Tnorm"]);
```

Run the animation to see the initially Gaussian temperature profile diffuse away.

6.1.5   Deja Vu

The postmodern reply to the modern consists of recognizing that the past, since it cannot really be destroyed, because its destruction leads to silence, must be revisited...
Umberto Eco, Italian novelist, *Reflections on the Name of the Rose, (1983)*

If this example seems familiar, it is. We revisit Recipe **01-1-6: Mr. Dirac's Famous Function** and, instead of using the Green's function method, Fourier transform the relevant differential equation and illustrate how contour integration can be used to evaluate the inverse transform.

Consider a very long uniform horizontal beam glued to an elastic foundation which exerts a Hooke's law restoring force (spring constant K) on the beam. Subject to a steady point force F exerted at $x = 0$, the static displacement $\psi(x) = (F/K)\, G(x)$ of the beam satisfies the following ODE,

$$\alpha^4 \frac{d^4 G}{dx^4} + G = \delta(x), \tag{6.7}$$

with $\alpha = (YI/K)^{1/4}$, Y being Young's modulus and I the beam's moment of inertia. Determine $G(x)$ for all x.

After loading the integral transform package, the following `interface` command is used to set $j = \sqrt{-1}$, to avoid confusion with the moment of inertia.

```
>  restart: with(inttrans): interface(imaginaryunit=j):
```
Equation (6.7) is entered in *ode* and then Fourier transformed in *eq*.

```
>  ode:=alpha^4*diff(G(x),x$4)+G(x)=Dirac(x);
```

$$ode := \alpha^4 \left(\frac{d^4}{dx^4}\, G(x) \right) + G(x) = \mathrm{Dirac}(x)$$

```
>  eq:=fourier(ode,x,k);
```

$$eq := \mathrm{fourier}(G(x),\, x,\, k)\,(\alpha^4\, k^4 + 1) = 1$$

eq is solved in *eq2* for the Fourier transform of $G(x)$.

```
>  eq2:=solve(eq,fourier(G(x),x,k));
```

$$eq2 := \frac{1}{\alpha^4\, k^4 + 1}$$

With Maple's convention for the Fourier transform, the inverse Fourier transform is given by $G(x) = \int_{-\infty}^{\infty} e^{jkx} (eq2/(2\pi))\, dk$. The integrand is now entered.

```
>  integrand:=exp(j*k*x)*eq2/(2*Pi);
```

$$integrand := \frac{1}{2}\, \frac{e^{(k\,x\,j)}}{(\alpha^4\, k^4 + 1)\, \pi}$$

The four complex k zeros of the integrand's denominator are determined.

```
>  sol:={solve(denom(integrand)=0,k)};
```

$$sol := \left\{ \frac{\frac{\sqrt{2}}{2} + \frac{1}{2}j\sqrt{2}}{\alpha}, \frac{\frac{1}{2}j\sqrt{2} - \frac{\sqrt{2}}{2}}{\alpha}, \frac{-\frac{\sqrt{2}}{2} - \frac{1}{2}j\sqrt{2}}{\alpha}, \frac{-\frac{1}{2}j\sqrt{2} + \frac{\sqrt{2}}{2}}{\alpha} \right\}$$

The integrand has four simple poles, the first two entries in *sol* corresponding to poles in the upper-half of the complex k-plane, the last two entries to poles in the lower-half k-plane. The integral in $G(x)$ is evaluated by performing a closed contour integration along the real k-axis from $-R$ to $+R$ and then around a semi-circle of radius R (where $k = R\,e^{j\theta}$), the limit $R \to \infty$ being taken. Noting that on the semi-circle the exponential term in the integrand takes the form $e^{jR e^{j\theta} x} = e^{jR\cos(\theta)x}\,e^{-R\sin(\theta)x}$, the semi-circle must be taken in the upper-half k-plane (where $\sin(\theta) > 0$) for $x > 0$ for the semi-circular contribution to vanish as $R \to \infty$. In this case, the two poles in the upper-half plane will contribute to the integral.

For $x < 0$, on the other hand, the semi-circle must be taken in the lower-half plane where $\sin(\theta) < 0$. The two poles in the lower-half k-plane then contribute.

A functional operator is formed for evaluating the residue of the integrand for the ith simple pole.

```
> res:=i->simplify(evalc(residue(integrand,k=sol[i]))):
```

Using the functional operator, the sequence of four residues is explicitly calculated, the first two residues corresponding to the two poles in the upper-half plane, the last two residues to the two poles in the lower-half plane.

```
> eq3:={seq(res(i),i=1..4)};
```

$eq3 :=$

$$\left\{ -\frac{1}{16}\frac{e^{\left(-\frac{x\sqrt{2}}{2\alpha}\right)}\sqrt{2}\left(\cos(\frac{x\sqrt{2}}{2\alpha}) - \sin(\frac{x\sqrt{2}}{2\alpha}) + \sin(\frac{x\sqrt{2}}{2\alpha})j + \cos(\frac{x\sqrt{2}}{2\alpha})j\right)}{\alpha\pi}, \right.$$

$$-\frac{1}{16}\frac{e^{\left(-\frac{x\sqrt{2}}{2\alpha}\right)}\sqrt{2}\left(-\cos(\frac{x\sqrt{2}}{2\alpha}) + \sin(\frac{x\sqrt{2}}{2\alpha}) + \sin(\frac{x\sqrt{2}}{2\alpha})j + \cos(\frac{x\sqrt{2}}{2\alpha})j\right)}{\alpha\pi},$$

$$\frac{1}{16}\frac{e^{\left(\frac{x\sqrt{2}}{2\alpha}\right)}\sqrt{2}\left(\cos(\frac{x\sqrt{2}}{2\alpha}) + \sin(\frac{x\sqrt{2}}{2\alpha}) - \sin(\frac{x\sqrt{2}}{2\alpha})j + \cos(\frac{x\sqrt{2}}{2\alpha})j\right)}{\alpha\pi},$$

$$\left. \frac{1}{16}\frac{e^{\left(\frac{x\sqrt{2}}{2\alpha}\right)}\sqrt{2}\left(-\cos(\frac{x\sqrt{2}}{2\alpha}) - \sin(\frac{x\sqrt{2}}{2\alpha}) - \sin(\frac{x\sqrt{2}}{2\alpha})j + \cos(\frac{x\sqrt{2}}{2\alpha})j\right)}{\alpha\pi} \right\}$$

The last two residues, containing $e^{x\sqrt{2}/(2\alpha)}$, are selected in *eq4*, while the first two residues, containing $e^{-x\sqrt{2}/(2\alpha)}$, are selected in *eq5*.

```
> eq4:=select(has,eq3,exp(x*sqrt(2)/(2*alpha))):
> eq5:=select(has,eq3,exp(-x*sqrt(2)/(2*alpha))):
```

For $x < 0$ (to the left of the applied force at the origin), then $G(x)$ is obtained by summing the two residues in *eq4*, multiplying by $2j\pi$, applying the complex evaluation command, and simplifying. The minus sign is inserted because for the semi-circle in the lower-half k-plane, the contour is taken clockwise, rather than counter-clockwise. The result, assigned the name *GL*, is the same as obtained previously in Recipe **01-1-6**.

```
> GL:=-simplify(evalc(2*j*Pi*(eq4[1]+eq4[2])));
```

$$GL := \frac{1}{4} \frac{e^{\left(\frac{x\sqrt{2}}{2\alpha}\right)} \sqrt{2}\,(-\sin(\frac{x\sqrt{2}}{2\alpha}) + \cos(\frac{x\sqrt{2}}{2\alpha}))}{\alpha}$$

The solution for $x > 0$ is similarly obtained by using *eq5*. The contour is counter-clockwise in this case, so a minus sign is not inserted. The result *GR* is identical with that previously obtained using the Green's function method.

```
> GR:=simplify(evalc(2*j*Pi*(eq5[1]+eq5[2])));
```

$$GR := \frac{1}{4} \frac{e^{\left(-\frac{x\sqrt{2}}{2\alpha}\right)} \sqrt{2}\,(\cos(\frac{x\sqrt{2}}{2\alpha}) + \sin(\frac{x\sqrt{2}}{2\alpha}))}{\alpha}$$

6.2 Laplace Transforms

The *Laplace transform* of a function $f(t)$ is defined to be

$$\mathcal{L}(f(t)) = F(s) = \int_0^\infty e^{-st} f(t)\,dt, \tag{6.8}$$

where $s = x + iy$ is a complex variable. The inverse transform is given by the *Bromwich integral formula*,

$$\mathcal{L}^{-1}(F) = f(t) = \frac{1}{2\pi i} \int_{c-i\infty}^{c+i\infty} e^{st} F(s)\,ds, \quad t > 0 \tag{6.9}$$

and $f(t) = 0$ for $t < 0$. The integration in (6.9) is to be performed along a line $s = c + iy$ in the complex plane. The real number c is chosen so that $s = c$ lies to the right of all poles, branch points, or essential singularities. With Maple, the `laplace` command is used to calculate the Laplace transform, and `invlaplace` for the inverse transform.

A couple of the more important properties of Laplace transforms are:

- The Laplace transform of the nth derivative of $f(t)$ is given by

$$\mathcal{L}\left(\frac{d^n f}{dt^n}\right) = s^n\, F(s) - \sum_{k=0}^{n-1} \left(\frac{d^k f}{dt^k}\right)_{t=0} s^{n-k-1}.$$

- Defining the convolution of the functions $f_1(t)$ and $f_2(t)$ to be

$$C(t) = \int_0^t f_1(t-y)\, f_2(y)\, dy, \quad \text{then} \quad \mathcal{L}(C(t)) = F_1(s)\, F_2(s),$$

where $F_1(s)$ and $F_2(s)$ are the Laplace transforms of $f_1(t)$ and $f_2(t)$.

6.2.1   Jennifer Consults Mr. Spiegel

*Histories are more full of examples of the fidelity of dogs
than of friends.*
Alexander Pope, English satirical poet, (1688–1744).

To illustrate the use of Laplace transforms in solving ODEs, Jennifer has created a recipe for solving the following problem taken from Murray Spiegel's *Advanced Mathematics*. Solve the fourth order inhomogeneous ODE

$$y''''(t) + 2\,y''(t) + y(t) = \sin t, \quad y(0) = 1,\; y'(0) = -2,\; y''(0) = 3,\; y'''(0) = 0.$$

After loading the integral transform package, Jennifer lets the Laplace transform of $y(t)$ be represented by $F(s)$ for notational convenience.

```
>   restart: with(inttrans): LT:=laplace(y(t),t,s)=F(s):
```
The ODE is now entered.

```
>   ode:=diff(y(t),t$4)+2*diff(y(t),t$2)+y(t)=sin(t);
```

$$ode := (\frac{d^4}{dt^4}\,y(t)) + 2\,(\frac{d^2}{dt^2}\,y(t)) + y(t) = \sin(t)$$

She then takes the Laplace transform of *ode* and substitutes *LT*,

```
>   eq:=subs(LT,laplace(ode,t,s));
```

$$eq := s^4\,F(s) - (\mathrm{D}^{(3)})(y)(0) - s\,(\mathrm{D}^{(2)})(y)(0) - s^2\,\mathrm{D}(y)(0) - s^3\,y(0)$$
$$+ 2\,s^2\,F(s) - 2\,\mathrm{D}(y)(0) - 2\,s\,y(0) + F(s) = \frac{1}{s^2+1}$$

the result being expressed in terms of $y(0)$ and the first three derivatives at $t=0$. Maple has simply applied the rule for taking the Laplace transform of a derivative to the fourth and second order derivatives. The last term, $1/(s^2+1)$, in *eq* is the Laplace transform of $\sin(t)$. If proceeding by hand, one would look up this result in a table of Laplace transforms.

The initial conditions are now entered in *ic*, using the differential operator,

```
>   ic:=(y(0)=1,D(y)(0)=-2,D(D(y))(0)=3,D(D(D(y)))(0)=0):
```
and *eq* is evaluated in *eq2* with the initial conditions.

```
>   eq2:=eval(eq,{ic});
```

$$eq2 := s^4\,F(s) + 4 - 5\,s + 2\,s^2 - s^3 + 2\,s^2\,F(s) + F(s) = \frac{1}{s^2+1}$$

Jennifer then solves *eq2* for $F(s)$.

```
>   eq3:=solve(eq2,F(s));
```

$$eq3 := \frac{-6\,s^2 - 3 + 6\,s^3 + 5\,s - 2\,s^4 + s^5}{s^6 + 3\,s^4 + 3\,s^2 + 1}$$

The solution, Y, of the original ODE then follows on applying the inverse Laplace transform to *eq3*. If proceeding by hand, as in Spiegel, one would laboriously express *eq3* as a sum of partial fractions and then use a table to look up the inverse transforms of the various terms.

```
>  Y:=invlaplace(eq3,s,t);
```

$$Y := -\frac{1}{8}\sin(t)\,(21+t^2-16\,t) + \frac{1}{8}\cos(t)\,(8+5\,t)$$

Exactly the same answer can be obtained more easily by using the `dsolve` command with the option `method=laplace`, as illustrated in Y2.

```
>  Y2:=dsolve({ode,ic},y(t),method=laplace);
```

$$Y2 := y(t) = -\frac{1}{8}\sin(t)\,(21+t^2-16\,t) + \frac{1}{8}\cos(t)\,(8+5\,t)$$

Finally, Jennifer completes the recipe by plotting Y over the time interval $t=0$ to 100, the solution being shown in Figure 6.5.

```
>  plot(Y,t=0..100,thickness=2,labels=["t","y"],numpoints=200);
```

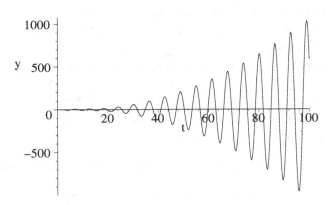

Figure 6.5: Time-dependent solution of the ODE.

As time increases, the oscillations grow indefinitely in amplitude, never settling down to a steady-state solution.

6.2.2   Jennifer's Heat Diffusion Problem

Gossip is a sort of smoke that comes from the dirty ...pipes of those who diffuse it: it proves nothing but the bad taste of the smoker.
George Eliot, English novelist, (1819–80)

The Laplace transform approach can be used to solve the diffusion equation with specified boundary conditions and an initial profile. Jennifer will illustrate this by solving the following temperature diffusion problem and animating it over the time interval $t=0$ to 0.05 seconds,

$$T_t = 4\,T_{xx},\quad T(0,t)=T(3,t)=0,\quad T(x,0)=10\sin(2\pi x)-6\sin(4\pi x)+2\sin(6\pi x).$$

After loading the integral transform and plots packages,

```
>  restart: with(inttrans): with(plots):
```

Jennifer enters the relevant PDE and the initial temperature profile $T(x, 0)$.

```
>  pde:=diff(T(x,t),t)=4*diff(T(x,t),x,x);
```

$$pde := \frac{\partial}{\partial t} T(x, t) = 4 \left(\frac{\partial^2}{\partial x^2} T(x, t) \right)$$

```
>  T(x,0):=10*sin(2*Pi*x)-6*sin(4*Pi*x)+2*sin(6*Pi*x);
```

Then *pde* is Laplace transformed in *eq* with respect to the time variable t, the initial condition being automatically substituted. For notational simplicity the Laplace transform of $T(x, t)$ is replaced with $F(x)$, the resulting ODE for $F(x)$ being displayed in *eq2*.

```
>  eq:=laplace(pde,t,s);
>  eq2:=subs(laplace(T(x,t),t,s)=F(x),eq);
```

$$eq2 := s F(x) - 10 \sin(2 \pi x) + 6 \sin(4 \pi x) - 2 \sin(6 \pi x) = 4 \left(\frac{d^2}{dx^2} F(x) \right)$$

Since the boundary conditions are $T(0, t) = T(3, t) = 0$, their Laplace transforms are equal to zero. So *eq2* is analytically solved for $F(x)$ subject to $F(0) = 0$ and $F(3) = 0$.

```
>  sol:=dsolve({eq2,F(0)=0,F(3)=0},F(x));
```

$$sol := F(x) = \left((160 \, s \, \pi^2 + 2 \, s^2 + 2048 \, \pi^4) \sin(6 \pi x) - 13824 \left(\left(\frac{s}{16} + \pi^2 \right) \sin(4 \pi x) \right. \right.$$

$$\left. \left. - \frac{20}{3} \sin(2 \pi x) \left(\frac{s}{64} + \pi^2 \right) \right) \left(\frac{s}{144} + \pi^2 \right) \right) \Big/ (s^3 + 224 \, s^2 \, \pi^2 + 12544 \, s \, \pi^4 + 147456 \, \pi^6)$$

Finally, applying the inverse Laplace transform to the right-hand side of *sol* yields the temperature profile $T(x, t)$ for $t \geq 0$, which is animated over the time interval $t = 0$ to 0.05.

```
>  T(x,t):=invlaplace(rhs(sol),s,t);
```

$$T(x, t) := 10 \sin(2\pi x) \, e^{(-16 \pi^2 t)} - 6 \sin(4\pi x) \, e^{(-64 \pi^2 t)} + 2 \sin(6\pi x) \, e^{(-144 \pi^2 t)}$$

```
>  animate(T(x,t),x=0..3,t=0..0.05,frames=100,numpoints=500,
   thickness=2,labels=["x","T"]);
```

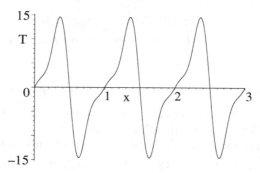

Figure 6.6: Initial frame of the animation.

The initial frame of the animation is shown in Figure 6.6, with 100 frames
being taken and 500 plotting points used to produce a smooth figure. The
temperature distribution becomes effectively zero throughout the entire range
$x=0$ to 3 on completion of the animation.

6.2.3   Daniel Strikes Yet Again: Mr. Laplace Appears

Propaganda is a soft weapon; hold it in your hands too long,
and it will move about like a snake, and strike the other way.
Jean Anouilh, French playwright, (1910–1987)

In **Example 4.1.3**, we imagined a scenario where young Daniel struck a light,
elastic, initially horizontal, string of length L fixed at both ends. Our approach
to determining the subsequent motion of the string was to build up a solution
using a Fourier series approach. In the present recipe, a similar struck string
is considered but now solved for the transverse displacement $U(x,t)$ using the
Laplace transform method. In particular, we take $L = \pi$ m, wave speed $c = 2$
m/s, and an initial transverse velocity $\partial U(x,0)/\partial t = 3\sin(2x) - 2\sin(5x)$.
After solving for $U(x,t)$, the motion of the string is animated over the time
interval $t=0$ to $T=5$ s.

After loading the integral transform and plots packages,

> `restart: with(inttrans): with(plots):`

the values of c, L, and T are entered.

> `c:=2: L:=Pi: T:=5:`

To simplify the notation, the `addtable` command is used to let $F(x)$ represent
the Laplace transform of $U(x,t)$ with respect to t in subsequent outputs.

> `addtable(laplace,U(x,t),F(x),t,s):`

The wave equation for $U(x,t)$ is entered in *pde*, the value of c being automatically substituted.

> `pde:=diff(U(x,t),x,x)=(1/c^2)*diff(U(x,t),t,t);`

$$pde := \frac{\partial^2}{\partial x^2}\, U(x,\,t) = \frac{1}{4}\left(\frac{\partial^2}{\partial t^2}\, U(x,\,t)\right)$$

The Laplace transform of *pde* with respect to time t is taken in *eq*.

> `eq:=laplace(pde,t,s);`

$$eq := \frac{d^2}{dx^2}\, F(x) = \frac{1}{4}s^2\, F(x) - \frac{1}{4}\,D_2(U)(x,\,0) - \frac{1}{4}\,s\,U(x,\,0)$$

The output in *eq* is given in terms of the initial profile ($U(x,0)$) and the initial
transverse velocity ($D_2(U)(x,0)$), with D_2 indicating a time derivative). These
are supplied in the initial condition, *ic*, which is substituted into *eq* in *eq2*.

> `ic:=U(x,0)=0,D[2](U)(x,0)=3*sin(2*x)-2*sin(5*x);`

$$ic := U(x,\,0) = 0,\ D_2(U)(x,\,0) = 3\sin(2x) - 2\sin(5x)$$

> `eq2:=subs(ic,eq);`

$$eq2 := \frac{d^2}{dx^2} F(x) = \frac{1}{4} s^2 F(x) - \frac{3}{4} \sin(2x) + \frac{1}{2} \sin(5x)$$

The ODE *eq2* is analytically solved for $F(x)$, subject to the two boundary conditions, $F(0)=0$ and $F(L)=0$

```
>  sol:=dsolve({eq2,F(0)=0,F(L)=0},F(x));
```

$$sol := F(x) = -2(16\cos(x)^4 s^2 + 256\cos(x)^4 - 192\cos(x)^2 - 12\cos(x)^2 s^2$$
$$-3s^2\cos(x) - 300\cos(x) + s^2 + 16)\sin(x)/(s^4 + 116s^2 + 1600)$$

The solution U of the wave equation follows on taking the inverse Laplace transform of the right-hand side of *sol*.

```
>  U:=invlaplace(rhs(sol),s,t);
```

$$U := \frac{1}{10} \sin(x) \left(15\cos(x)\sin(4t) + 2\sin(10t)(4\cos(x)^2 - (4\cos(x)^2 - 1)^2)\right)$$

U can be simplified by applying the **combine** command with the **trig** option.

```
>  U:=combine(U,trig);
```

$$U := \frac{1}{10}\cos(5x+10t) - \frac{1}{10}\cos(-5x+10t) - \frac{3}{8}\cos(2x+4t) + \frac{3}{8}\cos(-2x+4t)$$

The motion of the string is now animated over the time interval $t=0$ to T, with 100 frames being taken and 500 plotting points used.

```
>  animate(U,x=0..L,t=0..T,frames=100,numpoints=500,
      thickness=2,labels=["x","U"]);
```

You will have to run the animation to see how the string vibrates.

6.2.4   Infinite-medium Green's Function

Private information is practically the source
of every large modern fortune.
Oscar Wilde, Anglo-Irish playwright, author, (1854–1900)

In this recipe we consider the 3-dimensional infinite-medium diffusion equation (diffusion coefficient d) with a point source of unit magnitude located at $\vec{r} = (x, y, z) = 0$ and turned on at the instant $t = 0$. The relevant PDE for the Green's function $G(\vec{r}, t)$ is of the form

$$\frac{\partial G}{\partial t} - d\nabla^2 G = \delta(\vec{r})\,\delta(t). \tag{6.10}$$

Prior to the switching on of the source $G = 0$ everywhere. Our goal is to determine $G(x, y, z, t)$ for $t > 0$ and animate the solution.

In addition to the usual integral transform and plots packages that have appeared regularly in the recipes of this chapter, the VectorCalculus package is also loaded because the **Laplacian** command will be used to enter ∇^2 in (6.10).

```
>  restart: with(inttrans): with(plots): with(VectorCalculus):
```

Eq. (6.10) is entered in Cartesian coordinates, noting that $\delta(\vec{r}) = \delta(x)\,\delta(y)\,\delta(z)$.

```
>   pde:=diff(G(x,y,z,t),t)-d*Laplacian(G(x,y,z,t),'cartesian'
    [x,y,z])=Dirac(x)*Dirac(y)*Dirac(z)*Dirac(t);
```

$$pde := (\frac{\partial}{\partial t} G(x, y, z, t)) - d((\frac{\partial^2}{\partial x^2} G(x, y, z, t)) + (\frac{\partial^2}{\partial y^2} G(x, y, z, t))$$

$$+(\frac{\partial^2}{\partial z^2} G(x, y, z, t))) = \text{Dirac}(x) \, \text{Dirac}(y) \, \text{Dirac}(z) \, \text{Dirac}(t)$$

The Green's function $G(x, y, z, t)$ is equal to 0 up to the instant that the source is switched on. The Laplace transform of the first time derivative of G will be expressed in terms of $G(x, y, z, 0)$, which is set equal to 0.

```
>   G(x,y,z,0):=0:
```

A functional operator F is formed for calculating the 1-dimensional Fourier transform of an expression e with respect to an arbitrary variable v, the result being given in terms of a second variable k. A similar functional operator K is introduced to perform the 1-dimensional inverse Fourier transform of e from k-space back to v-space.

```
>   F:=(e,v,k)->fourier(e,v,k):  K:=(e,k,v)->invfourier(e,k,v):
```

Using F, a 3-dimensional Fourier transform of *pde* is taken in the spatial coordinates x, y, and z. This is implemented by nesting 3 one-dimensional transforms, the result being expressed in terms of kx, ky, and kz. A Laplace transform in time is then performed, and the result factored.

```
>   eq1:=factor(laplace(F(F(F(pde,x,kx),y,ky),z,kz),t,s));
```

$$eq1 := \text{laplace(fourier(fourier(fourier}(G(x, y, z, t), x, kx), y, ky), z, kz), t, s)$$
$$(d\,kz^2 + s + d\,kx^2 + d\,ky^2) = 1$$

eq1 is then solved for the Laplace-Fourier transform of $G(x, y, z, t)$.

```
>   eq2:=solve(eq1,laplace(F(F(F(G(x,y,z,t),x,kx),y,ky),z,kz),t,s));
```

$$eq2 := \frac{1}{s + d\,kz^2 + d\,ky^2 + d\,kx^2}$$

Applying the inverse Laplace transform to *eq2* puts us back into *t*-space.

```
>   eq3:=invlaplace(eq2,s,t);
```

$$eq3 := e^{(-(d\,kz^2 + d\,ky^2 + d\,kx^2)\,t)}$$

Finally, by performing three 1-dimensional inverse Fourier transforms on *eq3*, the desired Green's function solution G to Equation (6.10) is obtained. It is necessary to assume that both d and t are greater than zero for the transforms to be explicitly calculated.

```
>   G:=K(K(K(eq3,kx,x),ky,y),kz,z) assuming d>0,t>0;
```

$$G := \frac{1}{8} \frac{e^{(-\frac{x^2 + y^2 + z^2}{4\,t\,d})}}{t\,d\,\pi^{(3/2)}\,\sqrt{t\,d}}$$

G is expressed in terms of the radial distance r by making the algebraic substitution $x^2 + y^2 + z^2 = r^2$.

```
>  G:=algsubs(x^2+y^2+z^2=r^2,G);
```

$$G := \frac{1}{8}\frac{e^{\left(-\frac{r^2}{4td}\right)}}{td\,\pi^{(3/2)}\,\sqrt{td}}$$

As expected for a point source, G has spherical symmetry, depending only on the radial distance. The above form of G is valid for $t > 0$, with $G=0$ for $t < 0$. Taking the diffusion coefficient $d=1$, G is animated over the radial range $r=0$ to 2.5 and time interval $t=0.1$ to 3, the opening frame of the animation being shown in Figure 6.7.

```
>  animate(eval(G,d=1),r=0..2.5,t=0.1..3,frames=100,
   numpoints=200,thickness=2,labels=["r","G"]);
```

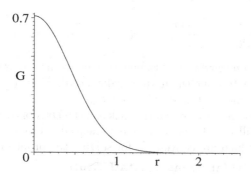

Figure 6.7: Initial frame of animation of the infinite-medium Green's function.

As discussed in Wallace [Wal84], Green's functions can be similarly derived for semi-infinite, and even finite media, with specified boundary conditions. The Green's functions can in turn be used to solve problems involving distributed time-dependent sources with the same boundary conditions. The interested reader is referred to Wallace's text.

6.2.5 Our Field of Dreams

A man is not old until regrets take the place of dreams.
John Barrymore, American actor, (1882–1942)

One summer I worked, along with other students, at a uranium mine on the Arctic circle to earn sufficient money to pay my college expenses for the next year. The students and the permanent workers would often play baseball under the midnight sun on a field which consisted of processed mine tailings. On and around this "field of dreams", the students would chat about what they planned to do in life. From this field of dreams came future engineers, chemists, and at least one physicist. This recipe is dedicated to those friends of long ago.

Radioactive gas is diffusing at a steady rate into the atmosphere from leveled mine tailings. Treating the ground and atmosphere as semi-infinite media, the X-axis is taken vertically upwards with the ground-atmosphere boundary at $X = 0$. Letting d be the diffusion constant and λ the decay rate of the radioactive gas (only one species is considered), the relevant modified diffusion equation for the gas concentration C and boundary condition (*Fick's law*) are of the form

$$\frac{\partial C(X,T)}{\partial T} = d\,\frac{\partial^2 C(X,T)}{\partial X^2} - \lambda\, C(X,T), \quad -d\,\frac{\partial C(0,T)}{\partial X} = k, \qquad (6.11)$$

with T the time and k a positive constant (units of kg/(m². s)). The various constants may be removed from the equations by setting $t = \lambda T$, $x = \sqrt{\lambda/d}\, X$, and $c(x,t) = (\sqrt{d\lambda}/k)\, C(X,T)$. Then Equation (6.11) becomes

$$\frac{\partial c(x,t)}{\partial t} = \frac{\partial^2 c(x,t)}{\partial x^2} - c(x,t), \quad \frac{\partial c(0,t)}{\partial x} = -1. \qquad (6.12)$$

Assuming that the concentration of radioactive gas in the atmosphere is initially zero, our goal is to determine the normalized concentration $c(x,t)$ in the region $x > 0$ at arbitrary time $t > 0$ and animate the result.

In addition to the integral transform package, the DEtools package is loaded. Using the latter will enable us to seek exponential solutions to the Laplace transformed PDE which are convenient for solving the problem.

```
>   restart: with(inttrans): with(DEtools):
```
The normalized diffusion equation in (6.12) is entered in *pde*.

```
>   pde:=diff(c(x,t),t)=diff(c(x,t),x,x)-c(x,t);
```

$$pde := \frac{\partial}{\partial t}\, c(x,\, t) = (\frac{\partial^2}{\partial x^2}\, c(x,\, t)) - c(x,\, t)$$

The addtable command is used to simplify the notation. The Laplace transform of $c(x,t)$ will be given as $F(x)$.

```
>   addtable(laplace,c(x,t),F(x),t,s):
```
The initial normalized concentration, $c(x,0)$, of radioactive gas in the atmosphere is set equal to zero, and the Laplace transform is applied to *pde*.

```
>   c(x,0):=0: eq:=laplace(pde,t,s);
```

$$eq := s\, F(x) = (\frac{d^2}{dx^2}\, F(x)) - F(x)$$

The expsols command is used to obtain exponential solutions for $F(x)$ in *eq*.

```
>   sol:=expsols(eq,F(x));
```

$$sol := [e^{(\sqrt{s+1}\, x)},\ e^{(-\sqrt{s+1}\, x)}]$$

As $x \to \infty$, the concentration of radioactive gas in the atmosphere must go to zero. Therefore the exponential solution with the minus sign (the second result in *sol*) is selected and multiplied by an arbitrary constant B.

```
>   sol2:=F(x)=B*sol[2];
```

$$sol2 := F(x) = B\,e^{(-\sqrt{s+1}\,x)}$$

The boundary condition at $x=0$ in (6.12) is applied as follows. The right-hand side of *sol2* is differentiated with respect to x and evaluated at $x = 0$. This result is equated to the Laplace transform of -1.

> `bc:=eval(diff(rhs(sol2),x),x=0)=laplace(-1,t,s);`

$$bc := -B\sqrt{s+1} = -\frac{1}{s}$$

The Laplace transformed boundary condition, *bc*, is solved for B and the form of the transformed concentration, $F(x)$, displayed.

> `B:=solve(bc,B); sol2;`

$$B := \frac{1}{\sqrt{s+1}\,s} \qquad F(x) = \frac{e^{(-\sqrt{s+1}\,x)}}{\sqrt{s+1}\,s}$$

To facilitate the calculation of the inverse transform, let's make a variable change, substituting $s = y - 1$ on the rhs of *sol2* and multiplying the result by e^{-t} to keep the Bromwich integral (6.9) the same.

> `F2:=exp(-t)*subs(s=y-1,rhs(sol2));`

$$F2 := \frac{e^{(-t)}\,e^{(-\sqrt{y}\,x)}}{\sqrt{y}\,(y-1)}$$

Taking the inverse Laplace transform of *F2* and assuming that $x > 0$ produces the concentration c, the answer given in terms of the *complimentary error function*, $\operatorname{erfc}(z) = 1 - \operatorname{erf}(z)$.

> `c:=invlaplace(F2,y,t) assuming x>0;`

$$c := -\frac{1}{2}\operatorname{erfc}(\frac{x+2\,t}{2\sqrt{t}})\,e^{x} + \frac{1}{2}\operatorname{erfc}(-\frac{-x+2\,t}{2\sqrt{t}})\,e^{(-x)}$$

The concentration c is now animated over the time interval $t = 0$ to 10, 100 frames being taken.

> `plots[animate](c,x=0..10,t=0..10,frames=100,thickness=2,`
>   `numpoints=200,color=blue,labels=["x","c"]);`

Execute the recipe to see how the concentration of radioactive gas in the atmosphere builds up from zero to the steady-state concentration, where the diffusion of radioactive gas molecules is balanced by their radioactive decay.

6.3   Bromwich Integral and Contour Integration

In performing inverse Laplace transforms in the previous section, we bypassed evaluating the Bromwich integral by simply using the `invlaplace` command. This approach is equivalent to consulting tables of Laplace transforms and their inverses in the pre-computer age. Occasionally, Maple may stumble, unless carefully guided, in performing the inverse Laplace transform, so it is instructive to look at a few examples of evaluating the Bromwich integral.

Recall that the inverse Laplace transform $\mathcal{L}^{-1}(F(s)) = f(t)$ is given by,

$$f(t) = \frac{1}{2\pi i} \int_{c-i\infty}^{c+i\infty} e^{st} F(s)\, ds, \quad t > 0, \text{ and } f(t) = 0, \quad t < 0. \tag{6.13}$$

The integration involved in determining $f(t)$ for $t > 0$ is to be performed along a line $s = c + iy$ in the complex s-plane, c being chosen so that $s = c$ lies to the right of all the poles of $F(s)$.

The central idea is to replace the integral in (6.13) with the closed contour integral $\frac{1}{2\pi i} \oint_\Gamma e^{st} F(s)\, ds$, and apply Cauchy's residue theorem. The basic Bromwich contour Γ is shown on the left of Figure 6.8, consisting of a vertical "leg" $A \to B$ between $s = c - iL$ and $s = c + iL$, and a circular arc $B \to C \to A$, centered on the origin O, of radius R.

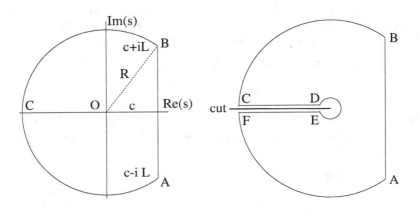

Figure 6.8: Left: Basic contour. Right: Contour modified with branch cut.

For $L \to \infty$, the integral contribution $A \to B$ will be the Bromwich integral. By Cauchy's residue theorem, assuming that the circular arc $(B \to C \to A)$ contribution approaches 0 as $R \to \infty$, the Bromwich integral will be equal to the sum of the residues of $e^{st} F(s)$ at the poles of $F(s)$ inside Γ.

A sufficient condition for $\int_{B \to C \to A} e^{st} F(s)\, ds \to 0$ as $R \to \infty$ is that $|F(s)| < \text{constant}/R^k$, with $k > 0$. This condition always holds if $F(s) = P(s)/Q(s)$, where $P(s)$ and $Q(s)$ are polynomials and the degree of P is less than that of Q. The sufficiency condition still holds if $F(s)$ has branch points.

However, then the basic Bromwich contour must be modified. For example, if $F(s)$ has a branch point at the origin, the contour is modified as shown on the right of Figure 6.8, a branch cut being introduced from $-\infty$ to 0. The contour is then given by $A \to B \to C \to D \to E \to F \to A$, the horizontal legs $C \to D$ and $E \to F$ ultimately being shrunk onto the $\text{Re}(s)$ axis and the small circular arc $D \to E$ shrunk onto the origin.

6.3.1 Spiegel's Transform Problem Revisited

When one pays a visit it is for the purpose of wasting
other people's time, not one's own.
Oscar Wilde, Anglo-Irish playwright, author, (1854–1900)

In Recipe **06-2-1**, we successfully solved a fourth order inhomogeneous ODE with specified initial conditions, suggested by Spiegel, using the `laplace` and `invlaplace` commands in the integral transform package. In this recipe, we shall solve the inverse Laplace transform of the previously obtained transform

$$F(s) = (s^5 - 2s^4 + 6s^3 - 6s^2 + 5s - 3)/(s^6 + 3s^4 + 3s^2 + 1),$$

by evaluating the Bromwich integral using contour integration. In this case, $F(s) \equiv P(s)/Q(s)$ involves the ratio of two polynomials, the largest exponent of s in P being one less than in Q. On the circular arc (radius R) of the Bromwich contour Γ, $s = R e^{i\theta}$, so $|F(s)| \sim 1/R$ for sufficiently large R. Applying the sufficiency condition, the circular arc contribution will be 0 for $R \to \infty$. So, taking the vertical leg to the right of all poles of $F(s)$, the Bromwich integral is equal to the sum of the residues of $e^{st} F(s)$ at the poles of $F(s)$ inside Γ.

$F(s)$ is now entered and the Bromwich integrand $e^{st} F(s)$ formed in f.

```
>  restart:
>  F(s):=(s^5-2*s^4+6*s^3-6*s^2+5*s-3)/(s^6+3*s^4+3*s^2+1);
```

$$F(s) := \frac{s^5 - 2s^4 + 6s^3 - 6s^2 + 5s - 3}{s^6 + 3s^4 + 3s^2 + 1}$$

```
>  f:=exp(s*t)*F(s);
```

$$f := \frac{e^{(s\,t)} (s^5 - 2s^4 + 6s^3 - 6s^2 + 5s - 3)}{s^6 + 3s^4 + 3s^2 + 1}$$

The denominator of f is factored in d, revealing that $F(s)$ has third-order poles at $s^2 + 1 = 0$, i.e., at $s = \pm i$.

```
>  d:=factor(denom(f));
```

$$d := (s^2 + 1)^3$$

Setting the denominator of f equal to 0, and solving for s yields the explicit locations ($\pm I$) of the poles, needed for evaluating the residues of $F(s)$. Each root is reproduced three times, again indicating that each pole is of third order.

```
>  pole:=solve(denom(f)=0,s);
```

$$pole := -I, I, -I, I, -I, I$$

Creating a functional operator r to evaluate the residue at an arbitrary pole,

```
>  r:=v->residue(f,s=pole[v]):
```

the residues of the two poles are determined in *r1* and *r2*.

```
>  r1:=r(1); r2:=r(2);
```

$$r1 := \frac{\frac{1}{8} I \left(8t - 4It - \frac{27}{4} + 8I - \frac{t^2}{2} + \frac{3}{4} I (2t - 16 + 5I)\right)}{e^{(t\,I)}}$$

$$r2 := \frac{1}{8} I(-8t\,e^{(t\,I)} - 4I\,t\,e^{(t\,I)} + 8I\,e^{(t\,I)} + \frac{1}{2}t^2\,e^{(t\,I)} + \frac{27}{4}\,e^{(t\,I)}$$

$$- \frac{3}{4}I\,e^{(t\,I)}\,(-2t + 16 + 5I))$$

The inverse Laplace transform, ILT, (Bromwich integral), then is equal to the sum of the two residues. Applying the complex evaluation command to the sum and simplifying yields the following result:

```
>  ILT:=simplify(evalc(r1+r2));
```

$$ILT := \cos(t) + \frac{5}{8}t\cos(t) - \frac{21}{8}\sin(t) - \frac{1}{8}t^2\sin(t) + 2t\sin(t)$$

The answer is further simplified by collecting sine and cosine terms.

```
>  ILT:=collect(%,[sin,cos]);
```

$$ILT := (2t - \frac{21}{8} - \frac{1}{8}t^2)\sin(t) + (1 + \frac{5t}{8})\cos(t)$$

As a check, the inverse Laplace transform of $F(s) = f/e^{st}$ is calculated with the invlaplace command, yielding the same analytic result.

```
>  check:=inttrans[invlaplace](f/exp(s*t),s,t);
```

$$check := -\frac{1}{8}\sin(t)\,(-16t + t^2 + 21) + \frac{1}{8}\cos(t)\,(8 + 5t)$$

6.3.2   Ms. Curious's Branch Point

Propaganda is that branch of the art of lying which consists in nearly deceiving your friends without quite deceiving your enemies.
F. M. Cornford, British author, poet, (1874–1943)

Ms. I. M. Curious has reached an important branch point in her educational career. Although, she has done very well to this point in her course work, she has to decide on whether to continue as a physics major and eventually seek an academic career, or would she better off to switch into the engineering physics program and go into industry. On seeking Jennifer's advice, Jennifer has temporarily deflected I. M.'s question by suggesting that, whatever decision I. M. makes, this week's homework is due tomorrow and no extensions will be allowed. Ironically, one of the questions on this week's quiz also deals with a branch point, albeit of the mathematical variety.

Specifically, I. M. is requested to calculate $\mathcal{L}^{-1}(e^{-a\sqrt{s}}/s)$, with $a > 0$, using contour integration. Then, employing this answer, determine $\mathcal{L}^{-1}(e^{-a\sqrt{s}}/\sqrt{s})$.

Here is the recipe that I. M. has created. She begins by entering $F(s) = e^{-a\sqrt{s}}/s$, which has a branch point at $s = 0$, and then forming the Bromwich integrand $f = e^{st}F(s)/(2\pi i)$.

```
>  restart:
>  F(s):=exp(-a*sqrt(s))/s; f:=exp(s*t)*F(s)/(2*Pi*I);
```

$$F(s) := \frac{e^{(-a\sqrt{s})}}{s} \qquad f := \frac{\frac{-1}{2} I e^{(st)} e^{(-a\sqrt{s})}}{s\pi}$$

Choosing the modified contour Γ on the right of Figure 6.8, I. M. notes that the only singular point in $F(s)$ is at $s=0$, which is outside the contour. Therefore, by Cauchy's theorem, the closed line integral $\oint_\Gamma f\,ds = 0$.

I. M. takes the radius of the outer circular arc ($B \to C$ and $F \to A$ in Figure 6.8) to be R and the radius of the inner arc ($D \to E$) to be ϵ. The limits $R \to \infty$ and $\epsilon \to 0$ are to be taken. As $R \to \infty$, the outer arc contribution $(\int_{B\to C} + \int_{F\to A}) f\,ds \to 0$, and $\int_{A\to B} f\,ds$ is just the Bromwich integral, BI. Then, $BI = -(\int_{C\to D} + \int_{D\to E} + \int_{E\to F}) f\,ds$, the three integrals to be evaluated.

To determine $IDE = \int_{D\to E} f\,ds$, I. M. sets $s = s1 = \epsilon e^{i\theta}$. To evaluate $ICD = \int_{C\to D} f\,ds$, I. M. notes that $\theta = \pi$ for this leg, so that $s = r e^{i\pi} = -r$ and $s2 \equiv \sqrt{s} = \sqrt{r} e^{i\pi/2} = i\sqrt{r}$. For $IEF = \int_{E\to F} f\,ds$, $\theta = -\pi$ so that $s = r e^{-i\pi} = -r$ still, but $s3 \equiv \sqrt{s} = \sqrt{r} e^{-i\pi/2} = -i\sqrt{r}$. I. M. now enters $s1$, $s2$, and $s3$.,

```
>  s1:=epsilon*exp(I*theta); s2:=sqrt(r)*exp(I*Pi/2);
   s3:=sqrt(r)*exp(-I*Pi/2);
```

$$s1 := \epsilon e^{(\theta I)} \qquad s2 := \sqrt{r} I \qquad s3 := -I\sqrt{r}$$

The integral IDE is given by $\lim_{\epsilon\to 0} \int_\pi^{-\pi} (f\,i\,s)|_{s=s1}\,d\theta$, which is now evaluated.

```
>  IDE:=int(limit(subs(s=s1,f*I*s),epsilon=0),theta=Pi..-Pi);
```

$$IDE := -1$$

To evaluate ICD and IEF, a functional operator G is introduced to calculate $\int_0^\infty (f)|_{\sqrt{s}=v, s=-r}\,dr$, where v is equal to $s2$ for ICD, and $s3$ for IEF. The inert form of the integral command is used here, as Maple is not able to actually perform the integrations.

```
>  G:=v->Int(subs({sqrt(s)=v,s=-r},f),r=0...infinity):
```

Using G, ICD and IEF take the following forms. For the latter, I. M. inserts a minus sign, since the integral IEF is in the opposite direction to ICD.

```
>  ICD:= G(s2); IEF:=-G(s3);
```

$$ICD := \int_0^\infty \frac{\frac{1}{2} I e^{(-rt)} e^{(-I a\sqrt{r})}}{r\pi}\,dr \qquad IEF := -\int_0^\infty \frac{\frac{1}{2} I e^{(-rt)} e^{(a\sqrt{r} I)}}{r\pi}\,dr$$

The Bromwich integral BI then is equal to $-(IDE + ICD + IEF)$. This sum is now converted to a trig form and the combine command applied.

```
>  BI:=-combine(convert(IDE+ICD+IEF,trig));
```

$$BI := -\int_0^\infty \frac{\cosh(r\,t)\sin(a\sqrt{r}) - \sinh(r\,t)\sin(a\sqrt{r})}{r\pi}\,dr + 1$$

The integrand of BI can be simplified by making the algebraic substitution $\cosh(rt) - \sinh(rt) = e^{-rt}$.

```
>  BI:=algsubs(cosh(r*t)-sinh(r*t)=exp(-r*t),BI);
```

$$BI := -\int_0^\infty \frac{\sin(a\sqrt{r})\,e^{(-rt)}}{r\,\pi}\,dr + 1$$

Applying the value command to BI yields the answer to the first part of the question, the result being expressed in terms of the error (erf) function or, on applying convert(BI,erfc), in terms of the complimentary error (erfc) function.

> BI:=value(BI); BI:=convert(BI,erfc);

$$BI := -\text{erf}(\frac{a}{2\sqrt{t}}) + 1 \qquad BI := \text{erfc}(\frac{a}{2\sqrt{t}})$$

So, $\mathcal{L}^{-1}(e^{-a\sqrt{s}}/s) = \text{erfc}(a/(2\sqrt{t}))$.

As a check, I. M. will derive exactly the same result by directly applying the inverse Laplace transform command. She mentally replaces $a\sqrt{s}$ with $\sqrt{y}$ and t with t/a^2 in the Bromwich integral, so that the function to be inverse transformed is $F2 = e^{-\sqrt{y}}/y$. Loading the integral transform package,

> F2:=exp(-sqrt(y))/y; with(inttrans):

$$F2 := \frac{e^{(-\sqrt{y})}}{y}$$

she inverse transforms $F2$ and applies the radical simplification command.

> check:=radsimp(invlaplace(F2,y,t/a^2));

$$check := \text{erfc}(\frac{a}{2\sqrt{t}})$$

The result is exactly the same as in BI.

To answer the second part of the question, I. M. notes that $e^{-a\sqrt{s}}/\sqrt{s} = -\frac{d}{da}(e^{-a\sqrt{s}}/s) = -\frac{dF(s)}{da}$. Thus, the second inverse Laplace transform, $ILT2$, follows on differentiating BI with respect to a and multiplying by -1.

> ILT2:=-diff(BI,a);

$$ILT2 := \frac{e^{(-\frac{a^2}{4t})}}{\sqrt{\pi}\,\sqrt{t}}$$

Once again I. M. confirms her answer by making the same variable transformation as above, forming the new function $F3 = e^{-\sqrt{y}}/(a\sqrt{y})$, and applying the inverse Laplace transform to $F3$ and simplifying.

> F3:=exp(-sqrt(y))/(a*sqrt(y));

$$F3 := \frac{e^{(-\sqrt{y})}}{a\sqrt{y}}$$

> check2:=radsimp(invlaplace(F3,y,t/a^2));

$$check2 := \frac{e^{(-\frac{a^2}{4t})}}{\sqrt{\pi t}}$$

The answer, of course, is the same as in $ILT2$.

6.3.3   Cooling That Weenie Rod

Write while the heat is in you. The writer who postpones the record-
ing of his thoughts uses an iron which has cooled to burn a hole with.
He cannot inflame the minds of his audience.
Henry David Thoreau, American author, philosopher, naturalist (1817–62)

Returning to our "tale" of the Northern weenie roast, having finished with
the long iron rods that they used for cooking their weenies, Russell places them
outdoors again to cool in the 0° temperature. This recipe is an idealization of
that cooling process.

   A very long circular rod of radius R, initially having a constant temperature
T_0 throughout, has a constant temperature of 0° applied to its surface for times
$t \geq 0$. Determine the temperature at an arbitrary point inside the rod after the
cooling process has begun. Animate the temperature profile inside the rod.

   If the rod were of finite length, one would use cylindrical coordinates (r, θ, z)
where r is the radial distance from the cylinder axis, θ the angle around the cir-
cumference of the cylinder, and z the distance along the cylinder axis. Treating
the rod as being infinitely long (which might have presented a "slight" problem
for the weenie roast!) removes the z-dependence from the problem. Since the
rod initially has a constant temperature throughout and the boundary condition at the surface has no angular dependence, the temperature T inside the
rod depends only on the radial distance r, i.e., $T = T(r,t)$. $T(r,t)$ then satisfies
the following heat diffusion equation (a^2 being the heat diffusion coefficient) for
$0 < r < R$ and $t > 0$,

$$\frac{\partial T}{\partial t} = a^2 \left(\frac{\partial^2 T}{\partial r^2} + \frac{1}{r} \frac{\partial T}{\partial r} \right), \quad T(R,t) = 0, \ T(r,0) = T_0. \tag{6.14}$$

The problem can be made dimensionless by setting $u \equiv T/T_0$, $x \equiv r/R$, and
$\tau \equiv a^2 t/R^2$, so that

$$\frac{\partial u}{\partial \tau} = \frac{\partial^2 u}{\partial x^2} + \frac{1}{x} \frac{\partial u}{\partial x}, \quad u(1,\tau) = 0, \ u(x,0) = 1. \tag{6.15}$$

   After loading the plots and integral transform packages, the normalized heat
flow equation given in (6.15) is entered in *pde*.

```
>  restart: with(plots): with(inttrans):
>  pde:=diff(u(x,tau),tau)=diff(u(x,tau),x,x)+diff(u(x,tau),x)/x;
```

$$pde := \frac{\partial}{\partial \tau} u(x, \tau) = \left(\frac{\partial^2}{\partial x^2} u(x, \tau) \right) + \frac{\frac{\partial}{\partial x} u(x, \tau)}{x}$$

For later notational convenience, the `addtable` command is used to replace the
Laplace transform of $u(x,\tau)$, with respect to τ, with the symbol $F(x,s)$.

```
>  addtable(laplace,u(x,tau),F(x,s),tau,s):
```

The initial condition $u(x, 0) = 1$ is entered, and *pde* Laplace transformed with respect to τ in *de*. The initial condition is automatically substituted into *de*.

> u(x,0):=1; de:=laplace(pde,tau,s);

$$u(x,\, 0) := 1$$

$$de := s\, F(x,\, s) - 1 = (\frac{\partial^2}{\partial x^2}\, F(x,\, s)) + \frac{\dfrac{\partial}{\partial x}\, F(x,\, s)}{x}$$

The spatially dependent ODE *de* is analytically solved for $F(x, s)$ in *sol*, the answer involving zeroth-order Bessel functions of the first and second kinds.

> sol:=dsolve(de,F(x,s));

$$sol := F(x,\, s) = \text{BesselJ}(0,\, \sqrt{-s}\, x)\, \_F2(s) + \text{BesselY}(0,\, \sqrt{-s}\, x)\, \_F1(s) + \frac{1}{s}$$

Since $Y_0(\sqrt{-s}\, x)$ diverges at $x = 0$ (the central axis of the rod), it is removed from the right-hand side of *sol*, $F(x, s)$ then taking the form shown in *sol2*.

> sol2:=remove(has,rhs(sol),BesselY);

$$sol2 := \text{BesselJ}(0,\, \sqrt{-s}\, x)\, \_F2(s) + \frac{1}{s}$$

From the boundary condition $u(1, \tau) = 0$, one must have $F(x = 1, s) = 0$. This transformed boundary condition (*bc*) is now applied to *sol2*.

> bc:=eval(sol2,x=1)=0;

$$bc := \text{BesselJ}(0,\, \sqrt{-s})\, \_F2(s) + \frac{1}{s} = 0$$

The boundary condition *bc* is then solved for the arbitrary function $\_F2(s)$ which, because it is assigned, is automatically substituted into *sol2*, or $F(x, s)$.

> _F2(s):=solve(bc,_F2(s)); F(x,s):=sol2;

$$\_F2(s) := -\frac{1}{\text{BesselJ}(0,\, \sqrt{-s})\, s}$$

$$F(x,\, s) := -\frac{\text{BesselJ}(0,\, \sqrt{-s}\, x)}{\text{BesselJ}(0,\, \sqrt{-s})\, s} + \frac{1}{s}$$

To determine $u(x, \tau)$, we must calculate the inverse Laplace transform of $F(x, s)$. As you may verify, using the inverse Laplace transform command in the integral transform package will not work here because, in addition to the simple pole at $s = 0$, $F(x, s)$ has an infinite number of simple poles associated with the zeros of $J_0(\sqrt{-s})$. This implies that $u(x, \tau)$ will consist of an infinite series. To determine the series form, the Bromwich integral must be evaluated.

Choosing the "basic" contour Γ on the left of Figure 6.8, all the poles lie inside Γ in the limit that the radius R of the circular arc goes to infinity. Since the circular arc contribution to the line integral vanishes in this limit, the inverse Laplace transform will be the sum of the residues of the poles inside Γ. We will now calculate the residues.

For the pole at $s = 0$, applying the **residue** command to $e^{s\tau}F(x, s)$ yields 0.

```
>  residue(exp(s*tau)*F(x,s),s=0);
```
$$0$$

Now the term $J_0(\sqrt{-s})$ in the denominator of the first term of $F(x,s)$ has simple zeros at, say, $\lambda_1, \lambda_2, \ldots$ Thus the integrand corresponding to the first term has simple poles at $s = s_n = -\lambda_n^2$, $n = 1, 2, \ldots$ The first 15 values of s_n are calculated in S using the BesselJZeros command, and S is assigned.

```
>  S:=[seq(s[n]=-(BesselJZeros(0,n))^2,n=1..15)]; assign(S):
```

$$S := [s_1 = -\text{BesselJZeros}(0, 1)^2, \; s_2 = -\text{BesselJZeros}(0, 2)^2, \; \cdots$$

Care must be taken in evaluating the residues at the zeros of $J_0(\sqrt{-s})$. The residue of the integrand at $s = s_n$ is given by

$$\lim_{s \to s_n} (s - s_n) \left(-\frac{e^{s\tau} J_0(\sqrt{-s}\,x)}{s \, J_0(\sqrt{-s})} \right) = - \left(\frac{e^{s\tau} J_0(\sqrt{-s}\,x)}{s \frac{d}{ds}\left(J_0(\sqrt{-s})\right)} \right)_{s=s_n},$$

where use has been made of L'Hospital's rule. A functional operator U is created to implement this latter form for the residue of the integrand at an arbitrary zero of J_0.

```
>  U:=sn->simplify(-eval(exp(s*tau)*BesselJ(0,sqrt(-s)*x)
          /(s*diff(BesselJ(0,sqrt(-s)),s)),s=sn));
```

$$U := sn \to \text{simplify} \left(-\frac{e^{(s\tau)} \text{BesselJ}(0, \sqrt{-s}\,x)}{s \left(\frac{d}{ds}\text{BesselJ}(0, \sqrt{-s})\right)} \bigg|_{s\,=\,sn} \right)$$

Making use of U, the formal series solution for u is given out to 15 terms.

```
>  u:=Sum(U(s[n]),n=1..15);
```

$$u := \sum_{n=1}^{15} \left(\frac{2 \, e^{(s_n \tau)} \text{BesselJ}(0, \sqrt{-s_n}\,x)}{\text{BesselJ}(1, \sqrt{-s_n}) \sqrt{-s_n}} \right)$$

The explicit series form of the normalized temperature distribution is now determined in $u2$, only the first few terms being displayed here in the text.

```
>  u2:=evalf(value(u));
```

$$u2 := 1.601974697 \, e^{(-5.783185964\,\tau)} \text{BesselJ}(0., 2.404825558\,x)$$
$$- 1.064799259 \, e^{(-30.47126234\,\tau)} \text{BesselJ}(0., 5.520078110\,x)$$
$$+ 0.8513991928 \, e^{(-74.88700679\,\tau)} \text{BesselJ}(0., 8.653727913\,x)$$
$$- 0.7296452400 \, e^{(-139.0402844\,\tau)} \text{BesselJ}(0., 11.79153444\,x) \; + \cdots$$

For $\tau > 0$, the terms quickly drop off in magnitude, but for $\tau = 0$ many terms must be kept to reduce the "ringing" associated with the initial step-function temperature profile.

The series solution $u2$ is animated over the time interval $\tau = 0.001$ to 2. The animation is started slightly past $\tau = 0$ to eliminate the worst of the ringing. The initial frame of the animation is shown in Figure 6.9.

```
>   animate(u2,x=0..1,tau=0.001..2,numpoints=100,frames=50,
    labels=["x=r/R","u=T/T0"]);
```

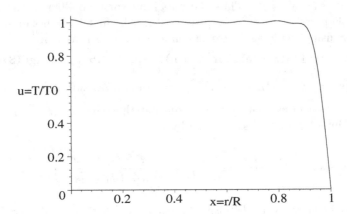

Figure 6.9: Initial frame of animated temperature distribution inside the rod.

By the time the animation reaches $\tau = 2$, the normalized temperature profile inside the rod has effectively decayed to 0.

6.4   Other Transforms

While the Fourier and Laplace transforms are the most recognized and useful of integral transforms, there exist other transforms which can be applied in particular geometries. We will conclude this chapter with an example involving the *Hankel transform* applied to a problem having cylindrical symmetry.

The Hankel transform $(F_m(k))$ of order $m > -1/2$ of a function $f(r)$, and the inverse transform, are given by

$$F_m(k) = \int_0^\infty f(r)\, J_m(kr)\, r\, dr, \qquad f(r) = \int_0^\infty F_m(k)\, J_m(kr)\, k\, dk, \qquad (6.16)$$

where J_m is the mth order Bessel function of the first kind. The reader is referred to standard mathematical physics texts for a discussion of the properties of Hankel transforms. However, it should be mentioned that there is no simple convolution theorem for the Hankel transform.

The Maple commands for performing the transform operations in equation (6.16) are hankel(f(r),r,k,m) and hankel(F$_m$(k),k,r,m), respectively. The integral transform package must be loaded to use these commands. In some cases, as the following recipe illustrates, it may be easier to make use of the basic defining relations (6.16) than Maple's Hankel transform commands.

6.4.1 Meet the Hankel Transform

Reason transformed into prejudice is the worst form of prejudice, because reason is the only instrument for liberation from prejudice. Allan Bloom, American educator, author, (1930–1992)

We wish to determine the subsequent transverse vibrations of a thin, infinite, elastic membrane which is initially at rest and has an initial displacement $\psi(r, \theta, t = 0) = A/\sqrt{1 + r^2/a^2}$. The solution $\psi(r, \theta, t > 0)$ will then be animated in three dimensions, taking $A = 1$, $a = 10$, and wave speed $c = 1$.

The plots, integral transform, and VectorCalculus packages are loaded, being required for the **animate**, **hankel**, and **Laplacian** commands, respectively.

```
> restart: with(plots): with(inttrans): with(VectorCalculus):
```

It is assumed that $c > 0$, $t > 0$, $A > 0$, $a > 0$, and $r > 0$. The initial condition for the shape is also entered.

```
> assume(c>0,t>0,A>0,a>0,r>0): ic:= A/sqrt(1+r^2/a^2);
```

$$ic := \frac{A}{\sqrt{\dfrac{r^2}{a^2} + 1}}$$

Since the initial shape has no angular (θ) dependence, the subsequent membrane displacement must also be independent of θ, i.e., $\psi = \psi(r, t)$. This also suggests that we set $m = 0$ when applying the Hankel transform and its inverse. The relevant wave equation for $\psi(r, t)$ is now entered in polar coordinates in *pde*.

```
> pde:=expand(Laplacian(psi(r,t),'polar'[r,theta]))
      -(1/c^2)*diff(psi(r,t),t,t)=0;
```

$$pde := \frac{\frac{\partial}{\partial r}\psi(r,\,t)}{r} + \left(\frac{\partial^2}{\partial r^2}\psi(r,\,t)\right) - \frac{\frac{\partial^2}{\partial t^2}\psi(r,\,t)}{c^2} = 0$$

Then *pde* is Hankel transformed with respect to the radial coordinate r, with $m = 0$, and multiplied by $-c^2$. A time-dependent ODE results in *eq*.

```
> eq:=-c^2*hankel(pde,r,k,0);
```

$$eq := \left(\frac{\partial^2}{\partial t^2}\,\text{hankel}(\psi(r,\,t),\,r,\,k,\,0)\right) + k^2\,\text{hankel}(\psi(r,\,t),\,r,\,k,\,0)\,c^2 = 0$$

To simplify the notation, the Hankel transform of $\psi(r, t)$ with respect to r with $m = 0$ is replaced with $F(t)$ in *eq*.

```
> eq2:=subs(hankel(psi(r,t),r,k,0)=F(t),eq);
```

$$eq2 := \left(\frac{d^2}{dt^2}\,F(t)\right) + k^2\,F(t)\,c^2 = 0$$

Since $\partial\psi(r, t)/\partial t = 0$ at $t = 0$, *eq2* is analytically solved using **dsolve** for $F(t)$, subject to the initial condition $dF(t)/dt = 0$ at $t = 0$.

```
> sol:=dsolve({eq2,D(F)(0)=0},F(t));
```

$$sol := F(t) = \_C2\cos(c\,k\,t)$$

At $t = 0$, one has from *eq2*, $\_C2 = F(0)$. So we need to calculate the Hankel transform (with $m=0$) of the initial membrane shape. If the `hankel` command was used, $\_C2$ would be expressed as a combination of generalized hypergeometric functions, which are difficult to reduce to a simpler form. This then would lead to further difficulty in performing the inverse Hankel transform. Instead, let's determine $\_C2$ by using the basic defining relation (6.16) and calculating $\int_0^\infty ic\, r\, J_0(kr)\, dr$. The result is then simplified.

```
>  _C2:=simplify(int(ic*r*BesselJ(0,k*r),r=0..infinity));
```

$$\_C2 := \frac{A\, e^{(-a\,k)}\, a}{k}$$

With the form of $\_C2$ automatically substituted, $F(t)$ is given by the rhs of *sol*.

```
>  F(t):=rhs(sol);
```

$$F(t) := \frac{A\, e^{(-a\,k)}\, a\, \cos(c\,k\,t)}{k}$$

Applying the inverse Hankel transform command to $F(t)$ would again generate an answer in terms of generalized hypergeometric functions which is difficult to simplify and express in real form. So, the basic defining relations will be used. First, however, it's necessary to convert $F(t)$ to an exponential form.

```
>  F2:=simplify(convert(F(t),exp));
```

$$F2 := \frac{1}{2}\, \frac{A\, a\, \left(e^{(k\,(-a+c\,t\,I))} + e^{(-k\,(a+c\,t\,I))}\right)}{k}$$

Then ψ is obtained by calculating the integral $\int_0^\infty F2\, k\, J_0(kr)\, dk$.

```
>  psi:=int(F2*k*BesselJ(0,k*r),k=0..infinity);
```

$$\psi := \frac{A\,a}{2r\sqrt{1 + \dfrac{(a - ct\,I)^2}{r^2}}} + \frac{A\,a}{2r\sqrt{1 + \dfrac{(a + ct\,I)^2}{r^2}}}$$

The displacement ψ is expressed in terms of $I \equiv \sqrt{-1}$. As the initial profile was real, ψ should be real, however, not complex. That it is real may be confirmed by applying the complex evaluation command to ψ and simplifying.

```
>  psi2:=simplify(evalc(psi));
```

$$\psi2 := Aa\sqrt{2}\sqrt{\sqrt{r^4 + 2r^2a^2 - 2r^2c^2t^2 + a^4 + 2a^2c^2t^2 + c^4t^4} + r^2 + a^2 - c^2t^2}$$
$$/\left(2\sqrt{r^4 + 2r^2a^2 - 2r^2c^2t^2 + a^4 + 2a^2c^2t^2 + c^4t^4}\right)$$

To animate the vibrations, the given parameter values are substituted into $\psi2$, which is also converted to Cartesian coordinates by setting $r = \sqrt{x^2 + y^2}$.

```
>  psi3:=subs({A=1,a=10,c=1,r=sqrt(x^2+y^2)},psi2);
```

Finally, $\psi3$ is animated with the `animate` command. A triangular grid style is used to smooth out the wave form.

```
>  animate(plot3d,[psi3,x=-100..100,y=-100..100],t=0..90,
   frames=100,axes=boxed,gridstyle=triangular,orientation
   =[25,65],tickmarks=[3,3,2],labels=["x","y","psi"]);
```

Execute the recipe to see what happens.

6.5 Supplementary Recipes

06-S01: Verifying the Convolution Theorem
Using the symmetric factor convention, verify that the convolution theorem is satisfied for the functions $f1 = A\,e^{-a^2 x^2}$ and $f2 = A\,e^{-a\,|x|}$ with $a > 0$, $A > 0$.

06-S02: Bandwidth Theorem
An approximately monochromatic plane wave packet in one dimension has the instantaneous form $u(x,0) = f(x)\,e^{ik_0 x}$, with $f(x)$ the envelope function and k_0 the central wave number. Consider:

(a) $f_1 = A\,e^{-a^2 x^2}$; (b) $f_2 = A\,e^{-a|x|}$; (c) $f3 = A$ for $|x| < a$, 0 for $|x| > a$.

For each of the $f(x)$, explicitly evaluate the root mean square deviations from the means, i.e., $\Delta x = \sqrt{<x^2> - <x>^2}$ and $\Delta k = \sqrt{<k^2> - <k>^2}$. The means $<\ >$ are defined in terms of the respective intensities $|u(x,0)|^2 = |f(x)|^2$ and $|F(k)|^2$, respectively, where $F(k)$ is the Fourier transform of $f(x)$. Show that in each case the *bandwidth theorem* (the optical analogue of the *uncertainty principle*) $\Delta x\,\Delta k \geq 1/2$ is satisfied.

06-S03: Solving an Integral Equation
Solve the following *integral equation* for $f(x)$, and plot the solution:

$$\int_0^\infty f(x)\,\cos(\alpha\,x)\,dx = \{1 - \alpha \ \text{for}\ 0 \leq \alpha \leq 1,\ 0\ \text{for}\ \alpha > 1\}.$$

06-S04: Verifying Parseval's Theorem
For $f(x) = x^6\,\sin(x)\,e^{-2x^2}$, calculate the Fourier transform $F(k) = \mathcal{F}(f(x))$ using the symmetric convention. Plot the intensities $|f(x)|^2$ and $|F(k)|^2$. Verify that Parseval's theorem is satisfied.

06-S05: Heat Diffusion in a Copper Rod
Suppose that the initial temperature distribution inside an infinitely long, thin, insulated, copper rod is $T = T0$ for $-x0 \leq x \leq x0$ and zero otherwise. By Fourier transforming the heat diffusion equation with respect to x, solving the resulting time-dependent ODE with the initial condition, and performing the inverse transform, determine the temperature distribution inside the rod for arbitrary time $t > 0$. For copper, the thermal conductivity $K = 386$ W/m·K°, the density $\rho = 8900$ kg/m^3, and the specific heat capacity $C = 390$ J/kg· K°. If $x0 = 1$ m and $T0 = 100°$, animate the temperature profile over the time interval $t = 0$ to the time it takes for the temperature at the origin to drop to $50°$. Determine the temperature at $x = 2$ m at the time the temperature at the origin has dropped to $50°$.

06-S06: Solving Another Integral Equation
Using the convolution theorem, solve the following integral equation for $y(x)$:

$$\int_{-\infty}^\infty (y(u)/((x-u)^2 + a^2))\,du = 1/(x^2 + b^2)^4, \quad 0 < a < b.$$

Verify the solution by direct substitution of $y(u)$ back into the integral equation. Taking $a = 1$ and $b = 2$, plot $y(x)$ over the range $x = -5$ to $+5$.

06-S07: Free Vibrations of an Infinite Beam

Consider a horizontal beam which is initially at rest ($\partial \psi(x,0)/\partial t = 0$) with the shape $\psi(x,0) = A e^{-b^2 x^2}$, A and b being positive constants. When released, the transverse vibrations of the beam are governed by $\psi_{xx} + (1/a^2)\psi_{tt} = 0$, with a positive and the subscripts denoting partial derivatives. By Fourier transforming the beam PDE with respect to x, solving the resulting time-dependent ODE with the initial conditions, and performing the inverse transform, determine $\psi(x, t)$ for $t > 0$. Taking $A = b = a = 1$, animate the motion of the beam.

06-S08: A Potential Problem

The potential $V(x,y)$ (in volts) in the region $x > 0$, $y > 0$ has the boundary conditions $V(0, y) = 0$ and $V(x, 0) = 1$. Using an appropriate Fourier transform approach, determine $V(x, y)$. Plot the equipotentials in 0.1 volt increments.

06-S09: Solving an ODE

Use the Laplace transform method to solve the ODE

$$y'''(t) + 8 y(t) = 32 t^3 - 16 t, \quad y(0) = y'(0) = y''(0) = 0.$$

Verify the solution by directly solving the ODE using the dsolve command with the Laplace transform option. Plot $y(t)$ over the range $t = 0$ to 1.6, and determine qualitatively and quantitatively the time at which the first minimum in the solution curve occurs.

06-S10: Impulsive Force

A unit mass attached to a spring and lying on a flat rough table experiences an impulsive force $F(t) = 5 t^3 \sin(t)$ for $t \leq \pi$ and $= 0$ for $t \geq \pi$, the ODE being

$$y''(t) + 2 y'(t) + y(t) = F(t), \quad y(0) = 0, \ y' = 1.$$

Use the Laplace transform method to determine the displacement $y(t)$ of the mass. Confirm your result with the dsolve command, using the Laplace transform option. Plot the solution over the range t=0 to 10. Determine the maximum displacement and the time at which it occurs.

06-S11: Bromwich Integral

Consider $F(s) = 1/(s^6 - 6s^5 + 9s^4 + 8s^3 - 24s^2 + 16)$. Identify the singular points of $F(s)$. Calculate $\mathcal{L}^{-1}(F(s))$ by replacing the Bromwich integral with an appropriate closed contour integral and using Cauchy's residue theorem. Verify the inverse Laplace transform by directly using the invlaplace command.

06-S12: Branch Point

Consider $F(s) = \sqrt{s}/(s - 1)$. Identify the singular points. Plot a modified Bromwich contour which can be used to evaluate $\mathcal{L}^{-1}(F(s))$. Use this contour to explicitly carry out the evaluation. Verify your answer by directly using the invlaplace command.

Chapter 7

Calculus of Variations

Although one could solve variational calculus problems with computer algebra by completely mimicking a hand calculation, in all the recipes of this chapter we will use the VariationalCalculus package to ease the work.

7.1   Euler–Lagrange Equation

Among all functions $y(x)$ with fixed points $y(x_0) = y_0$ and $y(x_1) = y_1$ at two distinct points A and B respectively, find the $y(x)$ which gives an *extremum* (minimum or maximum) or *stationary value* to the integral

$$I[y] = \int_{x_0}^{x_1} F(x, y, y') \, dx. \tag{7.1}$$

The value of the integral I depends on what functional form is chosen for $y(x)$. In a given problem, F is a known function of x, y, and $y' \equiv \frac{dy}{dx}$. The y which gives an extremum value to I is found by solving the *Euler–Lagrange equation*,

$$\frac{\partial F}{\partial y} - \frac{d}{dx} \left(\frac{\partial F}{\partial y'} \right) = 0. \tag{7.2}$$

In physical problems, it is usually evident whether y minimizes or maximizes I.

If F does not explicitly depend on x, Equation (7.2) may be written as

$$\frac{d}{dx} \left(F - y' \frac{\partial F}{\partial y'} \right) = 0, \tag{7.3}$$

so that a *first integral* of Equation (7.2) is $F - y'(\partial F/\partial y') = $ constant.

7.1.1   Betsy's In A Hurry

Work expands so as to fill the time available for its completion.
C. Northcote Parkinson, English writer, *Parkinson's Law (1958)*

One of the oldest variational examples is the *brachistochrone*[1] problem proposed by John Bernoulli in 1696 and independently solved by him and his brother James, Gottfried Leibniz, and Isaac Newton.

---

[1]brachistos $\equiv$ shortest, chronos $\equiv$ time

Consider a tiny bug (Betsy Bug) of mass m starting from rest and sliding under the influence of gravity along a smooth greased wire from some fixed point A to another fixed point B somewhere (but not directly) below A. What should the shape $y(x)$ of the wire be between A and B so that Betsy's time of descent is a minimum?

Let's take the point A to be at the origin (i.e., $x_0 = 0$, $y_0 = 0$) and measure y downwards. Neglecting friction and air resistance, Betsy's speed v when she passes through an arbitrary lower point $P(x, y)$ is obtained by equating her increase in kinetic energy to the decrease in potential energy, i.e., $\frac{1}{2} m v^2 = m g y$, or $v = \sqrt{2gy}$, where g is the acceleration due to gravity. But $v = ds/dt$ where $ds = \sqrt{(dx)^2 + (dy)^2} = \sqrt{1 + (y')^2}\, dx$ is an element of arc length along the path at P and dt is the time it takes for Betsy to traverse ds. Combining these results, the time T of descent from A to B(where $x = x_1, y = y_1$) is

$$T = \int_0^{x_1} \frac{\sqrt{1 + (y')^2}\, dx}{\sqrt{2gy}}.$$

The $y(x)$ which minimizes T follows on solving (7.2) with $F = \sqrt{1 + (y')^2}/\sqrt{2gy}$.

In this recipe, we will derive Betsy's path in the parametric form

$$x = \frac{a}{2}(\theta - \sin\theta), \qquad y = \frac{a}{2}(1 - \cos\theta),$$

with θ a parameter which is equal to 0 at A. Then, taking $\theta = \theta 1$ at B, we will show that the minimum time of descent is $T = \sqrt{a/2g}\,\theta 1$. Finally, choosing $x_1 = 5$ m, $y_1 = 2$ m, Betsy's path will be plotted and T evaluated.

If the form of F is specified, the `EulerLagrange` command will calculate the left-hand side of (7.2), and even produce the first integral if possible. To use this command, the VariationalCalculus package must be first loaded.

```
>  restart: with(VariationalCalculus):
```

The relevant F for the brachistochrone problem is entered. The factor $1/\sqrt{2g}$ is omitted since it will cancel out of the Euler–Lagrange equation, (7.2).

```
>  F:=sqrt(1+diff(y(x),x)^2)/sqrt(y(x));
```

$$F := \frac{\sqrt{1 + (\frac{d}{dx}y(x))^2}}{\sqrt{y(x)}}$$

The `EulerLagrange` command, with F, x, and $y(x)$ given as arguments, is applied to F and simplified, generating two results.

```
>  eq:=simplify(EulerLagrange(F,x,y(x)));
```

$$eq := \left\{ \frac{1}{\sqrt{1 + (\frac{d}{dx}y(x))^2}\,\sqrt{y(x)}} = K_1,\ -\frac{1}{2}\frac{1 + (\frac{d}{dx}y(x))^2 + 2\,(\frac{d^2}{dx^2}y(x))\,y(x)}{y(x)^{(3/2)}\,(1 + (\frac{d}{dx}y(x))^2)^{(3/2)}} \right\}$$

Since F doesn't explicitly depend on x here, the first integral is generated in the first expression in the above output, the integration constant being K_1. The second expression is just the left-hand side of the resulting ODE generated by the Euler–Lagrange equation. We will work with the first integral and now select it in *eq2*. The last argument, `[1]`, in the command line removes the curly (Maple set) brackets which would otherwise enclose the first integral expression.

```
>  eq2:=select(has,eq,K[1])[1];
```

$$eq2 := \cfrac{1}{\sqrt{1+(\cfrac{d}{dx}\,y(x))^2}\,\sqrt{y(x)}} = K_1$$

The `dsolve` command is used to analytically solve *eq2* for $y(x)$. The `parametric` option is specified, since a parametric form of the solution is desired.

```
>  sol:=dsolve(eq2,y(x),parametric);
```

$$sol := \left[y(\_T) = \frac{1}{K_1{}^2\,(1+\_T^2)}, \right.$$

$$\left. x(\_T) = \frac{-\_T - \arctan(\_T) - \arctan(\_T)\,\_T^2 + \_C1\,K_1{}^2 + \_C1\,K_1{}^2\,\_T^2}{K_1{}^2\,(1+\_T^2)} \right]$$

The mathematical forms of x and y are given in terms of the parameter $\_T$. The quantity $\_C1$ is a second integration constant. A new parameter θ is introduced by setting $\_T = \cos(\theta/2)/\sin(\theta/2)$. The constant K_1 is set equal to $1/\sqrt{a}$.

```
>   _T:=cos(theta/2)/sin(theta/2): K[1]:=1/sqrt(a):
```

Let's tackle x first. The right-hand side of the second expression in *sol* is symbolically simplified with the trig option, the above assignments having been automatically substituted.

```
>  x:=simplify(rhs(sol[2]),trig,symbolic);
```

$$x := -\cos(\frac{\theta}{2})\,a\sin(\frac{\theta}{2}) - \arctan\left(\frac{\cos(\frac{\theta}{2})}{\sin(\frac{\theta}{2})}\right)a + \_C1$$

Applying the `combine` command with the trig option reduces the first term in x to one of the terms desired in the final form of x.

```
>  x:=combine(x,trig);
```

$$x := -\frac{1}{2}\,a\sin(\theta) - \arctan\left(\frac{\cos(\frac{\theta}{2})}{\sin(\frac{\theta}{2})}\right)a + \_C1$$

We must choose the constant $\_C1$ in such a way that the last two terms in the above output reduce to $a\,\theta/2$. To do this, recall that the parameter θ must have the value 0 when $x=0$. For $\theta=0$, the first term in the above output is 0, while the second term yields $-a\arctan(\infty)$, or $-a\pi/2$. Thus, to make $\theta=0$ at $x=0$, we must choose $\_C1 = a\pi/2$. This substitution is now made in x. We further substitute $\cos(\theta/2) = \cot(\theta/2)\sin(\theta/2)$.

```
> x:=subs({_C1=a*Pi/2,cos(theta/2)=cot(theta/2)*sin(theta/2)},x);
```

$$x := -\frac{1}{2} a \sin(\theta) - \arctan(\cot(\frac{\theta}{2})) a + \frac{a\pi}{2}$$

Applying the combine command with the trig option, and then making the substitution $\operatorname{arccot}(\cot(\theta/2)) = \theta/2$ produces the desired final form for x.

> x:=combine(x,trig);

$$x := -\frac{1}{2} a \sin(\theta) + a \operatorname{arccot}(\cot(\frac{\theta}{2}))$$

> x:=subs(arccot(cot(theta/2))=theta/2,x);

$$x := -\frac{1}{2} a \sin(\theta) + \frac{a\theta}{2}$$

Next, we tackle y by selecting the first expression from the right-hand side of *sol* and simplifying with the trig and symbolic options.

> y:=simplify(rhs(sol[1]),trig,symbolic);

$$y := a \sin(\frac{\theta}{2})^2$$

Applying combine, with the trig option, yields the final desired form of y.

> y:=combine(y,trig);

$$y := \frac{a}{2} - \frac{1}{2} a \cos(\theta)$$

With the parametric forms of x and y determined, the minimum time of descent can be calculated. First note that the time of descent

$$T = \int_0^{x_1} \frac{\sqrt{1 + \left(\frac{dy(x)}{dx}\right)^2}}{\sqrt{2\,g\,y(x)}}\,dx = \int_0^{\theta_1} \frac{\sqrt{1 + \left(\frac{dy(\theta)/d\theta}{dx(\theta)/d\theta}\right)^2} \left(\frac{dx(\theta)}{d\theta}\right)}{\sqrt{2\,g\,y(\theta)}}\,d\theta,$$

where θ_1 is the value of θ at the end point B. The latter integrand is now calculated and simplified assuming that $\theta > 0$.

> integrand:=simplify(sqrt(1+(diff(y,theta)/diff(x,theta))^2)
             *diff(x,theta)/sqrt(2*g*y)) assuming theta>0;

$$integrand := \frac{1}{2} \frac{\sqrt{2}\,a}{\sqrt{g\,a}}$$

The minimum time of descent, assigned the name T, follows on integrating the integrand from $\theta = 0$ to θ_1.

> T:=int(integrand,theta=0..theta1);

$$T := \frac{1}{2} \frac{\sqrt{2}\,a\,\theta_1}{\sqrt{g\,a}}$$

The value of θ_1 depends on the coordinates x_1, y_1 of the point B, and must be determined numerically. This is now done for the specified values $x_1 = 5$, $y_1 = 2$.

> x1:=x=5; y1:=y=2;

$$x_1 := -\frac{1}{2} a \sin(\theta) + \frac{a\theta}{2} = 5 \qquad y_1 := \frac{a}{2} - \frac{1}{2} a \cos(\theta) = 2$$

The above pair of equations is solved numerically for the constant a and the value of θ at B, i.e., $\theta 1$.

```
>  sol2:=fsolve({x1,y1},{a,theta});
```

$$sol2 := \{\theta = 3.819665136,\ a = 2.248728127\}$$

The values of $\theta 1$ and a are now expressed separately.

```
>  theta1:=eval(theta,sol2); a:=eval(a,sol2);
```

$$\theta 1 := 3.819665136 \qquad a := 2.248728127$$

Betsy's path then is given by the following expressions for x and y. Taking $g = 9.8$ m/s^2, her time of descent is also calculated.

```
>  x:=x; y:=y; g:=9.8: T:=evalf(T);
```

$$x := -1.124364064\sin(\theta) + 1.124364064\,\theta$$

$$y := 1.124364064 - 1.124364064\cos(\theta)$$

$$T := 1.293795782$$

Along the path given by the above forms of x and y, Betsy takes about 1.29 seconds to travel from the origin (A) to B. This is the minimum time possible between these points, any other path producing a longer time. If you don't believe it, try calculating the time for descent along, e.g., a straight line path between A and B.

Finally, Betsy's path is plotted from $\theta = 0$ to $\theta 1$ with constrained scaling. The resulting picture is shown in Figure 7.1. Mathematically, the path is a portion of an inverted *cycloid*. Another example of a cycloidal path is the trajectory traced out by a point on the rim of a wheel rolling without slipping.

```
>  plot([x,-y,theta=0..theta1],scaling=constrained,
   thickness=2,labels=["x","y"]);
```

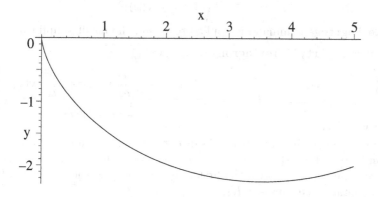

Figure 7.1: Betsy's path which minimizes the time of descent.

7.1.2   Fermat's Principle

The light that radiates from the great novels time can never dim....
Milan Kundera, Czech author, critic, (1929–)

Fermat's principle states that a ray of light in a medium with a variable refractive index n will follow the path which requires the shortest traveling time.
Consider a medium which has a refractive index $n(x,y) = e^{ay}$, with x the horizontal and y the vertical coordinate, respectively. Noting that the speed of light
is $v = c/n$, where c is the vacuum speed of light, determine the general equation
for the light ray path. Taking $a = 1$, determine the light ray path between the
points $(-1, 1)$ and $(1, 1)$ and plot it.

Now the light ray speed $v = ds/dt = c/n$, where $ds = \sqrt{(dx)^2 + (dy)^2} = \sqrt{1 + (y')^2}\,dx$ is an element of arclength and dt is a time interval. So, since
$n = e^{ay}$, the total time to travel between $x = -1$ and $+1$ is given by

$$T = \frac{1}{c} \int_{-1}^{1} e^{ay} \sqrt{1 + (y')^2}\,dx.$$

Omitting the constant factor $1/c$ which cancels out, the Euler–Lagrange equation will be solved with $F = e^{ay} \sqrt{1 + (y')^2}$. As in the previous recipe, since F
doesn't explicitly depend on x, a first integral must exist.

After loading the VariationalCalculus package, the refractive index n is entered and F formed.

```
>   restart: with(VariationalCalculus):
>   n:=exp(a*y(x)); F:=n*sqrt(1+diff(y(x),x)^2);
```

$$n := e^{(a\,y(x))}$$

$$F := e^{(a\,y(x))} \sqrt{1 + (\frac{d}{dx}\,y(x))^2}$$

The EulerLagrange command is applied to F and the result simplified.

```
>   eq:=simplify(EulerLagrange(F,x,y(x)));
```

$$eq := \left\{ \frac{e^{(a\,y(x))}}{\sqrt{1 + (\frac{d}{dx}\,y(x))^2}} = K_1,\; -\frac{e^{(a\,y(x))}\,(-a - a\,(\frac{d}{dx}\,y(x))^2 + (\frac{d^2}{dx^2}\,y(x)))}{(1 + (\frac{d}{dx}\,y(x))^2)^{(3/2)}} \right\}$$

In addition to generating the left-hand side of the Euler–Lagrange equation (second term in *eq*), the first integral has been generated (first term) as expected.
Once again, we will work with the first integral, selecting the expression from
eq which contains the constant K_1.

```
>   eq2:=select(has,eq,K[1])[1];
```

$$eq2 := \frac{e^{(a\,y(x))}}{\sqrt{1 + (\frac{d}{dx}\,y(x))^2}} = K_1$$

Then, *eq2* is analytically solved for $y(x)$ using the `dsolve` command and assuming that K_1 is real. If this assumption is not made, a much more formidable form of the solution will occur. Further, even with the assumption included, there occasionally may appear more possible forms of $y(x)$ in the output of *sol*. If this occurs, either select the solution which corresponds to the one chosen here or re-execute the work sheet to get an identical output.

> `sol:=dsolve(eq2,y(x)) assuming K[1]::real;`

$$sol := y(x) = \frac{1}{2}\frac{\ln(K_1{}^2)}{a}, \; y(x) = \frac{\ln(\frac{\sqrt{\tan(a\,(x - \_C1))^2 + 1}\,K_1}{\tan(a\,(x - \_C1))})}{a},$$

$$y(x) = \frac{\ln(-\frac{\sqrt{\tan(a\,(x - \_C1))^2 + 1}\,K_1}{\tan(a\,(x - \_C1))})}{a}$$

Noting that the second and third solutions in *sol* turn out to be equivalent, the rhs of the third solution is chosen and converted to sines and cosines.

> `y:=convert(rhs(sol[3]),sincos);`

$$y := \frac{\ln\left(-\frac{\sqrt{\frac{\sin(a\,(x - \_C1))^2}{\cos(a\,(x - \_C1))^2} + 1}\,K_1\cos(a\,(x - \_C1))}{\sin(a\,(x - \_C1))}\right)}{a}$$

The form of y is considerably simplified in *y2* by applying the radical simplification command followed by `combine` with the trig option.

> `y2:=combine(radsimp(y),trig);`

$$y2 := \frac{\ln(-\frac{K_1}{\sin(a\,x - a\,\_C1)})}{a}$$

With an analytic form for the light ray path determined, the given parameter values are now entered. Here h is the vertical coordinate of the end points of the light ray path and $x = \pm L$ will be the end points' horizontal coordinates.

> `a:=1: h:=1: L:=1:`

The light ray must pass through the points $(x = L,\ y = h)$ and $(-L, h)$. These boundary conditions are entered in *bc1* and *bc2*, respectively.

> `bc1:=eval(y2,x=L)=h; bc2:=eval(y2,x=-L)=h;`

$$bc1 := \ln(\frac{K_1}{\sin(-1 + \_C1)}) = 1 \qquad bc2 := \ln(\frac{K_1}{\sin(1 + \_C1)}) = 1$$

Then *bc1* and *bc2* are numerically solved for K_1 and $\_C1$ and the solution assigned. K_1 and $\_C1$ are automatically substituted into *y2* which is displayed.

> `sol2:=fsolve({bc1,bc2},{K[1],_C1}); assign(sol2): y2:=y2;`

$$sol2 := \{\_C1 = -7.853981634,\ K_1 = -1.468693940\}$$

$$y2 := \ln\left(\frac{1.468693940}{\sin(x + 7.853981634)}\right)$$

The solution $y2$ is now plotted with constrained scaling,

```
> plot(y2,x=-L..L,thickness=2,tickmarks=[3,2],view=
    [-L..L,0..h],scaling=constrained,labels=["x","y"]);
```

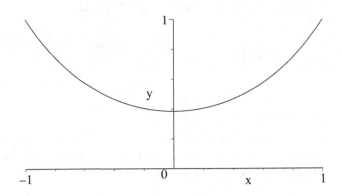

Figure 7.2: Path traced by light ray between the points $(-1, 1)$ and $(1, 1)$.

the light ray path being shown in Figure 7.2. The path joining the two points is curved rather than a straight line.

7.1.3   Betsy's Other Path

Adulthood is the ever-shrinking period between childhood and old age. It is the apparent aim of modern industrial societies to reduce this period to a minimum.
Thomas Szasz, American psychiatrist, *The Second Sin, "Social Relations" (1973)*

Suppose that we want to find the function $y(x)$ which minimizes or maximizes $I[y] = \int_{x_0}^{x_1} F(x, y, y') \, dx$, with y fixed at x_0, but with the other end point x_1 free to lie anywhere along the curve $G(x, y) = 0$. According to Mathews and Walker [MW71], $y(x)$ may still be found by solving the Euler–Lagrange equation, but subject to the subsidiary condition

$$\left(F - y'\frac{\partial F}{\partial y'}\right)\frac{\partial G}{\partial y} - \frac{\partial F}{\partial y'}\frac{\partial G}{\partial x} = 0. \tag{7.4}$$

For the brachistochrone, $F = \sqrt{(1 + (y')^2)/y}$ and the condition (7.4) reduces to $y' = (\partial G/\partial y)/(\partial G/\partial x)$. I.e., the curve of quickest descent must intersect $G(x, y) = 0$ at right angles.

As a simple illustration of applying this latter condition, suppose that Betsy bug now slides down a smooth wire from the origin (starting at rest, and with y measured downwards) to the parabola $y = m\, x^2 + c$, where the parabola passes through the two points $(x=0, y=50)$ and $(30,0)$ cm. Determine the path which minimizes Betsy's time of descent to the parabola. Taking $g=980$ cm/s^2, how long does it take Betsy to reach the parabola and through what distance does she drop? Plot Betsy's path and the parabola, using constrained scaling.

After loading the plots package, the parabolic equation G is entered.

```
>   restart: with(plots):
>   G:=Y-m*X^2-c=0;
```

$$G := Y - m\,X^2 - c = 0$$

Evaluating G at the two given points and solving yields the values of m and c.

```
>   sol:=solve({eval(G,{X=0,Y=50}),eval(G,{X=30,Y=0})},{m,c});
```

$$sol := \{c = 50,\ m = \frac{-1}{18}\}$$

Assigning sol, the explicit form of the parabola G then is as follows.

```
>   assign(sol): G:=G;
```

$$G := Y + \frac{X^2}{18} - 50 = 0$$

The quantity $(\partial G/\partial Y)/(\partial G/\partial X)$ is calculated and evaluated at the (still unknown) end point $X = x_1$.

```
>   s:=eval(lhs(diff(G,Y))/lhs(diff(G,X)),X=x1);
```

$$s := \frac{9}{x1}$$

Since F is the same as in Recipe **07-1-1**, the path of quickest descent must have the same mathematical form as obtained earlier. In parametric form, the equations describing the x and y coordinates of the path are as follows:

```
>   x:=(a/2)*(theta-sin(theta)); y:=(a/2)*(1-cos(theta));
```

$$x := \frac{1}{2}\,a\,(\theta - \sin(\theta)) \qquad y := \frac{1}{2}\,a\,(1 - \cos(\theta))$$

The slope of the path, dy/dx or $(dy/d\theta)/(dx/d\theta)$, must be equal to s at some unknown value $\theta1$ of the parameter θ. This boundary condition is now entered.

```
>   bc||1:=eval(diff(y,theta)/diff(x,theta),theta=theta1)=s;
```

$$bc1 := \frac{\sin(\theta1)}{1 - \cos(\theta1)} = \frac{9}{x1}$$

As two additional conditions, x and y evaluated at $\theta = \theta1$ must be equal to $x1$ and $y1$. Finally, in $bc4$ we evaluate G at $X = x1$ and $Y = y1$.

```
>   bc||2:=x1=eval(x,theta=theta1);
```

$$bc2 := x1 = \frac{1}{2}\,a\,(\theta1 - \sin(\theta1))$$

```
>   bc||3:=y1=eval(y,theta=theta1);
```

$$bc3 := y1 = \frac{1}{2}\,a\,(1 - \cos(\theta1))$$

```
>   bc||4:=eval(G,{X=x1,Y=y1});
```

$$bc4 := y1 + \frac{x1^2}{18} - 50 = 0$$

The 4 boundary conditions are numerically solved for $x1$, $y1$, a, and $\theta1$.

```
>  sol2:=fsolve({seq(bc||i,i=1..4)},{x1,y1,a,theta1},x1=0..30);
```

$sol2 := \{a = 26.49236117, x1 = 22.15176281, \theta1 = 2.369761726, y1 = 22.73885580\}$

Assigning $sol2$, the parametric equations describing Betsy's path are as follows:

```
>  assign(sol2): x:=x; y:=y;
```

$$x := 13.24618058\,\theta - 13.24618058\sin(\theta)$$

$$y := 13.24618058 - 13.24618058\cos(\theta)$$

Entering the value of g, the distance h through which Betsy drops and the (minimum) time T it takes to reach the parabola are calculated. h is given by the numerical value of $y1$, while for T we borrow the result $T = \sqrt{a/(2g)}\,\theta$ from recipe **07-1-1** and evaluate it at $\theta = \theta1$.

```
>  g:=980: h:=evalf(y1); T:=evalf(sqrt(a/(2*g))*theta1);
```

$$h := 22.73885580 \qquad T := 0.2755097538$$

Betsy drops through 22.7 cm and takes about 0.28 seconds to reach the parabola. In **gr1** and **gr2**, Betsy's path and the parabola are plotted, respectively.

```
>  gr1:=plot([x,-y,theta=0..theta1],color=blue,thickness=2):
>  gr2:=plot(lhs(G)-Y,X=0..30,thickness=2,linestyle=3):
```

The two graphs are superimposed with the **display** command, constrained scaling being used. The resulting picture is shown in Figure 7.3, the solid curve being Betsy's path, the dashed curve the parabola.

```
>  display({gr1,gr2},scaling=constrained,labels=["x","y"]);  ·
```

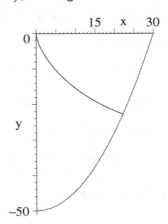

Figure 7.3: Betsy's path which minimizes the time of descent to the parabola.

Since constrained scaling has been used, it may be seen from the figure that Betsy's path does indeed intersect the parabola at right angles.

7.2 Subsidiary Conditions

In differential calculus, we may require a function $\phi(x, y)$ to be a *conditional* minimum or maximum subject to a subsidiary condition $y = y_1(x)$ or, equivalently $g(x, y) = 0$. How is a conditional extremum calculated?

A straightforward way is to replace y in $\phi(x, y)$ with $y_1(x)$ and calculate

$$\frac{d\phi(x, y_1(x))}{dx} = 0, \quad \text{or using the chain rule,} \quad \frac{\partial \phi}{\partial x} + \frac{\partial \phi}{\partial y_1} \frac{\partial y_1}{\partial x} = 0. \tag{7.5}$$

An alternate procedure is the *method of Lagrange multipliers*. One extremizes $\phi + \lambda g$, subject to $g(x, y) = 0$. λ is called the *Lagrange multiplier*. More explicitly, one must solve the simultaneous equations

$$\frac{\partial}{\partial x} (\phi + \lambda g) = 0, \quad \frac{\partial}{\partial y} (\phi + \lambda g) = 0, \quad g = 0. \tag{7.6}$$

In variational calculus, we may require an integral $I[y] = \int_{x_0}^{x_1} F(x, y, y') \, dx$ to be a minimum or maximum subject to an integral subsidiary condition $N = \int_{x_0}^{x_1} G(x, y, y') = C$, where C is a known constant. In this case, one sets $\mathcal{F} = F + \lambda G$ and solves the Euler–Lagrange equation

$$\frac{\partial \mathcal{F}}{\partial y} - \frac{d}{dx} \left(\frac{\partial \mathcal{F}}{\partial y'} \right) = 0. \tag{7.7}$$

Extensions can be made to handle more than one subsidiary condition. In the following recipes, examples from differential and variational calculus are given.

7.2.1 Ground State Energy

Authority has always attracted the lowest elements in the human race. All through history mankind has been bullied by scum...Every government is a parliament of whores. The trouble is, in a democracy the whores are us.
P. J. ORourke, American journalist, (1947–)

The quantum mechanical ground state energy of a particle of mass m in a box (rectangular parallelepiped) with sides a, b, c, is given ([Wie73]) by

$$E = \frac{\hbar^2 \pi^2}{2m} \left(\frac{1}{a^2} + \frac{1}{b^2} + \frac{1}{c^2} \right),$$

where $\hbar = h/(2\pi)$, h being Planck's constant.

What shape must the box have to minimize the energy E, subject to the constraint that the volume $V = abc$ of the box is constant? What is the minimum energy? Solve this problem by (a) the "direct" method, (b) the method of Lagrange multipliers. Here is Ms. I. M. Curious's solution.

(a) I. M. enters the energy expression E, setting $A \equiv (\hbar^2 \pi^2)/(2m)$.

```
>   restart: E:=A*(1/a^2+1/b^2+1/c^2);
```

$$E := A \left(\frac{1}{a^2} + \frac{1}{b^2} + \frac{1}{c^2} \right)$$

The volume $V = a\,b\,c$ is constant, so I. M. eliminates the variable c by setting $c = V/(ab)$. The new form of E is then displayed.

```
>  c:=V/(a*b): E:=E;
```

$$E := A\left(\frac{1}{a^2} + \frac{1}{b^2} + \frac{a^2\,b^2}{V^2}\right)$$

To determine the values of a and b which will extremize E, she sets $\partial E/\partial a = 0$ and $\partial E/\partial b = 0$ in *eq1* and *eq2*, respectively.

```
>  eq1:=diff(E,a)=0; eq2:=diff(E,b)=0;
```

$$eq1 := A\left(-\frac{2}{a^3} + \frac{2\,a\,b^2}{V^2}\right) = 0 \qquad eq2 := A\left(-\frac{2}{b^3} + \frac{2\,a^2\,b}{V^2}\right) = 0$$

For the extremum to be a minimum, I. M. notes that one must have [Ste87]

$$\frac{\partial^2 E}{\partial a^2} > 0, \quad \text{and} \quad \left(\frac{\partial^2 E}{\partial a^2}\right)\left(\frac{\partial^2 E}{\partial b^2}\right) - \left(\frac{\partial^2 E}{\partial a \partial b}\right)^2 > 0.$$

The left-hand sides of these two conditions are now entered in *eq3a* and *eq3b*.

```
>  eq3a:=diff(E,a,a);
   eq3b:=diff(E,a,a)*diff(E,b,b)-diff(E,a,b)^2;
```

$$eq3a := A\left(\frac{6}{a^4} + \frac{2\,b^2}{V^2}\right) \qquad eq3b := A^2\left(\frac{6}{a^4} + \frac{2\,b^2}{V^2}\right)\left(\frac{6}{b^4} + \frac{2\,a^2}{V^2}\right) - \frac{16\,A^2\,a^2\,b^2}{V^4}$$

Clearly $\partial^2 E/\partial a^2$ is positive, but it is not yet clear whether the second expression in the above output is also positive. I. M. now solves *eq1* and *eq2* for a and b.

```
>  sol:=solve({eq1,eq2},{a,b});
```

$$sol := \{a = \text{RootOf}(-V + \_Z^3), b = \text{RootOf}(-V + \_Z^3)\}, \;....$$

Four possible solutions for a and b are generated in *sol*, all expressed as RootOf. These four solutions are now converted to radical forms in *sol2*.

```
>  sol2:=seq(convert(sol[i],radical),i=1..4);
```

$$sol2 := \{a = V^{(1/3)}, b = V^{(1/3)}\}, \{a = -(-V)^{(1/3)}, b = (-V)^{(1/3)}\},$$
$$\{a = -V^{(1/3)}, b = V^{(1/3)}\}, \{a = (-V)^{(1/3)}, b = (-V)^{(1/3)}\}$$

Clearly, the first solution in *sol2* is the desired one, which I. M. now assigns. The values of a, b, c, and the minimum energy, *Emin*, immediately follow.

```
>  assign(sol2[1]): a:=a; b:=b; c:=c; Emin:=E;
```

$$a := V^{(1/3)} \quad b := V^{(1/3)} \quad c := V^{(1/3)} \quad Emin := \frac{3\,A}{V^{(2/3)}}$$

To minimize the energy, the box must be cubical. That the energy is a minimum follows on displaying *eq3a* and *eq3b*, which are both positive.

```
>  eq3a:=eq3a; eq3b:=eq3b;
```

$$eq3a := \frac{8\,A}{V^{(4/3)}} \qquad eq3b := \frac{48\,A^2}{V^{(8/3)}}$$

(b) Now, I. M. uses the method of Lagrange multipliers. The energy expression E is entered along with $g \equiv abc - V$. The subsidiary condition is $g = 0$.

```
>   restart: E:=A*(1/a^2+1/b^2+1/c^2); g:=a*b*c-V;
```

$$E := A\left(\frac{1}{a^2} + \frac{1}{b^2} + \frac{1}{c^2}\right) \qquad g := abc - V$$

I. M. forms a functional operator f to differentiate $E + \lambda g$ with respect to an arbitrary variable v, and then set the result equal to zero.

```
>   f:=v->diff(E+lambda*g,v)=0:
```

Then f is used, with v equal to a, b, and c in *eq1*, *eq2*, and *eq3*, respectively. The relation $g=0$ is given in *eq4*.

```
>   eq||1:=f(a); eq||2:=f(b); eq||3:=f(c); eq||4:=g=0;
```

$$eq1 := -\frac{2A}{a^3} + \lambda bc = 0 \qquad eq2 := -\frac{2A}{b^3} + \lambda ac = 0$$

$$eq3 := -\frac{2A}{c^3} + \lambda ab = 0 \qquad eq4 := abc - V = 0$$

The sequence of four equations is solved for a, b, c, and λ, and each of the solutions converted from RootOf forms to radical notation.

```
>   sol:=solve({seq(eq||i,i=1..4)},{a,b,c,lambda}):
```

```
>   sol2:=seq(convert(sol[i],radical),i=1..4);
```

$$sol2 := \{b = V^{(1/3)}, \; c = V^{(1/3)}, \; a = V^{(1/3)}, \; \lambda = \frac{2A}{V^{(5/3)}}\}, \; ....$$

The first solution in *sol2* is assigned and a, b, c, and *Emin* displayed.

```
>   assign(sol2[1]): a:=a; b:=b; c:=c; Emin:=E;
```

$$a := V^{(1/3)} \quad b := V^{(1/3)} \quad c := V^{(1/3)} \quad Emin := \frac{3A}{V^{(2/3)}}$$

As I. M. expected, the answer agrees with that obtained with the direct method.

7.2.2 Erehwon Hydro Line

The fundamental concept in social science is Power, in the same sense in which Energy is the fundamental concept in physics.
Bertrand Russell, British philosopher, mathematician, (1872–1970)

Erehwon Hydro has suspended a power line, of length $L = 1.5$ km and linear mass density $\epsilon = 1000$ kg/km, across a deep, wide, gorge. Relative to the origin at the gorge bottom, the two towers from which the cable is suspended are at $(-a/2, b)$ and $(a/2, b)$, where $a = 1.25$ km and $b = 1$ km. Assuming that the equilibrium shape of the cable is such as to minimize the potential energy, determine the cable shape and plot it. Take the gravitational acceleration $g = 9.8/1000$ km/s^2. What is the distance between the lowest point in the cable and the gorge bottom?

If an arclength element $ds = \sqrt{1 + (y')^2}\, dx$ of cable at a point x is a distance $y(x)$ above the gorge bottom, the potential energy of the whole cable is

$$V = \int \epsilon \, ds \, g \, y = \int_{-a/2}^{a/2} \epsilon \, g \, y \, \sqrt{1 + (y')^2} \, dx.$$

After loading the VariationalCalculus package,

> `restart: with(VariationalCalculus):`

the integrand $F = \epsilon g \, y \sqrt{1 + (y')^2}$ of the potential energy V is entered.

> `F:=epsilon*g*y(x)*sqrt(1+diff(y(x),x)^2);`

$$F := \varepsilon \, g \, y(x) \sqrt{1 + (\frac{d}{dx} y(x))^2}$$

The subsidiary condition is that the length of the cable must be L, i.e.,

$$\int ds = \int_{-a/2}^{a/2} \sqrt{1 + (y')^2} \, dx = L.$$

The integrand $G = \sqrt{1 + (y')^2}$ of this constraint relation is entered,

> `G:=sqrt(1+diff(y(x),x)^2);`

$$G := \sqrt{1 + (\frac{d}{dx} y(x))^2}$$

and the combination $FF = F + \lambda G$ formed, where λ is the Lagrange multiplier.

> `FF:=simplify(F+lambda*G);`

$$FF := \sqrt{1 + (\frac{d}{dx} y(x))^2} \, (\varepsilon \, g \, y(x) + \lambda)$$

The `EulerLagrange` command is applied to FF and simplified. Since FF doesn't explicitly depend on x, in addition to the lhs of the Euler–Lagrange equation, a first integral is generated, the integration constant being K_1.

> `eq:=simplify(EulerLagrange(FF,x,y(x)));`

$$eq := \left\{ \frac{\varepsilon \, g \, y(x) + \lambda}{\sqrt{1 + (\frac{d}{dx} y(x))^2}} = K_1, \right.$$

$$\left. \frac{\varepsilon \, g + \varepsilon \, g \, (\frac{d}{dx} y(x))^2 - (\frac{d^2}{dx^2} y(x)) \, \varepsilon \, g \, y(x) - (\frac{d^2}{dx^2} y(x)) \, \lambda}{(1 + (\frac{d}{dx} y(x))^2)^{(3/2)}} \right\}$$

The `select` command is used to extract the first integral relation.

> `eq2:=select(has,eq,K[1])[1];`

$$eq2 := \frac{\varepsilon \, g \, y(x) + \lambda}{\sqrt{1 + (\frac{d}{dx} y(x))^2}} = K_1$$

The ODE in *eq2* is analytically solved for $y(x)$, assuming that $\epsilon > 0$, $g > 0$.

> `sol:=dsolve(eq2,y(x)) assuming epsilon>0,g>0;`

Two equivalent solutions (not shown here) are produced in *sol*, the rhs of the second solution being chosen, simplified, and assigned the name Y.

```
> Y:=simplify(rhs(sol[2]));
```

$$Y := \frac{1}{2} \frac{(K_1{}^2 + e^{\left(-\frac{2\varepsilon g(x - \_C1)}{K_1}\right)} - 2 e^{\left(-\frac{\varepsilon g(x - \_C1)}{K_1}\right)} \lambda) e^{\left(\frac{\varepsilon g(x - \_C1)}{K_1}\right)}}{\varepsilon g}$$

The solution Y is converted to a trig form, and the **combine** command applied.

```
> Y2:=combine(convert(Y,trig));
```

$$Y2 := \frac{1}{2}(K_1{}^2 \cosh(\frac{\varepsilon g(x - \_C1)}{K_1}) + K_1{}^2 \sinh(\frac{\varepsilon g(x - \_C1)}{K_1})$$
$$+ \cosh(\frac{\varepsilon g(x - \_C1)}{K_1}) - \sinh(\frac{\varepsilon g(x - \_C1)}{K_1}) - 2\lambda) / (\varepsilon g)$$

The cosh and sinh terms are collected in *Y2*.

```
> Y3:=collect(Y2,[cosh,sinh]);
```

$$Y3 := \frac{1}{2} \frac{(1 + K_1{}^2) \cosh(\frac{\varepsilon g(x - \_C1)}{K_1})}{\varepsilon g} + \frac{1}{2} \frac{(K_1{}^2 - 1) \sinh(\frac{\varepsilon g(x - \_C1)}{K_1})}{\varepsilon g} - \frac{\lambda}{\varepsilon g}$$

To evaluate the constants K_1, $\_C1$, and λ, three conditions are needed. Since the cable will hang symmetrically, the slope at the center of the cable $(x = 0)$ must be zero. This is entered as boundary condition *bc1*. At the end point $x = a/2$, the vertical coordinate is b. This is entered in *bc2*.

```
> bc1:=eval(diff(Y3,x),x=0)=0: bc2:=eval(Y3,x=a/2)=b:
```

The two boundary conditions are solved for λ and $\_C1$.

```
> sol2:=solve({bc1,bc2},{lambda,_C1}): assign(sol2):
```

After assigning *sol2*, the cable shape is given by *Y4*.

```
> Y4:=expand(Y3);
```

$$Y4 := \frac{\cosh(\frac{\varepsilon g x}{K_1}) K_1}{\varepsilon g} - \frac{\cosh(\frac{1}{2} \frac{a \varepsilon g}{K_1}) K_1}{\varepsilon g} + b$$

The constant K_1 remains to be evaluated. The constraint $\int_{-a/2}^{a/2} G\, dx = L$ is applied in *bc3*. To perform the integration, it is assumed that $a > 0$, $\epsilon > 0$, $g > 0$, and $K_1 > 0$.

```
> bc3:=simplify(int(eval(G,y(x)=Y4),x=-a/2..a/2))=L
        assuming a>0,epsilon>0,g>0,K[1]>0;
```

$$bc3 := \frac{e^{(-1/2 \frac{a\varepsilon g}{K_1})} K_1 (-1 + e^{(\frac{a\varepsilon g}{K_1})})}{\varepsilon g} = L$$

bc3 is a transcendental equation, which must be solved numerically for K_1. The given parameter values are entered.

```
> epsilon:=1000: L:=1.5: a:=1.25: b:=1: g:=9.8/1000:
```

Then *bc3* is numerically solved for K_1, the value being labeled *K1*. Evaluating *Y4* at $K_1 = K1$, the equilibrium shape of the cable is given by *Y5*.

```
> K1:=fsolve(bc3,K[1]=0..10); Y5:=eval(Y4,K[1]=K1);
```

$$K1 := 5.751883654$$

$$Y5 := 0.5869269033 \cosh(1.703789678\,x) + 0.0476433492$$

The equilibrium shape *Y5*, called a *catenary*, is plotted with constrained scaling, the resulting picture being shown in Figure 7.4.

```
> plot(Y5,x=-a/2..a/2,thickness=2,view=[-a/2..a/2,0..b],
  scaling=constrained,tickmarks=[3,3],labels=["x","y"]);
```

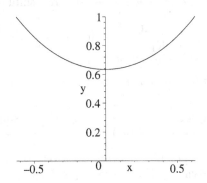

Figure 7.4: Equilibrium shape of the power line.

Evaluating *Y5* at $x = 0$, the lowest point on the cable

```
> height:=eval(Y5,x=0);
```

$$height := 0.6345702525$$

is about 0.635 km, or 635 m, above the gorge bottom.

7.3 Lagrange's Equations

In classical mechanics, the *Lagrangian L* is defined by $L = T - V$, where T is the kinetic energy and V the potential energy of the system of interest. If $L = L(q_i,\ \dot{q}_i,\ t)$ where q_i are the *generalized coordinates* (any set of coordinates completely specifying the state of the system), $\dot{q}_i$ the (generalized) velocity components, and t the time, then *Hamilton's principle* states that of all possible motions the actual motion of the system over a time interval t_0 to t_1 is the one for which $\int_{t_0}^{t_1} L(q_i,\ \dot{q}_i,\ t)\,dt$ is an extremum. Setting the variation of the integral equal to zero leads to *Lagrange's equations of motion*

$$\frac{\partial L}{\partial q_i} - \frac{d}{dt}\left(\frac{\partial L}{\partial \dot{q}_i}\right) = 0, \tag{7.8}$$

which are equivalent to Newton's law of motion. Since (7.8) is of the same structure as Eq. (7.2), we can also make use of the EulerLagrange command.

7.3.1 Daniel's Chaotic Pendulum

Out of chaos God made a world,
and out of high passions comes a people.
Lord Byron, English poet, (1788–1824)

Daniel's engineering father has given him a toy for his birthday which basically consists of a simple pendulum (a small mass m attached to the end of a light rod of length b), whose pivot point is attached to the rim of a wheel of radius a which rotates at a constant angular velocity ω. The pendulum is in the same vertical plane as the wheel and is free to rotate completely about its pivot point. All frictional effects are neglected. This recipe will determine the equation of motion for m, numerically solve the resulting nonlinear ODE, and animate the motion. The plots and VariationalCalculus packages are loaded.

```
>  restart: with(plots): with(VariationalCalculus):
```

If $\theta(t)$ is the angle that the pendulum rod makes with the vertical, the horizontal coordinate of m at time t is $x = a\cos(\omega t) + b\sin(\theta(t))$ which is entered.

```
>  x:=a*cos(omega*t)+b*sin(theta(t));
```

$$x := a\cos(\omega t) + b\sin(\theta(t))$$

The vertical coordinate of m at time t is $y = a\sin(\omega t) - b\cos(\theta(t))$.

```
>  y:=a*sin(omega*t)-b*cos(theta(t));
```

$$y := a\sin(\omega t) - b\cos(\theta(t))$$

The horizontal ($\dot{x}$) and vertical ($\dot{y}$) components of m's velocity are calculated.

```
>  xdot:=diff(x,t); ydot:=diff(y,t);
```

$$xdot := -a\sin(\omega t)\,\omega + b\cos(\theta(t))\,(\frac{d}{dt}\,\theta(t))$$

$$ydot := a\cos(\omega t)\,\omega + b\sin(\theta(t))\,(\frac{d}{dt}\,\theta(t))$$

The kinetic ($T = \frac{1}{2}m\,(\dot{x}^2 + \dot{y}^2)$) and potential ($V = m\,g\,y$) energy are determined.

```
>  T:=simplify(m*(xdot^2+ydot^2)/2); V:=m*g*y;
```

$$T := \frac{1}{2}m(a^2\,\omega^2 - 2\,a\sin(\omega t)\,\omega\,b\cos(\theta(t))\,(\frac{d}{dt}\,\theta(t))$$
$$+ 2\,a\cos(\omega t)\,\omega\,b\sin(\theta(t))\,(\frac{d}{dt}\,\theta(t)) + b^2\,(\frac{d}{dt}\,\theta(t))^2)$$

$$V := m\,g\,(a\sin(\omega t) - b\cos(\theta(t)))$$

The Lagrangian $L = T - V$ is formed, T being simplified by applying the `combine` command with the trig option.

```
>  L:=combine(T,trig)-V;
```

$$L := \frac{m\,a^2\,\omega^2}{2} - m\,a\,\omega\,b\,(\frac{d}{dt}\,\theta(t))\sin(\omega t - \theta(t)) + \frac{1}{2}\,m\,b^2\,(\frac{d}{dt}\,\theta(t))^2$$
$$- m\,g\,(a\sin(\omega t) - b\cos(\theta(t)))$$

The governing equation of motion, *ode*, is obtained by applying the `EulerLagrange` command as follows. Because L depends explicitly on t, a first integral is not generated with the `EulerLagrange` command.

```
> ode:=simplify(-EulerLagrange(L,t,theta(t))[1]/(m*b))=0;
```

$$ode := g\sin(\theta(t)) - a\,\omega^2\cos(\omega\,t - \theta(t)) + b\left(\frac{d^2}{dt^2}\,\theta(t)\right) = 0$$

Since *ode* is a nonlinear ODE, it must be solved numerically. We will consider a giant version of Daniel's toy taking $a=1$ m and $b=3$ m. The wheel is rotated at $\omega=2$ rads/s and $g=9.8$ m/s^2. The ODE is then displayed in *ode2*.

```
> a:=1: b:=3: omega:=2: g:=9.8: ode2:=ode;
```

$$ode2 := 9.8\sin(\theta(t)) - 4\cos(2\,t - \theta(t)) + 3\left(\frac{d^2}{dt^2}\,\theta(t)\right) = 0$$

For the initial condition, let's take $\theta(0) = \pi/6$ rads and $\dot{\theta}(0) = 0$.

```
> ic:=theta(0)=Pi/6,D(theta)(0)=0:
```

Then, *ode2* is numerically solved in `sol`, subject to the initial condition, for $\theta(t)$. The output is given as a listprocedure.

```
> sol:=dsolve({ode2,ic},theta(t),type=numeric,
        output=listprocedure):
```

Making use of *sol*, the next three command lines allow us to evaluate the horizontal and vertical coordinates of m at an arbitrary time t.

```
> Theta:=eval(theta(t),sol):
> X:=t->eval(x,theta(t)=Theta(t)):
> Y:=t->eval(y,theta(t)=Theta(t)):
```

The `circle` command, found in the plottools library package, is used to plot a thick red circle of radius a, centered on the origin, representing the wheel's rim.

```
> c:=plottools[circle]([0,0],a,color=red,style=line,
        thickness=2):
```

The `line` command (also in the plottools package) is used to create an arrow operator `l` to draw the pendulum rod as a thick green line at arbitrary time t.

```
> l:=t->plottools[line]([a*cos(omega*t),a*sin(omega*t)],
        [X(t),Y(t)],color=green,thickness=2):
```

The `pointplot` command is used to form an operator `p` to draw size 16 blue circles at the pivot point and at m at arbitrary time t.

```
> p:=t->pointplot({[a*cos(omega*t),a*sin(omega*t)],
        [X(t),Y(t)]},symbol=circle,symbolsize=16,color=blue):
```

The three graphs are superimposed in `gr||i` at a time $t=0.1\,i$, and a do loop used to generate plots for i from 0 to 200, i.e, for $t=0$ to 20 s.

```
> for i from 0 to 200 do
> t:=0.1*i:
> gr||i:=display({c,p(t),l(t)},labels=["x","y"]):
> end do:
```

The motion of the pendulum is animated by displaying the sequence of graphs, using the `insequence=true` option.

```
>  display(seq(gr||i,i=0..200),insequence=true);
```

A typical (at $t=0.1$ s) frame of the animation is shown in Figure 7.5. Click on the computer plot and on the start arrow to see the entire pendulum motion.

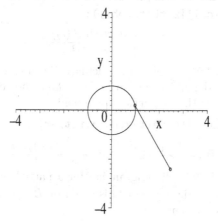

Figure 7.5: Typical frame of the animation of Daniel's chaotic pendulum.

For the choice of parameter values, the motion is quite irregular or "chaotic" in appearance. You should try other values and see what happens.

7.3.2   Van Allen Belts

He could not see a belt without hitting below it.
Margot Asquith, British socialite about a former prime minister, (1864–1945)

As an interesting and non-trivial example, in this recipe we will determine and animate the motion of a proton moving in the magnetic dipole field of the earth. The animated motion is characteristic of the behavior of charged particles trapped in the Van Allen radiation belts surrounding the earth. [Bra68]

This recipe uses the plots, VariationalCalculus, and VectorCalculus packages which will generate several warnings unless suppressed, e.g., by using the `interface` command to set the warnlevel to 0.

```
>  restart: interface(warnlevel=0):
```

```
>  with(plots): with(VariationalCalculus): with(VectorCalculus):
```

The magnetic field of the earth will be treated as a pure magnetic dipole oriented along the z-axis and spherical polar coordinates (r, θ, ϕ) used. The velocity vector $\vec{v}$ of the proton at arbitrary time t is entered.

```
>  v:=VectorField(<diff(r(t),t),r(t)*diff(theta(t),t),r(t)*
        sin(theta(t))*diff(phi(t),t)>,'spherical'[r,theta,phi]);
```

$$v := (\frac{d}{dt}\, \mathrm{r}(t))\, \bar{e}_r + \mathrm{r}(t)\, (\frac{d}{dt}\, \theta(t))\, \bar{e}_\theta + \mathrm{r}(t)\sin(\theta(t))\, (\frac{d}{dt}\, \phi(t))\, \bar{e}_\phi$$

From Griffiths [Gri99], the spherical polar components of the vector poten-
tial $\vec{A}$ at time t for a pure magnetic dipole are $A_r = 0$, $A_\theta = 0$, and $A_\phi = (\mu_0/4\pi)\, m\, \sin(\theta(t))/r^2(t)$, where μ_0 is the permeability of free space and m the
magnetic dipole moment (of the earth). The vector field $\vec{A}$ is now entered.

```
>  A:=VectorField(<0,0,(mu[0]/(4*Pi))*m*sin(theta(t))/r(t)^2>,
      'spherical'[r,theta,phi]);
```

$$A := \frac{1}{4}\, \frac{\mu_0\, m\, \sin(\theta(t))}{\pi\, r(t)^2}\, \overline{e}_\phi$$

As a check on the form of $\vec{A}$, let's calculate the magnetic field $\vec{B} = \nabla \times \vec{A}$.
Removing the time dependence from $\vec{A}$ and taking the curl yields the standard
expression [Gri99] for the magnetic dipole field.

```
>  B:=Curl(subs({theta(t)=theta,r(t)=r},A));
```

$$B := \frac{1}{2}\, \frac{\mu_0\, m\, \cos(\theta)}{r^3\, \pi}\, \overline{e}_r + \frac{1}{4}\, \frac{\sin(\theta)\, \mu_0\, m}{r^3\, \pi}\, \overline{e}_\theta$$

From Goldstein [GPS02], the Lagrangian for a particle of mass M and charge
q moving in the magnetic dipole field is given by $L=(1/2)\, M\, (\vec{v} \cdot \vec{v}) + q\, (\vec{A} \cdot \vec{v})$.
This Lagrangian is now entered.

```
>   L:=(1/2)*M*(v . v)+q*(A . v);
```

$$L := \frac{1}{2}\, M\, ((\frac{d}{dt}\, r(t))^2 + r(t)^2\, (\frac{d}{dt}\, \theta(t))^2 + r(t)^2\, \sin(\theta(t))^2\, (\frac{d}{dt}\, \phi(t))^2)$$

$$+ \frac{1}{4}\, \frac{q\, \mu_0\, m\, \sin(\theta(t))^2\, (\frac{d}{dt}\, \phi(t))}{\pi\, r(t)}$$

Applying the `EulerLagrange` command to L, and specifying the time-dependent
spherical polar coordinates, yields three equations of motion for r, θ, and ϕ, as
well as two constants of the motion. The lengthy output is suppressed here.

```
>  EL:=EulerLagrange(L,t,[r(t),theta(t),phi(t)]);
```

What do the constants of the motion tell us? First, let's select the expression
in EL containing the constant K_1.

```
>  eq1:=select(has,EL,K[1])[1];
```

$$eq1 := M\, r(t)^2\, \sin(\theta(t))^2\, (\frac{d}{dt}\, \phi(t)) + \frac{1}{4}\, \frac{q\, \mu_0\, m\, \sin(\theta(t))^2}{\pi\, r(t)} = K_1$$

The result $eq1$ can be used to determine the range of motion of the proton. For
given initial conditions, there are "forbidden" regions of space which the particle
cannot reach. The particle is "trapped" in an allowed region. This topic will
not be pursued here. The interested reader is referred to Bradbury. [Bra68]

Now, the expression in EL containing the constant K_2 is selected, simplified,
and simplified further in $eq2b$ with the substitution $\cos^2(\theta(t))=1 - \sin^2(\theta(t))$.

```
>  eq2:=simplify(select(has,EL,K[2])[1]);

>  eq2b:=algsubs(cos(theta(t))^2=1-sin(theta(t))^2,eq2);
```

$$eq2b := -\frac{1}{2} M ((\frac{d}{dt} \text{r}(t))^2 + \text{r}(t)^2 (\frac{d}{dt} \theta(t))^2 + \text{r}(t)^2 \sin(\theta(t))^2 (\frac{d}{dt} \phi(t))^2) = K_2$$

eq2b tells us that the speed of the particle remains constant with time. This is hardly surprising, as it is well-known that a magnetic field can do no work on a charged particle. Now, the motion of the proton in the magnetic dipole field must be determined by solving the equations of motion numerically. We remove the expressions involving the constants K_1 and K_2 from *EL*,

> eq3:=remove(has,EL,[K[1],K[2]]):

and write out the relevant equations in *eq4*, *eq5*, *eq6*. Only *eq4* is shown here.

> eq4:=expand(eq3[1]/M)=0; eq5:=expand(eq3[2]/M)=0;
  eq6:=expand(eq3[3]/M)=0;

$$eq4 := -2\,\text{r}(t) \sin(\theta(t))^2 (\frac{d}{dt} \phi(t)) (\frac{d}{dt} \text{r}(t))$$

$$- 2\,\text{r}(t)^2 \sin(\theta(t)) (\frac{d}{dt} \phi(t)) \cos(\theta(t)) (\frac{d}{dt} \theta(t)) - \text{r}(t)^2 \sin(\theta(t))^2 (\frac{d^2}{dt^2} \phi(t))$$

$$- \frac{1}{2} \frac{q\,\mu_0\,m\,\sin(\theta(t)) \cos(\theta(t)) (\frac{d}{dt} \theta(t))}{M\,\pi\,\text{r}(t)} + \frac{1}{4} \frac{q\,\mu_0\,m\,\sin(\theta(t))^2 (\frac{d}{dt} \text{r}(t))}{M\,\pi\,\text{r}(t)^2} = 0$$

To solve the three coupled nonlinear ODEs in *eq4*, *eq5*, and *eq6*, we enter the numerical values of the proton rest mass (*M0* in kilograms), the proton charge (*q* in Coulombs), the earth's magnetic dipole moment (*m* in Joules/Tesla), the permeability (μ_0) of free space, the mean radius of the earth (*RE* in meters), and the vacuum speed of light (*c* in meters/second).

> MO:=1.67*10^(-27); q:=1.60*10^(-19); m:=7.94*10^(22);
  mu[0]:=4*Pi*10^(-7); RE:=6.37*10^6; c:=3*10^8;

$$MO := 0.1670000000\,10^{-26} \quad q := 0.1600000000\,10^{-18} \quad m := 0.7940000000\,10^{23}$$

$$\mu_0 := \frac{\pi}{2500000} \quad RE := 0.637000000\,10^7 \quad c := 300000000$$

At time $t = 0$, let's take $r(0) = 2\,RE$ (radial distance twice the earth's radius), $\theta(0) = \pi/2$ rads, $\phi(0) = 0$, $\dot{r}(0) = 0$, $\dot{\phi}(0) = 0$, and $\dot{\theta}(0) = 19$ rads/sec.

> ic:=r(0)=2*RE,theta(0)=evalf(Pi/2),phi(0)=0,
  D(r)(0)=0,D(phi)(0)=0,D(theta)(0)=19;

$$ic := \text{r}(0) = 0.1274000000\,10^8, \ \theta(0) = 1.570796327, \ \phi(0) = 0,$$
$$\text{D}(r)(0) = 0, \ \text{D}(\phi)(0) = 0, \ \text{D}(\theta)(0) = 19$$

The mass of the proton is given by $M = M0/\sqrt{1 - \beta^2}$, with $\beta = v/c$. In the next two command lines, β and M are calculated.

> beta:=evalf(subs(ic,sqrt(D(r)(0)^2+(r(0)*D(theta)(0))^2
  +(r(0)*sin(theta(0))*D(phi)(0))^2)/c));

$$\beta := 0.8068666667$$

> M:=MO/sqrt(1-beta^2);

$$M := 0.2826993436\,10^{-26}$$

In this simulation, the proton is traveling at about 8/10 the (vacuum) speed of light, and its mass is about 5/3 times its rest mass. *eq4*, *eq5*, and *eq6* are expressed with the numerical values substituted, their outputs suppressed here.

> ```
eq4:=evalf(eq4); eq5:=evalf(eq5); eq6:=evalf(eq6);
```

The three ODES are numerically solved in `sol` subject to the initial condition,

> ```
sol:=dsolve({eq4,eq5,eq6,ic},{r(t),theta(t),phi(t)},
      type=numeric,output=listprocedure):
```

and $r(t)$, $\theta(t)$, $\phi(t)$ evaluated at arbitrary time t in R, Theta, and Phi.

> ```
R:=eval(r(t),sol): Theta:=eval(theta(t),sol):
Phi:=eval(phi(t),sol):
```

For plotting purposes, three functional operators are now introduced to convert from spherical polar coordinates to Cartesian coordinates at arbitrary time $t$.

> ```
x:=t->R(t)*sin(Theta(t))*cos(Phi(t)):
```

> ```
y:=t->R(t)*sin(Theta(t))*sin(Phi(t)):
```

> ```
z:=t->R(t)*cos(Theta(t)):
```

The total time (tt) for the animation is taken to be 3.5 seconds, $N = 100$ frames will be used, and the time step will be tt/N.

> ```
tt:=3.5: N:=100: step:=tt/N:
```

In `gr1`, the `spacecurve` command is used to plot the proton's trajectory from $t = 0$ to $t = tt$. The coordinates $x(t)$, $y(t)$, and $z(t)$ are normalized by dividing by the radius $RE$ of the earth. The trajectory is colored with the `zhue` option.

> ```
gr1:=spacecurve([x(t)/RE,y(t)/RE,z(t)/RE],t=0..tt,
      numpoints=2000,shading=zhue):
```

In `gr2`, the `plot3d` command is used in spherical coordinates to plot a bluish colored sphere of radius 1 to represent the earth.

> ```
gr2:=plot3d(1,theta=0..Pi,phi=0..2*Pi,coords=spherical,
      color=COLOR(RGB,0.1,0.5,0.8),style=patchnogrid):
```

In the following do loop, the proton's position is plotted at each time step and superimposed on the graphs `gr1` and `gr2`.

> ```
for n from 0 to N do
```

> ```
t:=step*n;
```

> ```
gr3||n:=pointplot3d([x(t)/RE,y(t)/RE,z(t)/RE],style=point,
        symbol=circle,symbolsize=16,color=red);
```

> ```
pl||n:=display({gr1,gr2,gr3||n});
```

> ```
end do:
```

Using the `display` command with the `insequence=true` option allows the proton's motion to be animated.

> ```
display(seq(pl||n,n=0..N),insequence=true,
    scaling=constrained,axes=framed,labels=["x","y","z"],
    orientation=[45,55],tickmarks=[3,3,3]);
```

On executing the last command line and clicking on the computer plot and on
```

the start arrow, the motion of the proton may be viewed. The opening frame
of the animation is shown in Figure 7.6, the proton represented by the small
circle at the bottom left of the figure.

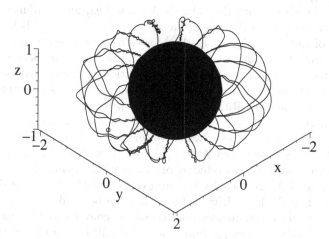

Figure 7.6: Path traced out by proton in the magnetic dipole field of the earth.

In the animation, the proton spirals around the magnetic dipole field lines, undergoing repeated reflections in the vicinity of the poles, and precessing around
the earth. The aurora occur in the vicinity of the turning points near the poles.

7.4 Rayleigh–Ritz Method

In Chapter 1, we noted that the simple harmonic oscillator, Bessel, Legendre,
and some other commonly occuring ODEs of physical interest, are all special
cases of the Sturm–Liouville (S-L) equation,

$$\frac{d}{dx}\left[p(x)\,\frac{dy}{dx}\right] - q(x)\,y = -\lambda\,w(x)\,y. \tag{7.9}$$

for particular choices of the real functions $p(x)$, $q(x)$, and $w(x)$. $w(x)$ is taken
to be non-negative over the range $x = a$ to b of interest. In boundary-value
problems where y satisfies the Sturm–Liouville boundary conditions (y or its
derivative vanish at a and b), y is the eigenfunction and λ the real eigenvalue.

The S-L equation can be formulated as a variational problem. Suppose that
we want to extremize the integral

$$I[y] = \int_a^b [p\,(y')^2 + q\,y^2]\,dx, \quad \text{subject to} \quad J[y] = \int_a^b w\,y^2\,dx = 1. \tag{7.10}$$

The form that y must take is determined by solving the Euler–Lagrange (E-L)
equation with $\mathcal{F} = p\,(y')^2 + q\,y^2 - \lambda\,w\,y^2$, where λ is the Lagrange multiplier.
Substituting $\mathcal{F}$ into the E-L equation just yields the S-L equation, (7.9). So the
function y which makes $I[y]$ an extremum subject to the subsidiary condition
on $J[y]$ is an eigenfunction of the S-L equation. The Lagrange multiplier λ is

the corresponding eigenvalue, while the subsidiary condition corresponds to the normalization condition on the eigenfunctions.

Now form the quantity $\Lambda[y] = I[y]/J[y]$. Extremizing Λ is exactly equivalent to extremizing I with the subsidiary condition on J. For S-L boundary conditions, it can be shown that the value of Λ for $y = y_n$, an eigenfunction of the S-L equation, is equal to the eigenvalue λ_n, i.e., $\Lambda[y_n] = \lambda_n$. This latter result is the basis of the *Rayleigh–Ritz* method for estimating eigenvalues.

For example, for atomic or molecular systems more complicated than the hydrogen atom, an exact analytical determination of the eigenvalues (energy levels) and eigenfunctions is not possible, so one must resort to approximate techniques. The Rayleigh–Ritz method is one such approach.

In this method one introduces a "trial function" ϕ, satisfying the boundary conditions. For the lowest eigenvalue λ_1 (ground state energy for molecular systems), it can be shown that $\Lambda[\phi] \geq \lambda_1$, i.e., Λ provides an upper bound to the lowest eigenvalue. By introducing one or more adjustable parameters into ϕ, the estimate of λ_1 can usually be improved, the estimate converging on λ_1 from above. The Rayleigh–Ritz method can also be used to estimate higher eigenvalues, but the estimate does not necessarily converge to the exact answer from above. In the recipes presented here, we will only estimate the lowest eigenvalue. The text *Mathematics in Physics and Engineering* by Irving and Mullineux [IM69] deals with estimating higher eigenvalues.

7.4.1   I. M. Estimates a Bessel Zero

To arrive at a just estimate of a renowned man's character
one must judge it by the standards of his time, not ours.
Mark Twain, American author, (1835–1910)

This recipe, provided by Ms. I.M. Curious, solves the following problem, which has often appeared in various guises on my mathematical physics exams.

The Bessel function of order 1 satisfies the equation

$$\frac{d^2 J_1}{dr^2} + \frac{1}{r}\frac{dJ_1}{dr} + (k^2 - \frac{1}{r^2})J_1 = 0.$$

Given that $J_1(r=0) = 0$, and noting that if we impose the boundary condition $J_1(r=1) = 0$, then k must be a zero of $J_1(kr)$, use the Rayleigh–Ritz method with no adjustable parameters to obtain an approximate value of the first zero of J_1. What is the percentage error in your answer when compared to the exact value? Improve your estimate by including one adjustable parameter. Generally, the agreement between the trial function and the exact solution is not nearly as good as between the eigenvalue estimate and the exact eigenvalue. Confirm that this is the case here by plotting the trial function with one adjustable parameter and the exact first-order Bessel function solution.

I. M. begins by entering the general form of $\Lambda[\phi(r)]$. The range of interest here is from $r=0$ to 1, so the integrals are taken to have these limits. From the structure of Λ, she notes that the trial function need not be normalized as the normalization constant will obviously cancel out.

```
>   restart:
>   Lambda:=int(p(r)*diff(phi(r),r)^2+q(r)*phi(r)^2,r=0..1)
        /int(w(r)*phi(r)^2,r=0..1);
```

$$\Lambda := \frac{\displaystyle\int_0^1 p(r)\,(\frac{d}{dr}\,\phi(r))^2 + q(r)\,\phi(r)^2\,dr}{\displaystyle\int_0^1 w(r)\,\phi(r)^2\,dr}$$

The first-order Bessel equation can be put into Sturm–Liouville form by choosing $p(r)=r$, $q(r)=1/r$, $w(r)=r$, and noting that $\lambda=k^2$.

```
>   p(r):=r: q(r):=1/r: w(r):=r:
```

I. M. chooses $\phi = r(1 - r)(1 + cr)$ as her[2] trial function, with c an adjustable parameter which she will set to 0 in the first part of the problem. The boundary conditions are clearly satisfied, i.e., $\phi = 0$ at $r=0$ and 1.

```
>   phi(r):=r*(1-r)*(1+c*r):
```

The trial function ϕ is automatically substituted into Λ, which on being simplified takes the following form.

```
>   Lambda:=simplify(Lambda);
```

$$\Lambda := \frac{14\,(7\,c^2 + 18\,c + 15)}{5\,c^2 + 16\,c + 14}$$

To answer the first part of the question, I. M. evaluates Λ with $c=0$.

```
>   Lambda0:=eval(Lambda,c=0);
```

$$\Lambda 0 := 15$$

Then the estimated value of k is obtained by taking the square root of $\Lambda 0$.

```
>   k0:=evalf(sqrt(Lambda0));
```

$$k0 := 3.872983346$$

The exact (numerical) k value is obtained by using the BesselJZeros command.

```
>   kexact:=BesselJZeros(1.0,1);
```

$$kexact := 3.831705970$$

I. M. now calculates the percentage error, $100\,(k0 - kexact)/kexact$, and finds it to be about 1.08%.

```
>   percenterror0:=((k0-kexact)/kexact)*100;
```

$$percenterror0 := 1.077258441$$

Now, she adjusts the parameter c to minimize the value of Λ. This is accomplished by calculating $d\Lambda/dc$ and setting the result equal to 0.

```
>   eq:=diff(Lambda,c)=0;
```

$$eq := \frac{14\,(14\,c + 18)}{5\,c^2 + 16\,c + 14} - \frac{14\,(7\,c^2 + 18\,c + 15)\,(10\,c + 16)}{(5\,c^2 + 16\,c + 14)^2} = 0$$

---

[2]Since the form of ϕ is deliberately not specified in the wording of the problem, I do get a variety of possible forms submitted by my students, sometimes even trial functions that do not satisfy the boundary conditions!

Then, *eq* is solved for c, and put into decimal form by applying the floating point evaluation command.

```
>  sol:=evalf(solve(eq));
```
$$sol := -0.3055081542, -1.785400936$$

Two values are generated for c. To decide on which value will minimize Λ, I. M. plots Λ over the range $c = -2$ to 1. The resulting picture is shown in Figure 7.7.

```
>  plot(Lambda,c=-2..1,labels=["c","Lambda"]);
```

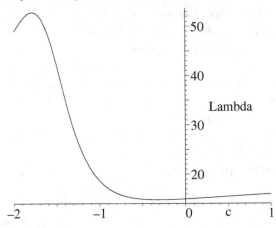

Figure 7.7: Λ versus c.

From the figure, the value $c \simeq -1.79$ must be ruled out as then $\Lambda \simeq 50 > \Lambda 0$. Λ should be evaluated with $c \simeq -0.31$. This is done in $\Lambda 1$, the number being slightly lower than $\Lambda 0 = 15$.

```
>  Lambda1:=eval(Lambda,c=sol[1]);
```
$$\Lambda 1 := 14.84137598$$

Then, $k1 = \sqrt{\Lambda 1}$ is calculated and the percentage error determined once again.

```
>  k1=sqrt(Lambda1);
```
$$k1 := 3.852450646$$

```
>  percenterror:=((k1-kexact)/kexact)*100;
```
$$percenterror := 0.5413952992$$

With one adjustable parameter included, the estimated value of k, i.e., $k1$, now only differs by 0.54% from the exact value. The corresponding trial function is given by $\phi 1$.

```
>  phi1:=eval(phi(r),c=sol[1]);
```
$$\phi 1 := r\,(1-r)\,(1-0.3055081542\,r)$$

To finish the problem, I. M. plots $\phi 1$ and the exact Bessel solution in Fig. 7.8.

```
>  plot([phi1,BesselJ(1,kexact*r)],r=0..1,color=[red,blue],
   labels=["r","phi"]);
```

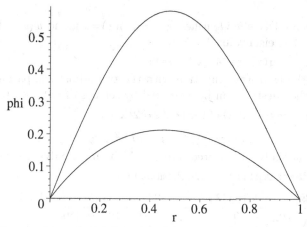

Figure 7.8: Comparison of trial function (lower curve) and exact solution (top).

She observes that although the estimated value of k is extremely good, the trial function $\phi 1$ doesn't do a very good job of representing the exact first-order Bessel function solution over the range $r = 0$ to 1. Part of the discrepancy could be due to the fact that $\phi 1$ and J_1 are not similarly normalized.

7.4.2 I. M. Estimates the Ground State Energy

*The only difference between a tax man and a taxidermist
is that the taxidermist leaves the skin.*
Mark Twain, American author, (1835–1910)

In the last example, the trial function contained one adjustable parameter. In the following problem, for which I. M. Curious will provide her recipe as a solution, there are two parameters which must be adjusted to minimize Λ.

Use the Rayleigh–Ritz procedure and the trial function $\phi = e^{-x^2}$ to estimate the ground state energy E_1 of a particle satisfying the one-dimensional Schrödinger equation (in appropriate units)

$$\frac{d^2\psi}{dx^2} - x^4\,\psi = -E\,\psi, \qquad -\infty < x < \infty.$$

Improve your estimate by considering $\phi = e^{-x^2}(1 + c1\,x^2 + c2\,x^4)$, with two adjustable parameters $c1$ and $c2$.

I. M. enters the general form of Λ, the range now from $x = -\infty$ to ∞.

```
>   restart: with(plots):
>   Lambda:=int(p(x)*diff(phi(x),x)^2+q(x)*phi(x)^2,x=-infinity
    ..infinity)/int(w(x)*phi(x)^2,x=-infinity..infinity);
```

$$\Lambda := \frac{\displaystyle\int_{-\infty}^{\infty} p(x)\,(\frac{d}{dx}\,\phi(x))^2 + q(x)\,\phi(x)^2\,dx}{\displaystyle\int_{-\infty}^{\infty} w(x)\,\phi(x)^2\,dx}$$

From the given ODE, she identifies $p(x)=1$, $q(x)=x^4$, and $w(x)=1$, which are now entered. The eigenvalue $\lambda = E$.

```
>   p(x):=1: q(x):=x^4: w(x):=1:
```

I. M. enters the trial wave function with the two adjustable parameters. The first part of the question can be answered by setting $c1=0$, $c2=0$.

```
>   phi(x):=exp(-x^2)*(1+c1*x^2+c2*x^4);
```

$$\phi(x) := e^{(-x^2)} \left(1 + c1\, x^2 + c2\, x^4\right)$$

$\phi(x)$ is automatically substituted into Λ, which on simplifying is as follows.

```
>   Lambda:=simplify(value(Lambda));
```

$$\Lambda := \frac{13995\, c2^2 + 10248\, c2\, c1 - 1248\, c2 + 3472\, c1^2 + 4864 - 128\, c1}{16\, (256 + 105\, c2^2 + 120\, c2\, c1 + 128\, c1 + 96\, c2 + 48\, c1^2)}$$

The estimate of the ground state energy with no adjustable parameters is determined in $\Lambda 0$.

```
>   Lambda0:=evalf(eval(Lambda,{c1=0,c2=0}));
```

$$\Lambda 0 := 1.187500000$$

So, I. M. estimates the ground state energy E_1 to be about 1.19 units. Now, she considers the situation when the parameters are non-zero. In this case, she must minimize Λ with respect to both parameters. This is done by imposing the conditions $\partial \Lambda / \partial c1 = 0$ and $\partial \Lambda / \partial c2 = 0$, which is done in $C1$ and $C2$.

```
>   C1:=simplify(diff(Lambda,c1))=0;
    C2:=simplify(diff(Lambda,c2))=0;
```

$$C1 := -2(-68352\, c2 + 18855\, c2^3 + 19200\, c2^2\, c1 - 24576\, c2\, c1$$
$$+\, 20976\, c2^2 + 2352\, c2\, c1^2 - 40960\, c1 - 14080\, c1^2 + 20480)$$
$$/(256 + 105\, c2^2 + 120\, c2\, c1 + 128\, c1 + 96\, c2 + 48\, c1^2)^2 = 0$$

$$C2 := 6(64000\, c2 + 6285\, c2^2\, c1 + 37600\, c2\, c1 + 15360\, c2^2$$
$$+\, 6400\, c2\, c1^2 + 19712\, c1 + 9728\, c1^2 + 784\, c1^3 - 8192)$$
$$/(256 + 105\, c2^2 + 120\, c2\, c1 + 128\, c1 + 96\, c2 + 48\, c1^2)^2 = 0$$

The formidable nonlinear algebraic equations $C1$ and $C2$ must be numerically solved for $c1$ and $c2$. To guide her in arriving at the correct solution, I. M. forms a functional operator `gr` to apply the `implicitplot` command to a specified algebraic equation A, the color of the curve being dictated by the choice of B. To obtain smooth curves, the number of plotting points is increased to 2000.

```
>   gr:=(A,B)->implicitplot(A,c1=-10..10,c2=-10..10,
    numpoints=2000,color=B):
```

A plot of $c2$ versus $c1$ satisfying the equation $C1$ follows on taking $A=C1$ in `gr`. The curve will be colored blue. Similarly, a green curve is generated for the choice $A=C2$. The two curves are superimposed with the `display` command, the resulting picture being shown in Figure 7.9.

```
>   display({gr(C1,blue),gr(C2,green)});
```

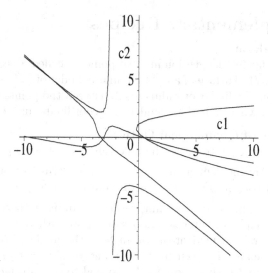

Figure 7.9: Possible values of $c1$ and $c2$ occur at intersection points.

Within the range of the plot, I. M. spots three intersection points. Applying the floating point solve command to $C1$ and $C2$ without specifying the search range for $c1$ and $c2$,

> s1:=fsolve({C1,C2},{c1,c2});

$$s1 := \{c1 = -7.524875339,\ c2 = 4.879077669\}$$

yields the intersection point in the second quadrant of Figure 7.9. To obtain the other two intersection points, I. M. now specifies the appropriate plotting ranges in the fsolve command in $s2$ and $s3$

> s2:=fsolve({C1,C2},{c1,c2},{c1=0..2,c2=-2..2});

$$s2 := \{c1 = 0.5221933646,\ c2 = -0.05771753990\}$$

> s3:=fsolve({C1,C2},{c1,c2},{c1=-5..0,c2=-2..2});

$$s3 := \{c2 = -0.2086941568,\ c1 = -3.169995132\}$$

An operator L is created to evaluate λ for a specified s, which I. M. then applies to $s1$, $s2$, and $s3$ in $\Lambda1$, $\Lambda2$, and $\Lambda3$.

> L:=s->eval(Lambda,s):

> Lambda1:=L(s1); Lambda2:=L(s2); Lambda3:=L(s3);

$$\Lambda1 := 16.71552532 \qquad \Lambda2 := 1.061092905 \qquad \Lambda3 := 7.535881788$$

The estimates $\Lambda1$ and $\Lambda3$ are rejected as both are larger than $\Lambda0 \simeq 1.19$. The estimate $\Lambda2 \simeq 1.06$ is lower than $\Lambda0$, so represents an improved estimate of the ground state energy. It would appear from Iam's recipe that her estimates are converging on the value $E_1 = 1$. It is left as an exercise for you to confirm whether this conclusion is correct or not.

7.5 Supplementary Recipes

07-S01: Geodesic
The curve of shortest length joining two points is called a geodesic. Show that the geodesic on the surface of a right circular cylinder of radius a is a helix. Plot the geodesic on a cylinder of radius $a=2$ between the points $(z=1/2,\ \theta=\pi/8)$ and $(5/2, \pi/2)$, where z is the cylinder axis coordinate and θ the polar angle.

07-S02: Laws of Geometrical Optics
Use Fermat's principle to prove the following geometrical optics relations:

(a) A light ray incident at an angle i to the normal on a planar mirror is reflected back into the same medium at an angle $r' = i$ to the normal.

(b) Consider a light ray traveling from a medium with refractive index n_1 through a planar interface into a medium with refractive index n_2. If the light ray is incident at an angle i, the angle of refraction r in the second medium is given by *Snell's law*: $n_1 \sin i = n_2 \sin r$. Both angles are measured with respect to the normal to the interface.

07-S03: Bending of Starlight
According to the theory of general relativity, the trajectory of starlight traveling in the spherically symmetric static field of the sun is such as to minimize the integral

$$I = \int \sqrt{(dr/\gamma)^2 + (r\,d\theta)^2/\gamma}$$

where (r, θ) are polar coordinates and $\gamma = 1-(2\,G\,ms)/(c^2\,r)$, G being the gravitational constant, ms the mass of the sun, and c the vacuum speed of light. Using the variational approach and setting $u = 1/r$, prove that the differential equation of the trajectory can be written in the form

$$\frac{d^2 u}{d\theta^2} + a\,u = b\,u^2,$$

where a and b remain to be identified. Taking $G=6.673\times10^{-11}$ N·m^2/kg^2, $ms=1.99\times10^{30}$ kg, $c=2.997\times10^8$ m/s, and the sun's radius $Rs=6.96\times10^8$ m, numerically solve the nonlinear ODE for $u(\theta)$, taking $u(0)=1/(10Rs)$ and $du(0)/d\theta=0$. Then use the `spacecurve` command to plot $(r(\theta)\cos(\theta),\ r(\theta)\sin(\theta))$ for $\theta = -\pi/2$ to $\pi/2$. Include the sun in your figure.

07-S04: Another Refractive Index
Using Fermat's principle, prove that light rays in a medium with refractive index $n(x, y)=1/y$ will follow a path which is the arc of a circle.

07-S05: Mirage
Assuming that the refractive index varies linearly with height in the following way, $n(x, y) = n_0\,(1 + \alpha\,y)$ with $n_0 > 0$ and $\alpha > 0$, use Fermat's principle to determine the angle θ by which a point P is apparently lowered when viewed from a point P' at the same height as P at a horizontal distance d. The parameter α is sufficiently small that you may take $\alpha\,d << 1$. This problem is relevant to the phenomenon of mirages observed in hot desert regions.

07-S06: A Constrained Extremum

Determine a function $y(x)$ for which $I = \int_0^\pi ((y')^2 - y^2)\, dx$ is an extremum, subject to $N = \int_0^\pi y\, dx = 1$, and $y(0) = 0$, $y(\pi) = 1$. Plot $y(x)$.

07-S07: Maximum Volume

Find the maximum value of the volume $V = x\, y\, z$, subject to the condition $x^2/a^2 + y^2/b^2 + z^2/c^2 = 1$, with a, b, and c positive. Use (a) the direct method, (b) the Lagrange multiplier method.

07-S08: Eigenvalue Estimate

Taking a trial function of the form $y = A_0 + A_1 x + A_2 x^2 + A_3 x^3$, estimate the smallest eigenvalue k of $y'' + k^2 y = 0$, subject to $y(0) = y(1) = 0$. Compare the estimate and the trial function with the exact results.

07-S09: Surface of Revolution

Consider a surface of revolution generated by revolving a curve $y(x)$ about the x-axis. The curve is required to pass through the fixed end points $(x=0, y=1)$ and $(x=1, y=2)$. Determine what shape $y(x)$ must have so that the area of the resulting surface will be a minimum. Plot the curve.

07-S10: Dido Wasn't a Dodo

Given a closed planar curve C of a fixed length L, what shape should the curve have to maximize the enclosed area A? The answer, known as the *isoperimetric theorem*, is a circle. Evidently, this theorem has been known from about 900 BC, and according to Virgil's *Aneid* was applied by Queen Dido in a practical way to establish the city of Carthage (now Tunisia) in North Africa. The local ruler, King Jambas, agreed to sell her all the land that she could enclose within a bull's hide. She cleverly had the hide cut into thin strips which were joined end to end to form a long "string" and made use of a variation on the isoperimetric theorem. She cleverly maximized A by selecting a straight portion of the Mediterranean coast and laying the string out in a semicircle with the string's two ends touching the coast.

Given that $A = \frac{1}{2} \oint_C (x\, dy - y\, dx)$ for a closed planar curve C, make use of the Euler–Lagrange equation to prove the isoperimetric theorem.

07-S11: Another Approach to the String Equation

In recipe **04-1-1**, the wave equation for small transverse vibrations of a light flexible horizontal string under tension and fixed on its ends was derived using Newton's second law of motion. Another approach to deriving the wave equation is to use the variational method.

It can be shown that a necessary condition for $y = y(x,t)$ to be an extremal of $\int_{t_1}^{t_2} \int_{x_1}^{x_2} F(x,t,y,y_x,y_t)\, dx\, dt$ is

$$\frac{\partial F}{\partial y} - \frac{\partial}{\partial x}\left(\frac{\partial F}{\partial y_x}\right) - \frac{\partial}{\partial t}\left(\frac{\partial F}{\partial y_t}\right) = 0.$$

Here $y_x \equiv \partial y/\partial x$, $y_t \equiv \partial y/\partial t$, x_1 and x_2 are fixed end points, and t_1 and t_2 are specified times. Making use of this extension of Lagrange's equation, derive the wave equation for small vibrations of a flexible string, fixed at its end points $x=0$ and a, and under tension τ.

07-S12: Betsy Bug's Ride

Betsy bug is perched on a bead which slides along a smooth wire bent in the shape of a stylized "W". I.e., the bead's vertical coordinate at time t is $z(t) = -c_1 r(t)^2 + c_2 r(t)^4$, with $c_1 > 0$, $c_2 > 0$, and $r(t)$ the radial distance from the vertical symmetry axis. The wire is rotating about it's symmetry axis with angular velocity ω radians/s, so the bead's horizontal coordinates at time t are $x(t) = r(t) \cos(\omega t)$, $y(t) = r(t) \sin(\omega t)$.

(a) Using the Lagrangian approach, find the equation of motion for $r(t)$.

(b) Show that there is critical frequency $\omega_{cr} = \sqrt{2g(2c_2 r(0)^2 - c_1)}$, where g is the acceleration due to gravity, for which the bead will not move relative to the wire.

(c) Taking $c_1 = 1$ m^{-1}, $c_2 = 1/2$ m^{-3}, gravitational acceleration $g = 9.8$ m/s^2, $r(0) = 2$ m, $\dot{r}(0) = 0$ m/s, determine the critical frequency.

(d) Taking $\omega = 0.1\, \omega_{cr}$, numerically solve the equation of motion for $r(t)$.

(e) Animate the motion of the bead and the wire.

Part III

THE DESSERTS

The reasonable man adapts himself to the world;
the unreasonable one persists in trying to adapt the
world to himself. Therefore, all progress depends
on the unreasonable man.
George Bernard Shaw, Anglo-Irish playwright, *Man and Superman*, (1903)

Perfection is achieved not when you have nothing more
to add, but when you have nothing left to take away.
Antoine de Saint-Exupery, French pilot and author, (1900–1944)

I may not have gone where I intended to go,
but I think I have ended up where I intended to be.
Douglas Adams, English humorist and science fiction writer, (1952–2001)

Chapter 8

NLODEs & PDEs of Physics

So far, this gourmet selection of computer algebra recipes has emphasized physical examples governed by linear ODEs and PDEs. However, much of modern scientific research involves phenomena described by nonlinear differential equations, i.e, equations which are not linear in the dependent variable(s). In the **Desserts**, I will give you a small taste of this intellectually delectable area of mathematical physics. If you enjoy the nonlinear recipes that follow, and crave more, you should consult [EM00] for a much deeper treatment of the subject.

This chapter illustrates a few of the basic analytic (exact and approximate) and graphical methods for solving nonlinear ODEs and PDEs (NLODEs and NLPDEs). Basic numerical methods will be covered in Chapter 9. A comprehensive survey of analytic and numerical approaches may be found in Daniel Zwillinger's *Handbook of Differential Equations* [Zwi89].

8.1 Nonlinear ODEs: Exact Methods

In Chapter 1, we encountered a simple nonlinear ODE, describing the motion of a falling badminton bird acted upon by a nonlinear drag force, which had an exact analytic solution. Only a handful of NLODEs of physical interest can be exactly solved using elementary techniques. Here are a few more examples of increasing mathematical complexity.

8.1.1 Jacob Bernoulli and the Nonlinear Diode

I recognize the lion by his paw.
Jacob Bernoulli, Swiss mathematician, after reading an anonymous solution to a problem that he realized was Newton's solution, (1654–1705)

What does Jacob Bernoulli who lived some 300 years ago have to do with a modern nonlinear diode? As you will shortly see, the governing circuit equation in the following recipe is the nonlinear *Bernoulli equation* which Jacob discovered in another context in 1690 and solved in 1696. Bernoulli's equation is a first order nonlinear ODE of the form

$$y' + f_1(x)\, y = f_2(x)\, y^n \tag{8.1}$$

which can be cast into a linear ODE, and thus solved, by introducing a new dependent variable $p = 1/y^{n-1}$. Bernoulli's equation is known to Maple and thus can be easily solved without having to carry out the details of the variable transformation. Now let's formulate the nonlinear diode problem.

A linear capacitor C is connected in series to a nonlinear diode, a circuit element which has a current (i)–voltage (v) relation of the form $i = a\,v + b\,v^n$, where a and b are positive constants and $n = 2, 3, 4, \cdots$ Given that the voltage across the capacitor at time $t = 0$ is $v(0) = V$, what is the voltage $v(t)$ for $t > 0$? Express the solution in dimensionless form in terms of a single parameter β, which remains to be identified. Plot the analytic result for $n = 2$ to 5 and $\beta = 2$.

To solve this problem, the PDEtools library package is loaded. This contains the **dchange** command which will be used to cast the governing circuit equation and the voltage into dimensionless form.

>   `restart: with(PDEtools):`

From the definition of capacitance, the charge q on the capacitor at time t is $q = C\,v(t)$. So, the current in the circuit is $i = -dq/dt = a\,v(t) + b\,v(t)^n$. These two basic relations are entered, the expression for q being automatically substituted into i.

>   `q:=C*v(t);`

$$q := C\,v(t)$$

>   `i:=-diff(q,t)=a*v(t)+b*v(t)^n;`

$$i := -C\,(\frac{d}{dt}\,v(t)) = a\,v(t) + b\,v(t)^n$$

The above ODE is clearly of Bernoulli's form with $y \equiv v$, $f_1 \equiv a/C$ and $f_2 \equiv -b/C$. It can be transformed into dimensionless form by introducing a new time variable $\tau = a\,t/C$ and a dimensionless voltage variable $y(\tau) = v(t)/V$.

>   `tr:={t=C*tau/a,v(t)=V*y(tau)};`

$$tr := \{t = \frac{C\,\tau}{a},\ v(t) = V\,y(\tau)\}$$

The above transformation is applied to i in the following **dchange** command.

>   `ode:=dchange(tr,i,[tau,y(tau)]);`

$$ode := -a\,V\,(\frac{d}{d\tau}\,y(\tau)) = a\,V\,y(\tau) + b\,(V\,y(\tau))^n$$

Then *ode* is divided by $-a\,V$ and the result expanded.

>   `ode:=expand(ode/(-a*V));`

$$ode := \frac{d}{d\tau}\,y(\tau) = -y(\tau) - \frac{b\,(V\,y(\tau))^n}{a\,V}$$

Introducing a dimensionless parameter β through the substitution $b = a\,\beta/V^{(n-1)}$ and simplifying with the **symbolic** option produces the desired dimensionless nonlinear ODE in *ode2*.

> `ode2:=simplify(subs(b=a*beta/V^(n-1),ode),symbolic);`

$$ode2 := \frac{d}{d\tau}\, y(\tau) = -y(\tau) - \beta\, y(\tau)^n$$

Note that this equation now only depends on one parameter, β, and that the initial condition at $\tau=t=0$ is $y(\tau)=1$. Instead of mimicking a hand calculation, and making the suggested dependent variable substitution, I will directly use the `dsolve` command. First let's enter the following `infolevel` command to give us information on what approach Maple takes in solving *ode2*.

> `infolevel[dsolve]:=2:`

Then, *ode2* is solved for $y(\tau)$, subject to the initial condition $y(0)=1$.

> `sol:=simplify(dsolve({ode2,y(0)=1},y(tau)));`
> *Methods for first order ODEs:*
> — *Trying classification methods* —
> *trying a quadrature*
> *trying 1st order linear*
> *trying Bernoulli*
> *<— Bernoulli successful*

$$sol := \mathrm{y}(\tau) = \left(-\beta + e^{((n-1)\,\tau)} + e^{((n-1)\,\tau)}\,\beta\right)^{\left(-\frac{1}{n-1}\right)}$$

The NLODE is identified as a Bernoulli equation and successfully solved. To plot the solution, let's enter the suggested value $\beta=2$.

> `beta:=2:`

The rhs of *sol* is now plotted for $n=2$, 3, 4, 5, a different color being assigned to each curve. The resulting graph is shown in Figure 8.1, the bottom curve corresponding to $n=2$, the next highest curve to $n=3$, and so on. Because $y(\tau) < 1$ as τ increases, the curves decrease less rapidly as n is increased.

> `plot([seq(rhs(sol),n=2..5)],tau=0..2,color=[red,green,blue,`
> `black],thickness=2,labels=["tau","y"],tickmarks=[2,2])`

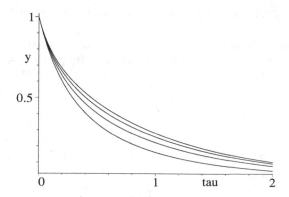

Figure 8.1: Normalized voltage as a function of time.

8.1.2   The Chase

I wouldn't think of asking you to lie;
you haven't the necessary diplomatic training.
A line spoken by one of the characters in the movie *Sea Chase (1955)*

A classic problem [Dav62] in the history of nonlinear ODEs involves the *curve of pursuit*. This is the trajectory generated by a point P which moves in such a way that its direction of motion is always towards a second point P', constrained to move along a prescribed path. This type of problem originated with Leonardo da Vinci in the 15th century, but the example in the following recipe is due to the French hydrographer Pierre Bouguer who published his solution in 1732. Bouguer's formulation is expressed in terms of one ship pursuing another, the latter moving along a straight line.

I will phrase the problem somewhat differently. Referring to the lhs of Figure 8.2, rascally Roger Rabbit is being pursued by ferocious Freddy Fox.

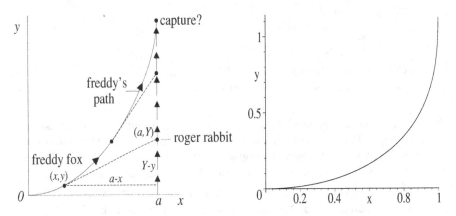

Figure 8.2: Left: Geometry for pursuit. Right: Freddy's calculated path.

Roger, who is initially at the point $x = a$, $y = 0$, runs at a constant speed along the vertical line $x = a$. Freddy Fox, who is initially at the origin O, pursues Roger by constantly aiming at Roger's current position. Freddy's constant speed is n times that of Roger's, with $n > 1$. The problem is to derive the equation $y(x)$ of Freddy's path (indicated schematically in the picture) and determine the theoretical point of capture. If $a = 1$ km and $n = 3/2$, will Roger escape capture by diving into his hole which is located 1195 meters from Roger's starting point. Plot Freddy's path up to the theoretical point of capture. So let the chase begin!

The assumption that $a > 0$, $x \geq 0$, $x < a$, and $n > 1$ is entered.

```
>  restart: assume(a>0,x>=0,x<a,n>1):
```

From the figure, the slope of Freddy's trajectory when he is at the point $(x, y(x))$ and Roger is at (a, Y) is given by $dy/dx = (Y - y(x))/(a - x)$.

```
> slope:=diff(y(x),x)=(Y-y(x))/(a-x);
```

$$slope := \frac{d}{dx} y(x) = \frac{Y - y(x)}{a - x}$$

The slope equation is solved for Y,

```
> Y:=solve(slope,Y);
```

$$Y := (\frac{d}{dx} y(x)) \, a - (\frac{d}{dx} y(x)) \, x + y(x)$$

and the x derivative of Y then taken, and factored.

```
> Yder:=factor(diff(Y,x));
```

$$Yder := (\frac{d^2}{dx^2} y(x)) \, (a - x)$$

The arclength along Freddy's path is $ds = \sqrt{(dx)^2 + (dy)^2}$ and his speed, ds/dt, is n times Roger's speed, dY/dt. So, $n \, (dY/dx) = \sqrt{1 + (dy/dx)^2}$ is entered.

```
> ode:=n*Yder=sqrt((diff(y(x),x))^2+1);
```

$$ode := n \, (a - x) \, (\frac{d^2}{dx^2} y(x)) = \sqrt{(\frac{d}{dx} y(x))^2 + 1}$$

Freddy's path is described by a second order NLODE which, amazingly, can be solved exactly. Although the `dsolve` command could be applied directly to *ode*, the subsequent manipulations to put the solution in a compact form are messy. Instead, let's mimic the approach that we would undertake by hand. Making the substitution $dy(x)/dx = p(x)$ reduces *ode* to a first order ODE in $p(x)$ displayed in *ode2*.

```
> ode2:=subs(diff(y(x),x)=p(x),ode);
```

$$ode2 := n \, (a - x) \, (\frac{d}{dx} p(x)) = \sqrt{p(x)^2 + 1}$$

Freddy's initial slope is zero, i.e., $p(0) = 0$. *ode2* is now analytically solved for $p(x)$, subject to the initial slope condition. The right-hand side of the solution is then taken and the result equated to dy/dx. Occasionally, on executing the worksheet, the form of the output will differ in *ode3* from what is reproduced here in the text. This will not affect the shape of Freddy's trajectory or the theoretical point of capture.

```
> ode3:=diff(y(x),x)=rhs(dsolve({ode2,p(0)=0},p(x)));
```

$$ode3 := \frac{d}{dx} y(x) = -\sinh(\frac{\ln(a - x) - \ln(a)}{n})$$

ode3 is analytically solved for $y(x)$, subject to the initial condition $y(0) = 0$. Again, the right-hand side of the solution is taken. The equation of Freddy's path is then displayed in the output of the following command line.

```
> y:=rhs(dsolve({ode3,y(0)=0},y(x)));
```

$$y := -\frac{n\,x\,(a-x)^{\left(\frac{1}{n}\right)}}{2\,(n+1)\,a^{\left(\frac{1}{n}\right)}} + \frac{a\,n\,(a-x)^{\left(\frac{1}{n}\right)}}{2\,(n+1)\,a^{\left(\frac{1}{n}\right)}} + \frac{\left(-\dfrac{a\,n}{2\,(n-1)} + \dfrac{n\,x}{2\,(n-1)}\right)a^{\left(\frac{1}{n}\right)}}{(a-x)^{\left(\frac{1}{n}\right)}}$$
$$-\frac{a\,n}{2\,(n+1)} + \frac{a\,n}{2\,(n-1)}$$

Next, let's substitute $x = a - X^n$ into y and expand the result.

>   `y:=expand(subs(x=a-X^n,y));`

$$y := \frac{n\,(X^n)^{\left(\frac{1}{n}\right)}X^n}{2\,(n+1)\,a^{\left(\frac{1}{n}\right)}} - \frac{a^{\left(\frac{1}{n}\right)}n\,X^n}{2\,(X^n)^{\left(\frac{1}{n}\right)}(n-1)} - \frac{a\,n}{2\,(n+1)} + \frac{a\,n}{2\,(n-1)}$$

The previous substitution is reversed by setting $X = (a-x)^{1/n}$ in y. Applying the **combine** command yields a compact equation describing Freddy's trajectory.

>   `y:=combine(subs(X=(a-x)^(1/n),y));`

$$y := \frac{(a-x)^{\left(\frac{1}{n}+1\right)}a^{\left(-\frac{1}{n}\right)}n}{2\,(n+1)} - \frac{(a-x)^{\left(1-\frac{1}{n}\right)}a^{\left(\frac{1}{n}\right)}n}{2\,(n-1)} - \frac{a\,n}{2\,(n+1)} + \frac{a\,n}{2\,(n-1)}$$

The theoretical point of capture, Yc, follows on evaluating y at $x = a$.

>   `Yc:=simplify(eval(y,x=a));`

$$Yc := \frac{a\,n}{n^2 - 1}$$

Roger would be captured at $Yc = (a\,n)/(n^2 - 1)$, unless he gets to his hole first. So, does he avoid capture? Substituting $a = 1$ and $n = 3/2$ into Yc and multiplying by 1000 to convert the result into meters,

>   `Yc:=evalf(subs({n=3/2,a=1},Yc))*1000;`

$$Yc := 1200.000000$$

yields a value for Yc of 1200 meters. Fortunately, for Roger, his hole was only 1195 meters from his starting point, so he survives for another day. To plot Freddy's path up to the point of theoretical capture, the values $n = 3/2$ and $a = 1$ are substituted into y, yielding the result shown in $y2$.

>   `y2:=subs({n=3/2,a=1},y);`

$$y2 := \frac{3\,(1-x)^{(5/3)}}{10} - \frac{3\,(1-x)^{(1/3)}}{2} + \frac{6}{5}$$

Freddy's trajectory $y2$ is then plotted up to the theoretical point of capture,

>   `plot(y2,x=0..1,thickness=2,numpoints=200,labels=["x","y"]);`

the result being shown on the right-hand side of Figure 8.2. Of course, if Roger's hole were not conveniently close, he would have had to take evasive action, leading to a much more difficult pursuit problem to solve.

8.1.3 Not As Hard As It Seems

Often, the less there is to justify a traditional custom,
the harder it is to get rid of it.
Mark Twain, American author, *Tom Sawyer, ch. 5, 1876*

When a spring is stretched from its equilibrium length by an amount x which is no longer small, Hooke's law must be modified so as to include nonlinear or *anharmonic* terms in x. For symmetric oscillations of a nonlinear spring about equilibrium, the next terms retained in the Taylor expansion of the force must be cubic in x. In this case, the force F required to stretch the nonlinear spring will be of the form $F = k_1 x + k_2 x^3 = k_1 (x + a x^3)$, with $k_1 > 0$ and $a \equiv k_2/k_1$. If $a > 0$, one has a *hard spring* because it is harder to stretch than a linear spring $(a=0)$. If $a < 0$, one has a *soft spring*. In this recipe, we will solve the equation of motion for the oscillations of a mass m attached to a (light) hard spring allowed to move horizontally on a smooth horizontal surface. Applying Newton's second law, the equation of motion of m at arbitrary time τ is

$$m \ddot{x}(\tau) + k_1 (x(\tau) + a x(\tau)^3) = 0 \tag{8.2}$$

or, on introducing a new time variable $t = \sqrt{k_1/m}\, \tau$,

$$\ddot{x}(t) + x(t) + a x(t)^3 = 0. \tag{8.3}$$

Unlike the situation for the linear spring where the solution is in terms of elementary functions (sine or cosine), the solution for the hard spring will involve another "special" function, the *Jacobian elliptic function*. As you will see, by using computer algebra, deriving the solution is not as hard as it seems.

Taking, say, $a=1$, the equation of motion (8.3) is entered in *ode*.

```
> restart: a:=1:
```

```
> ode:=diff(x(t),t,t)+x(t)+a*x(t)^3=0;
```

$$ode := (\frac{d^2}{dt^2} x(t)) + x(t) + x(t)^3 = 0$$

The dsolve command is applied to *ode*. Two *implicit* solutions are generated, with the time t given in integral form ($\_a$ is the integration variable) and two arbitrary constants, $\_C1$ and $\_C2$, present. The positive square root solution must be selected in order to produce a positive period for the oscillatory motion.

```
> sol:=dsolve(ode,x(t)); #choose positive root
```

$$sol := \int^{x(t)} \frac{2}{\sqrt{-4\_a^2 - 2\_a^4 + 4\_C1}}\, d\_a - t - \_C2 = 0,$$

$$\int^{x(t)} -\frac{2}{\sqrt{-4\_a^2 - 2\_a^4 + 4\_C1}}\, d\_a - t - \_C2 = 0$$

The positive square root solution (the first one here) is selected and differentiated with respect to t. This removes the constant $\_C2$.

```
> eq1:=diff(sol[1],t);
```

$$eq1 := \frac{2\left(\dfrac{d}{dt}\, x(t)\right)}{\sqrt{-4\, x(t)^2 - 2\, x(t)^4 + 4\_C1}} - 1 = 0$$

eq1 is solved in eq2 for $dx(t)/dt$, i.e., the speed.

> eq2:=solve(eq1,diff(x(t),t));

$$eq2 := \frac{1}{2}\, \sqrt{-4\, x(t)^2 - 2\, x(t)^4 + 4\_C1}$$

In order to perform the subsequent integrations, the dependent variable $x(t)$ is replaced in eq2 with a time-independent symbol, say y .

> eq3:=subs(x(t)=y,eq2);

$$eq3 := \frac{\sqrt{-4\, y^2 - 2\, y^4 + 4\_C1}}{2}$$

Let's call the amplitude of the oscillations A. When $y \equiv x(t) = A$, the speed of the mass m will be zero. This condition is expressed in eq4.

> eq4:=subs(y=A,eq3)=0;

$$eq4 := \frac{\sqrt{-4\, A^2 - 2\, A^4 + 4\_C1}}{2} = 0$$

eq4 is solved for the constant $\_C1$, which is automatically substituted into eq3 which is relabeled as eq5.

> _C1:=solve(eq4,_C1); eq5:=eq3;

$$\_C1 := A^2 + \frac{1}{2}\, A^4 \qquad eq5 := \frac{\sqrt{-4\, y^2 - 2\, y^4 + 4\, A^2 + 2\, A^4}}{2}$$

Since $dx(t)/dt \equiv dy/dt = eq5$, the period T (time for a complete oscillation) is given by $T = 2 \int_{-A}^{A} (1/eq5)\, dy$. This integral is entered. To accomplish the integration, the assumption that $A > 0$ must be included.

> T:=2*int(1/eq5,y=-A..A) assuming A>0;

$$T := \frac{4\, \sqrt{2}\, \mathrm{EllipticK}\!\left(\dfrac{\sqrt{2}\, A}{2\, \sqrt{1 + A^2}}\right)}{\sqrt{2 + 2\, A^2}}$$

Highlighting EllipticK in the output and opening the Help page reveals that the period T involves the *complete elliptic integral* of the first kind. The *incomplete elliptic integral of the first kind* is defined [AS72] by

$$F(\phi \backslash \alpha) = \int_0^\phi (1 - \sin^2 \alpha \, \sin^2 \theta)^{-1/2}\, d\theta, \tag{8.4}$$

or, on setting $m \equiv \sin^2 \alpha$, $y \equiv \sin \theta$, and $x \equiv \sin \phi$,

$$F(\phi \,|\, m) = \int_0^x [(1 - y^2)\,(1 - m\, y^2)]^{-1/2}\, dy. \tag{8.5}$$

The complete elliptic integral $K \equiv K(m)$ corresponds to setting $\phi = \pi/2$ in $F(\phi \backslash \alpha)$, or $x = 1$ in $F(\phi \,|\, m)$. The command EllipticF(sin(phi),sqrt(m)) generates $F(\phi \,|\, m)$, while EllipticK(sqrt(m)) produces $K(m)$. The period is now plotted over the range $A = 0$ to 5, the result being shown in Figure 8.3.

```
>  plot(T,A=0..5,labels=["A","T"],view=[0..5,0..2*Pi]);
```

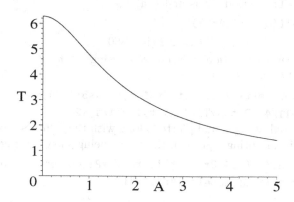

Figure 8.3: Period T of the hard spring versus amplitude A.

Unlike the situation for a linear spring, the period of the hard spring is amplitude dependent, decreasing with increasing values of A. To determine $x(t)$, the integral $t = \int_{-A}^{x}(1/eq5)\,dy$ is evaluated assuming that $A > 0$ and $x > -A$.

```
>  eq6:=t=int(1/eq5,y=-A..x) assuming A>0,x>-A;
```

$$eq6 := t = \frac{2\sqrt{2}\,\text{EllipticK}(\dfrac{\sqrt{2}\,A}{2\sqrt{1+A^2}})}{\sqrt{2+2\,A^2}} - \frac{\sqrt{2}\,\text{EllipticF}(\dfrac{\sqrt{A^2 - x^2}}{A}, \dfrac{\sqrt{2}\,A}{2\sqrt{1+A^2}})}{\sqrt{2+2\,A^2}}$$

eq6 is now solved for x, which produces positive and negative square root expressions, which are assigned the names *x1* and *x2*. Both solution branches will be needed to plot the complete oscillation of the hard spring.

```
>  eq7:=solve(eq6,x); x1:=eq7[1]; x2:=eq7[2];
```

$$x1 :=$$
$$\sqrt{1 - \text{JacobiSN}(\frac{1}{2}(t\sqrt{2+2\,A^2} - 2\sqrt{2}\,\text{EllipticK}(\frac{\sqrt{2}\,A}{2\sqrt{1+A^2}}))\sqrt{2}, \frac{\sqrt{2}\,A}{2\sqrt{1+A^2}})^2\,A},$$

$$x2 :=$$
$$-\sqrt{1 - \text{JacobiSN}(\frac{1}{2}(t\sqrt{2+2\,A^2} - 2\sqrt{2}\,\text{EllipticK}(\frac{\sqrt{2}\,A}{2\sqrt{1+A^2}}))\sqrt{2}, \frac{\sqrt{2}\,A}{2\sqrt{1+A^2}})^2\,A}$$

The "special" function JacobiSN appearing in the output is the *Jacobi elliptic sine function*. Setting $u \equiv F(\phi\backslash\alpha)$, ϕ is referred to as the "amplitude" of u, written as $\phi = \text{am}\,u$. The elliptic sine function is defined as $\text{sn}\,u = \sin(\text{am}\,u) = \sin\phi$. The Maple command for numerically calculating $\text{sn}\,u$ for a given m value is evalf(JacobiSN(u,sqrt(m))). The reader is referred to Abramowitz and Stegun [AS72] for the properties of the elliptic functions.

To plot the hard spring solution, let's choose a specific value of the amplitude, say $A=3$. The period $T1$ is first evaluated

```
>   T1:=evalf(eval(T,A=3));
```

$$T1 := 2.294401860$$

and then a piecewise function x is created, using $x1$ for $t < T1/4$, $x2$ for $T1/4 < t < 3T1/4$, and so on.

```
>   x:=piecewise(t<T1/4,x1,t<3*T1/4,x2,t<5*T1/4,x1,
        t<7*T1/4,x2,t<9*T1/4,x1,t<11*T1/4,x2):
```

Then x is evaluated at $A=3$ and plotted along with the solution $3\cos(t)$ to the corresponding linear spring equation, the result being shown in Figure 8.4.

```
>   plot([eval(x,A=3),3*cos(t)],t=0..2*Pi,color=[red,blue],
        linestyle=[1,3],labels=["t","x"]);
```

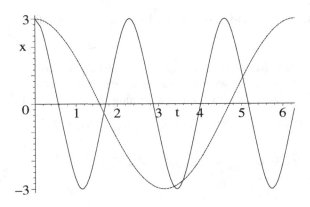

Figure 8.4: Solid curve: hard spring; Dashed curve: linear spring $(a=0)$.

As expected, the hard spring oscillates more rapidly than the linear spring.

8.2   Nonlinear ODEs: Graphical Methods

Two recipes are presented which illustrate how a second-order NLODE, or a system of two first-order equations, can be graphically solved and interpreted.

8.2.1   Joe and the Van der Pol Scroll

Sometimes one likes foolish people for their folly,
better than wise people for their wisdom.
Elizabeth Gaskell, English novelist, (1810–65)

This recipe is inspired by my reminiscences of a former student in my nonlinear physics class, whose identity I will protect by calling him "Joe". In class,

I had derived the "equation of motion" for a tunnel diode electrical circuit, the ODE taking the form of the nonlinear *Van der Pol (VdP) equation,*

$$\ddot{x} + \epsilon(x^2 - 1)\,\dot{x} + x = 0, \tag{8.6}$$

where x is related to the voltage change across the diode and the positive parameter ϵ depends on certain circuit parameters. From a mechanical viewpoint, the VdP equation follows on applying Newton's second law to a unit mass which experiences a Hooke's law restoring force, $F_{\text{Hooke}} = -x$, and a drag force, $F_{\text{drag}} = -\epsilon(x^2 - 1)\,\dot{x}$. For $\epsilon = 0$, Eq. (8.6) reduces to the simple harmonic oscillator equation which has undamped oscillatory solutions. For $\epsilon > 0$, the drag force has a rather peculiar property. For $x > 1$ and $\dot{x} > 0$, $F_{\text{drag}} < 0$, which tends to reduce the size of the oscillations, but for $x < 1$ and $\dot{x}$ still positive, $F_{\text{drag}} > 0$, and the oscillations tend to grow in amplitude. In the tunnel diode case, it is the latter feature which causes the diode to begin to spontaneously oscillate even if it is connected to a steady (non-oscillatory) power supply.

As a follow up to the class room derivation, I had then asked the class to solve the VdP equation for $\epsilon = 5$ for a few initial conditions of their own choosing, using a graphical/numerical technique called the *method of isoclines.* The basis of the isoclines method is as follows. Setting $y \equiv \dot{x}$, the second-order VdP equation can be reduced to a system of two coupled first-order ODEs, viz.,

$$\dot{x} = y, \quad \dot{y} = \epsilon(1 - x^2)\,y - x. \tag{8.7}$$

Since the time doesn't explicitly appear[1] in (8.7), it may be eliminated by forming the ratio $dy/dx = (\epsilon(1-x^2)\,y - x)/y$. But this ratio is just the slope tangent to the solution "trajectory" in the x-y plane (called the *phase plane*) at an instant in time. A *tangent field* picture can be created by drawing systematically spaced arrows in the x-y plane, the arrows oriented along the slope directions and the arrow heads pointing in the direction of increasing time. Given some initial condition, $x(0)$, $y(0)$, the solution curve (called a *phase-plane trajectory*) can be drawn in the x-y plane by following the arrows. A tangent field picture with one or more superimposed solution curves is referred to as a *phase-plane portrait* of the autonomous ODE or ODE system.

The method of isoclines was commonly used in the pre-computer age to systematically draw the tangent field arrows. In this method, curves corresponding to different constant slopes (the isoclines) were drawn and equally spaced arrows pointing in the slope direction placed on each isocline. Once the x-y plane was filled with a sufficiently fine grid of arrows, one could see how a solution would evolve with time from any point in the phase plane.

So, why do I remember Joe? It's because when Joe handed in his solution to the problem (a week late!), it was in the form of a large cylindrical scroll fastened with an elastic band. Removing the band, I began to unwind the scroll. To my amazement, I found that the scroll stretched across the width of my office. Joe had evidently chosen the wrong scale for his initial conditions and stubbornly kept splicing sheets together until he obtained a complete solution

---

[1]The equations are referred to as *autonomous.*

trajectory. To compound Joe's woes, he had not written a simple program to automate the process but had evidently used a calculator and then plotted the isoclines, arrows, and trajectories completely by hand. However, to his credit, the plot was basically correct and I didn't have the heart to penalize Joe for handing the problem in late.

The following recipe carries out the solution to Joe's problem painlessly and quickly. It makes use of the **phaseportrait** command which is in Maple's DEtools library, so this package is loaded. I will take $\epsilon = 5$.

```
>  restart: with(DEtools): epsilon:=5:
```

Equations (8.7) are entered in *ode1* and *ode2*.

```
>  ode1:=diff(x(t),t)=y(t);
```

$$ode1 := \frac{d}{dt} x(t) = y(t)$$

```
>  ode2:=diff(y(t),t)=epsilon*(1-x(t)^2)*y(t)-x(t);
```

$$ode2 := \frac{d}{dt} y(t) = 5 \left(1 - x(t)^2\right) y(t) - x(t)$$

Two initial conditions are considered. The phase-plane trajectory will start from near the origin for *ic1*, and reasonably far from the origin for *ic2*.

```
>  ic1:=x(0)=0.01,y(0)=0.01: ic2:=x(0)=-1,y(0)=6:
```

The following **phaseportrait** command is used to plot the tangent field[2] and the two trajectories corresponding to the above initial conditions. In the argument, the two ODEs are entered as a list, as are the dependent variables to be solved for. The time range is taken to be $t = 0$ to 20 for the solution curves and the initial conditions are entered as a list of lists. The plot range is taken to be $x = -2.5$ to 2.5, and $y = -10$ to 13. To obtain accurate solution curves, the numerical step size is taken to be 0.01. The **dirgrid** option controls the number of tangent arrows to be drawn. Here, I have chosen to plot $30 \times 30 = 900$ arrows. The default is 20×20. Various arrow styles are available. I have chosen **arrows=MEDIUM** and colored the arrows blue. Finally, the two trajectories are colored red and green. The result is shown on the left of Figure 8.5.

```
>  phaseportrait([ode1,ode2],[x(t),y(t)],t=0..20,[[ic1],[ic2]],
   x=-2.5..2.5,y=-10..13,stepsize=.01,dirgrid=[30,30],
   arrows=MEDIUM,color=blue,linecolor=[red,green]);
```

The tangent arrows provide a visual guide to how trajectories will evolve as the time increases. As time evolves, the trajectory corresponding to *ic1* unwinds in a spiral fashion from near the origin onto a closed loop, indicative of a cyclic solution. The trajectory corresponding to *ic2* winds onto the same closed loop. In fact, no matter what the initial conditions, all trajectories wind onto the closed loop, which is an example of a *limit cycle*. In fact, since all trajectories wind onto it as $t \to \infty$, it is referred to as a *stable* limit cycle.

By including the option **scene=[t,x(t)]**, the **phaseportrait** command can also be used to plot $x(t)$ versus t.

---

[2]Isoclines are not drawn.

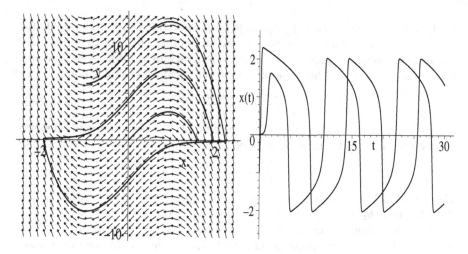

Figure 8.5: Left: trajectories winding onto limit cycle. Right: $x(t)$ versus t.

```
>  phaseportrait([ode1,ode2],[x(t),y(t)],t=0..30,[[ic1],[ic2]],
   x=-2.5..2.5,y=-10..13,stepsize=.01,scene=[t,x(t)],
   linecolor=[red,green]);
```

The result is shown on the right of Figure 8.5. After a transient interval, the two curves have identical shapes. The steady-state curves are characterized by periods of relatively slowly varying x, periodically interspersed with abrupt changes. These types of oscillations are referred to as *relaxation oscillations*. By increasing ϵ, the period of the oscillations can be increased.

8.2.2 Squid Munch (Slurp?) Herring

Man is the only animal that can remain on friendly terms with the victims he intends to eat until he eats them.
Samuel Butler, English author, (1835–1902)

Another pre-computer approach to qualitatively sketching possible solution curves of a second-order autonomous nonlinear ODE (or two coupled first-order NLODEs) in the phase-plane was to first locate all the *stationary* or *singular* points (where all derivatives vanish) of the nonlinear ODE, identify their "topological" nature, and use this knowledge to sketch the trajectories. More specifically, this topological approach is as follows.

Consider the pair of first-order autonomous ODEs,

$$\dot{x} = P(x,y), \qquad \dot{y} = Q(x,y), \tag{8.8}$$

where, in general, P and Q are nonlinear functions and we have taken time as the independent variable. For the Van der Pol oscillator of the last recipe, $P \equiv y$ and $Q \equiv -x + \epsilon(1 - x^2)y$. A stationary point (x_0, y_0) corresponds to

$\dot{x} = 0$ and $\dot{y} = 0$, or $P(x_0, y_0) = Q(x_0, y_0) = 0$. For the Van der Pol oscillator, there is only one stationary point, namely $(x_0 = 0,\ y_0 = 0)$, i.e., the origin. For any *ordinary* point outside a stationary point, we can write its coordinates as $x = x_0 + u$, $y = y_0 + v$. The slope of the trajectory at an ordinary point is

$$\frac{dy}{dx} = \frac{Q(x_0 + u, y_0 + v)}{P(x_0 + u, y_0 + v)}. \tag{8.9}$$

For an ordinary point close to a stationary point, the numerator and denominator on the rhs can be Taylor expanded about (x_0, y_0) in powers of u and v, so

$$\frac{dy}{dx} = \frac{dv}{du} = \frac{c\,u + d\,v + \cdots}{a\,u + b\,v + \cdots}, \tag{8.10}$$

where $a \equiv (\partial P/\partial x)_{x_0, y_0}$, $b \equiv (\partial P/\partial y)_{x_0, y_0}$, $c \equiv (\partial Q/\partial x)_{x_0, y_0}$, $d \equiv (\partial Q/\partial y)_{x_0, y_0}$. Provided that $bc - ad \neq 0$, the trajectories near the stationary point can be correctly described by retaining only the linear terms in u and v in (8.10). In this case, the stationary point is referred to as *simple*. In this approximation, equation (8.10) can be thought of as resulting from the pair of linear ODEs,

$$\dot{u} = a\,u + b\,v, \qquad \dot{v} = c\,u + d\,v, \tag{8.11}$$

which have solutions of the form $(u, v) \sim e^{\lambda t}$, with $\lambda = -\frac{p}{2} \pm \frac{1}{2}\sqrt{p^2 - 4q}$, where $p = -(a + d)$ and $q = ad - bc$. Detailed examination of the two λ roots reveals that there are only four types of simple stationary points, the *saddle*, *focal* or *spiral*, *nodal*, and *vortex* points. Which type occurs depends on the ranges of q, p, and $p^2 - 4q$ as indicated in Table 8.1.

Stationary Point	q=ad−bc	p=−(a+d)	p²−4q
saddle	< 0	≥ 0 and ≤ 0	> 0
higher order	= 0	≥ 0 and ≤ 0	≥ 0
stable focal		> 0	< 0
stable nodal		> 0	≥ 0
vortex or focal	> 0	= 0	< 0
unstable focal		< 0	< 0
unstable nodal		< 0	≥ 0

Table 8.1: Classification of stationary or singular points.

In the neighborhood of these simple stationary points, the trajectories have the schematic form shown in Figure 8.6, the arrows indicating increasing time. *Stable* focal and nodal points are shown, the trajectories approaching the stationary points as $t \to \infty$. For *unstable* focal and nodal points, the arrow directions are reversed. The origin of the phase-plane for the Van der Pol oscillator is an example of an unstable focal point, which can be seen in Figure 8.5.

For $q > 0$ and $p = 0$, note that either a vortex or focal point occurs. The reason for the "uncertainty" is that inclusion of quadratic (or higher) terms in

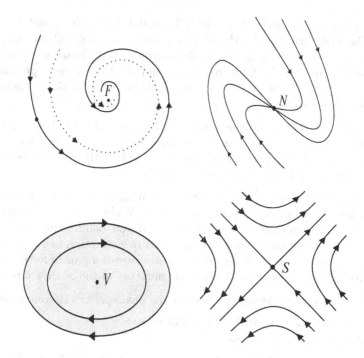

Figure 8.6: Curves near a focal (F), nodal (N), vortex (V), saddle (S) point.

the Taylor expansion may turn a closed loop (for the vortex) into a spiral. Instead of examining these higher-order terms, we quite often rely on a sufficient, but not necessary, *global* theorem due to Poincaré, which is built on symmetry considerations. *If $P(x, -y) = -P(x, y)$ and $Q(x, -y) = Q(x, y)$, then the stationary point is a vortex, not a focal point.*

For $q = 0$ and arbitrary p, the stationary point is no longer simple, and is referred to as a *higher-order* stationary point. Trajectories in the neighborhood of these points tend to be more complicated than in Figure 8.6.

In the pre-computer age, we would locate and identify the stationary points and attempt to draw the trajectories in the entire phase-plane by splicing the trajectories around each stationary point together. This procedure works well enough when the nonlinear ODE has only a few widely separated simple stationary points, with no higher-order points or limit cycles present. But in this computer age, we can simply use, e.g., the `phaseportrait` command. However, it's still important to know the locations and types of stationary points as this provides a deeper understanding of the solution curves. This is illustrated in the following mathematical biology example.

The major food source for squid is herring. If S and H are the numbers of squid and herring, respectively, per acre of seabed, the interaction between the two species can be modeled [Sco87] by the system (with time in years)

$$\dot{H} = k_1 H - k_2 H^2 - k_3 H S, \qquad \dot{S} = -k_4 S - k_5 S^2 + k_6 H S, \qquad (8.12)$$

with $k_1 = 1.1$, $k_2 = 10^{-5}$, $k_3 = 10^{-3}$, $k_4 = 0.9$, $k_5 = 10^{-4}$, and $k_6 = 2 \times 10^{-5}$. The first term on the rhs of $\dot{H}$ represents the natural growth of the herring population if the resources (food) were unlimited, the second term limits the growth because of the finite resources, and the third term represents the decrease in population due to interaction (i.e., being eaten) with the squid. You should be able to interpret the terms in the other equation.

(a) Locate and classify all the stationary points of the NLODEs. Plot the tangent field and show the trajectories for (i) $H(0) = 5, S(0) = 5$; (ii) $H(0) = 10000, S(0) = 1500$; (iii) $H(0) = 120000, S(0) = 500$. Take $t = 0$ to 50. Relate the results to the stationary points.

(b) Suppose that every squid was removed from the area occupied by the herring and from all surrounding areas. Would H increase indefinitely or would it approach an upper limit? If you believe the latter would occur, what is that number? If the squid free situation had persisted for many years, what are S and H two years later if a pair of fertile squid is introduced into the area? Round the numbers to the nearest integer.

The DEtools library package is needed for the **phaseportrait** command.

```
>  restart: with(plots): with(DEtools):
```
The coefficient values are entered,
```
>  k1:=1.1: k2:=10^(-5): k3:=10^(-3): k4:=0.9: k5:=10^(-4):
   k6:=2*10^(-5):
```
and functional operators P and Q introduced to generate the rhs of the NLODEs (8.12) for arbitrary H and S.
```
>  P:=(H,S)->k1*H-k2*H^2-k3*H*S; Q:=(H,S)->-k4*S-k5*S^2+k6*H*S;
```
$$P := (H, S) \to k1\, H - k2\, H^2 - k3\, H\, S$$
$$Q := (H, S) \to -k4\, S - k5\, S^2 + k6\, H\, S$$
Using P and Q, the necessary derivatives to calculate a, b, c, d are calculated. These expressions have to be evaluated at each stationary point.
```
>  a:=diff(P(H,S),H);   b:=diff(P(H,S),S);   c:=diff(Q(H,S),H);
   d:=diff(Q(H,S),S);
```
$$a := 1.1 - \frac{H}{50000} - \frac{S}{1000} \quad b := -\frac{H}{1000} \quad c := \frac{S}{50000} \quad d := -0.9 - \frac{S}{5000} + \frac{H}{50000}$$
The number of stationary points and their coordinates is determined by setting $P(H, S) = 0$ and $Q(H, S) = 0$ and solving for H and S.
```
>  sol:=solve({P(H,S)=0,Q(H,S)=0},{H,S});
```
$$sol := \{H = 0., S = 0.\}, \{H = 110000., S = 0.\}, \{H = 0., S = -9000.\},$$
$$\{S = 619.0476190, H = 48095.23810\}$$
The NLODEs have four stationary points at the locations indicated above. Note that the stationary point at $H = 0$, $S = -9000$ is in a non-physical portion of the phase-plane, since we must have $H \geq 0$ and $S \geq 0$.

Functional operators p and q are now formed to calculate $p = -(a + d)$ and $q = a\,d - b\,c$ for the ith stationary point.

```
>  p:=i->evalf(eval(-(a+d),sol[i])):
   q:=i->evalf(eval(a*d-b*c,sol[i])):
```

In the following do loop, p, q, and $r \equiv p^2 - 4q$ are calculated for each of the four singular points. The classification scheme of Table 8.1 is implemented by using a conditional if...then...elif...else...end if statement. elif is a contraction of "else if". This allows each singular point to be classified. For this conditional statement to work, numerical (not symbolic) values must be provided for the coefficients in the NLODEs.

```
>  for i from 1 to 4 do
>  sol[i]; p||i:=p(i); q||i:=q(i); r||i:=simplify(p||i^2-4*q||i);
>  if q||i<0 then s||i:=saddle;
   elif q||i>0 and p||i>0 and r||i<0  then s||i:=stablefocal;
   elif q||i>0 and p||i>0 and r||i>=0 then s||i:=stablenodal;
   elif q||i>0 and p||i<0 and r||i<0  then s||i:=unstablefocal;
   elif q||i>0 and p||i<0 and r||i>=0 then s||i:=unstablenodal;
   elif q||i>0 and p||i=0 then sing||i:=vortex_or_focal;
   else s||i:=higherorder; end if: s||i;
>  end do;
```

$$\{H = 0., \ S = 0.\} \quad p1 := -0.2000000000 \quad q1 := -0.9900000000$$
$$r1 := 4. \quad saddle$$
$$\{H = 110000., \ S = 0.\} \quad p2 := -0.2000000000 \quad q2 := -1.430000000$$
$$r2 := 5.760000000 \quad saddle$$
$$\{H = 0., \ S = -9000.\} \quad p3 := -11. \quad q3 := 9.090000000$$
$$r3 := 84.64000000 \quad unstablenodal$$
$$\{H = 48095.23810, \ S = 619.0476190\} \quad p4 := 0.5428571428 \quad q4 := 0.6252380953$$
$$r4 := -2.206258504 \quad stablefocal$$

So, the stationary points at $H = 0$, $S = 0$ and $H = 110,000$, $S = 0$ are saddle points, $H = 0$, $S = -9000$ is an unstable nodal point, and $H \simeq 48095$, $S \simeq 619$ is a stable focal point. This implies that all trajectories in the "physical region" $H \geq 0$, $S \geq 0$ will be attracted to the stable focal point as $t \to \infty$. I.e., if the equations are valid for all t (which they wouldn't be!), one would ultimately end up with about 48095 herring and 619 squid per acre of seabed. To confirm this, let's create a phase-plane portrait, first entering the NLODE system.

```
>  sys:=diff(H(t),t)=P(H(t),S(t)),diff(S(t),t)=Q(H(t),S(t));
```

$$sys := \frac{d}{dt} H(t) = 1.1\, H(t) - \frac{1}{100000} H(t)^2 - \frac{1}{1000} H(t)\, S(t),$$
$$\frac{d}{dt} S(t) = -0.9\, S(t) - \frac{1}{10000} S(t)^2 + \frac{1}{50000} H(t)\, S(t)$$

A functional operator PP is formed to generate the phase-plane portrait $S(t)$ vs. $H(t)$, as well as $H(t)$ vs. t and $S(t)$ vs. t, for the system of NLODEs.

```
>  PP:=(X,Y)->phaseportrait([sys],[H(t),S(t)],t=0..50,
      [[H(0)=5,S(0)=5],[H(0)=10000,S(0)=1500],
      [H(0)=120000,S(0)=500]],stepsize=.05,scene=[X,Y],
      arrows=MEDIUM,color=blue,dirgrid=[20,20],
      linecolour=[red,green,black],xtickmarks=3):
```

Which plot is produced depends on the choice of scene variables X and Y. The lists of equations and dependent variables are entered as arguments, along with the time range which is taken here to be 0 to 50 years. The initial population numbers are given as a list of lists. The numerical time stepsize is taken to be 0.05 years and medium blue arrows are chosen for the tangent field. The grid density for the tangent arrows is taken to be 20×20. The trajectories corresponding to the three initial conditions are colored red, green, and black, and the number of tickmarks along the horizontal axis controlled.

Then PP is used to produce the phase-plane portrait, as well as plots of $H(t)$ versus t and $S(t)$ versus t.

```
>  PP(H(t),S(t)); PP(t,H(t)); PP(t,S(t));
```

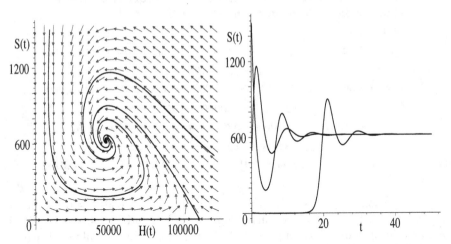

Figure 8.7: Phase-plane portrait (left) and S vs. t (right) for 3 initial conditions.

S vs. H is shown on the left of Figure 8.7, and S vs. t on the right. In the phase-plane picture on the left, one can clearly see that all three trajectories asymptotically wind onto the stable focal point, as was anticipated. The trajectory corresponding to $H(0)=5$, $S(0)=5$ initially travels horizontally very close to the $H(t)$ axis until it gets close to the saddle point at $S=0$, $H=110,000$. It then makes an abrupt turn, heading towards the stable focal point and winding onto it. The trajectory starting at $H(0)=10,000$, $S(0)=1500$ initially descends almost vertically until it begins to follow the tangent field in the vicinity of the saddle point at the origin. The fourth singular point at $H = 0$, $S = -9000$ plays no role in the "flow" of the trajectories. The phase-plane portrait doesn't show the times involved, but of course the other plot on the rhs of Figure 8.7 does.

If every squid was removed from the area occupied by the herring and from all surrounding areas, H would asymptotically approach an upper limit of 110,000 herring (the H value of the saddle point to which the trajectory would be attracted). To answer the last part of part (b), we numerically solve the system of NLODEs with the initial condition $H(0) = 110,000$, $S(0) = 2$, giving the output as a listprocedure.

```
> sol2:=dsolve({sys,H(0)=110000,S(0)=2},{H(t),S(t)},
    type=numeric,output=listprocedure):
```

The number of herring and squid at an arbitrary time t years later is now determined using sol2.

```
> H:=eval(H(t),sol2): S:=eval(S(t),sol2):
```

The number (per acre of seabed) of herring and squid two years later is now calculated, and (using the round command) found to be about 108,794 and 26.

```
> H2:=round(H(2)); S2:=round(S(2));
```

$$H2 := 108794 \qquad S2 := 26$$

8.3 Nonlinear ODEs: Approximate Methods

The *perturbation* and *Krylov–Bogoliubov* approximation methods, which may be applied when the nonlinear terms in the NLODEs are small, are illustrated, along with the *Ritz* trial function method which can be used when they are not.

8.3.1 Poisson's Method Isn't Fishy

Let a man get up and say, Behold, this is the truth, and instantly I perceive a sandy cat filching a piece of fish in the background. Look, you have forgotten the cat, I say.
Virginia Woolf, British novelist, (1882–1941)

Perturbation methods are applicable to NLODEs when the nonlinear terms are small. As a simple example, consider the motion of a unit mass moving in a viscous medium characterized by a Hooke's law restoring force $F_{Hooke} = -y$ and a drag force $F_{drag} = -a\,v - \epsilon\,v^3 = -a\,(dy/dt) - \epsilon\,(dy/dt)^3$ where y is the displacement from equilibrium, v is the velocity, a is a positive coefficient, and ϵ is a small positive parameter. The governing NLODE then is

$$\ddot{y} + a\dot{y} + \epsilon\dot{y}^3 + y = 0. \tag{8.13}$$

For $\epsilon = 0$, (8.13) reduces to the linearly damped simple harmonic oscillator equation. Our goal is to derive an approximate analytic solution when ϵ is small, but not zero, for the initial condition $y(0) = 1$, $\dot{y}(0) = 0$. Two values of a will be considered, first $a = 1/2$ which is entered below, and later, $a = 0$. The *Poisson perturbation method* is to assume a series solution in powers of ϵ, viz.,

$y(t) = y_0(t) + \epsilon\, y_1(t) + \epsilon^2\, y_2(t) + \cdots + \epsilon^N\, y_N(t)$. In our recipe, let's take $N = 2$, i.e., we will stop at second order in ϵ in our expansion.

```
> restart: with(plots): N:=2: a:=1/2:
```
A functional operator is introduced for generating the governing NLODE.
```
> ode:=y->diff(y(t),t,t)+a*diff(y(t),t)
         +epsilon*(diff(y(t),t))^3+y(t)=0;
```

$$ode := y \to (\frac{d^2}{dt^2}\, y(t)) + a\,(\frac{d}{dt}\, y(t)) + \epsilon\,(\frac{d}{dt}\, y(t))^3 + y(t) = 0$$

The Poisson perturbation expansion is now entered out to order N.
```
> Y(t):=sum(y[n](t)*epsilon^n,n=0..N);
```

$$Y(t) := y_0(t) + y_1(t)\,\epsilon + y_2(t)\,\epsilon^2$$

The perturbation expansion is automatically substituted into the NLODE in the following command line and the result expanded. The very lengthy output is suppressed with a line-ending colon.
```
> ode2:=expand(ode(Y)):
```
Powers of ϵ are now collected in *ode2*, all powers of order $\epsilon^{(N+1)}$ and higher being set equal to zero. The result is then simplified.
```
> eq1:=simplify(collect(ode2,epsilon),{epsilon^(N+1)=0}):
```
Since ϵ is arbitrary (but small), the coefficient of each power of ϵ must be equal to zero. This will generate a linear ODE for each order of the perturbation expansion. These equations are produced in the following do loop.
```
> for n from 0 to N do
> Eq[n]:=coeff(lhs(eq1),epsilon,n)=0;
> end do;
```

$$Eq_0 := (\frac{d^2}{dt^2}\, y_0(t)) + \frac{1}{2}\,(\frac{d}{dt}\, y_0(t)) + y_0(t) = 0$$

$$Eq_1 := (\frac{d^2}{dt^2}\, y_1(t)) + (\frac{d}{dt}\, y_0(t))^3 + y_1(t) + \frac{1}{2}\,(\frac{d}{dt}\, y_1(t)) = 0$$

$$Eq_2 := (\frac{d^2}{dt^2}\, y_2(t)) + 3\,(\frac{d}{dt}\, y_0(t))^2\,(\frac{d}{dt}\, y_1(t)) + \frac{1}{2}\,(\frac{d}{dt}\, y_2(t)) + y_2(t) = 0$$

The zeroth order equation, Eq_0, is just the linearly damped simple harmonic oscillator equation. It is now solved subject to the initial condition $y_0(0) = 1$, $\dot{y}_0(0) = 0$, and the solution is assigned.
```
> sol[0]:=dsolve({Eq[0],y[0](0)=1,D(y[0])(0)=0},y[0](t));
  assign(sol[0]):
```

$$sol_0 := y_0(t) = \frac{1}{15}\,\sqrt{15}\,e^{(-\frac{t}{4})}\,\sin(\frac{\sqrt{15}\,t}{4}) + e^{(-\frac{t}{4})}\,\cos(\frac{\sqrt{15}\,t}{4})$$

Since the zeroth-order solution satisfies the initial condition, all higher-order equations must be solved subject to $y_n(0) = \dot{y}_n(0) = 0$, where $n = 1, \cdots, N$.

This is done in the following do loop, using the Laplace transform option of the dsolve command. Each solution is assigned and the do loop ended.

```
>   for n from 1 to N do
>   sol[n]:=dsolve({Eq[n],y[n](0)=0,D(y[n])(0)=0},y[n](t),
            method=laplace); assign(sol[n]):
>   end do:
```

The lengthy perturbation expansion is now written out[3] in Y,

```
>   Y:=sum('epsilon^n*rhs(sol[n])','n'=0..N):
```

and ϵ and exponential terms sequentially collected.

```
>   Y:=collect(Y,[epsilon,exp]);
```

$$Y := (\frac{1}{15}\sin(\frac{\sqrt{15}\,t}{4})\sqrt{15}+\cos(\frac{\sqrt{15}\,t}{4}))\,e^{(-\frac{t}{4})}$$

$$+\epsilon\,((-\frac{4}{305}\cos(\frac{3\sqrt{15}\,t}{4})+\frac{116}{13725}\sin(\frac{3\sqrt{15}\,t}{4})\,\sqrt{15}+\frac{4}{5}\cos(\frac{\sqrt{15}\,t}{4})$$

$$+\frac{4}{75}\sin(\frac{\sqrt{15}\,t}{4})\,\sqrt{15})e^{(-\frac{3t}{4})}+(-\frac{48}{61}\cos(\frac{\sqrt{15}\,t}{4})+\frac{8}{305}\sin(\frac{\sqrt{15}\,t}{4})\,\sqrt{15})\,e^{(-\frac{t}{4})})$$

$$+\epsilon^2\,((\frac{127494}{161101}\cos(\frac{\sqrt{15}\,t}{4})-\frac{10818}{805505}\sin(\frac{\sqrt{15}\,t}{4})\,\sqrt{15})\,e^{(-\frac{t}{4})}$$

$$-(\frac{108}{61}\cos(\frac{\sqrt{15}\,t}{4})+\frac{84}{1525}\sin(\frac{\sqrt{15}\,t}{4})\,\sqrt{15}+\frac{32}{1525}\sin(\frac{3\sqrt{15}\,t}{4})\,\sqrt{15})\,e^{(-\frac{3t}{4})}$$

$$+(-\frac{648}{211975}\cos(\frac{5\sqrt{15}\,t}{4})-\frac{1544}{3179625}\sqrt{15}\sin(\frac{5\sqrt{15}\,t}{4})-\frac{486}{28975}\cos(\frac{3\sqrt{15}\,t}{4})$$

$$+\frac{2306}{86925}\sin(\frac{3\sqrt{15}\,t}{4})\,\sqrt{15}+\frac{28944}{28975}\cos(\frac{\sqrt{15}\,t}{4})+\frac{6896}{86925}\sin(\frac{\sqrt{15}\,t}{4})\,\sqrt{15})e^{(-\frac{5t}{4})})$$

Clearly, deriving the above solution by hand would be a tedious task. Let's now take a specific value for the parameter, say $\epsilon = 0.4$.

```
>   epsilon:=0.4:
```

The perturbation solution Y is plotted over the time interval $t=0$ to 20 in gr1.

```
>   gr1:=plot(Y,t=0..20,labels=["t","y"],thickness=2):
```

For comparison, the NLODE is now solved numerically, subject to the given initial condition,

```
>   X:=dsolve({ode(x),x(0)=1,D(x)(0)=0},x(t),numeric,
            output=listprocedure):
```

and plotted in gr2. A point style is chosen, with fifty size 12 black circles.

```
>   gr2:=odeplot(X,[t,x(t)],0..20,style=point,symbol=circle,
            symbolsize=12,numpoints=50,color=black):
```

The two graphs are superimposed with the display command,

```
>   display([gr1,gr2]);
```

---

[3]Note the enclosure of the summand and summation index in "right quotes" here. This is required for the sum to be carried out. See Maple's Help on sum for a discussion of this issue.

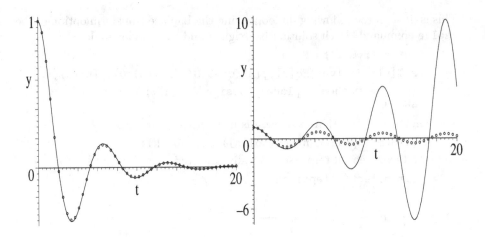

Figure 8.8: Numerical (circles) and perturbation (line) solutions for $a = 1/2$ (left) and $a=0$ (right).

the resulting picture being shown on the left of Figure 8.8. In this case, the perturbation expansion fits the numerical solution very well.

However, if the recipe is re-executed with $a = 0$, trouble occurs. Y then takes the following form, containing *secular terms* involving powers of t.

$$Y = \cos(t) + \epsilon \left(\frac{1}{32} \sin(3t) + \frac{9}{32} \sin(t) - \frac{3}{8} t \cos(t) \right)$$
$$+ \epsilon^2 \left(-\frac{9}{64} t \sin(t) + \frac{27}{128} t^2 \cos(t) - \frac{9}{256} t \sin(3t) - \frac{3}{1024} \cos(5t) + \frac{3}{1024} \cos(t) \right).$$

Referring to the right hand side picture in Figure 8.8, the perturbation solution agrees with the exact numerical result for short times, but eventually the secular terms cause the perturbation expansion to diverge from the exact solution, the oscillations in fact increasing with time, rather than decreasing. Going to higher order in the perturbation expansion will not "cure" the situation, as even higher powers of t then occur. Their are a variety of methods (see Zwillinger [Zwi89]) for dealing with the presence of secular terms in the perturbation solution of a NLODE. In the following recipe, we illustrate one method which can be applied to a NLODE having a periodic solution.

8.3.2   Lindstedt Saves the Day

I like to do all the talking myself.
It saves time, and prevents arguments.
Oscar Wilde, Anglo-Irish playwright, author, (1854–1900)

Consider a spring whose displacement from equilibrium satisfies the NLODE,

$$\ddot{y}(t) + y(t) + \epsilon\, y(t)^5 = 0, \tag{8.14}$$

with $\epsilon > 0$ and small. Our goal is to derive a second-order perturbation solution to equation (8.14), subject to the initial condition $y(0) = 1$, $\dot{y}(0) = 0$. The solution then will be compared to the exact numerical solution for $\epsilon = 1/2$. If the Poisson method is applied, secular terms will occur, which would destroy the expected periodic solution. So what do we do?

For $\epsilon = 0$, (8.14) is just the simple harmonic oscillator equation which has the periodic solution $\cos(t)$, with frequency 1. The *Lindstedt perturbation method* is to assume not only a perturbation expansion for y, but that the frequency changes slightly to a new frequency Ω, which can also be represented by a perturbation series. Introducing a new time variable $T = \Omega t$, and setting $X \equiv y$, equation (8.14) can be rewritten as

$$\Omega^2 \ddot{X}(T) + X(T) + \epsilon X(T)^5 = 0. \tag{8.15}$$

Lindstedt's procedure is to assume that

$$X(T) = x_0(T) + \epsilon x_1(T) + \epsilon^2 x_2(T) + \cdots, \quad \text{and} \quad \Omega = 1 + \epsilon \omega_1 + \epsilon^2 \omega_2 + \cdots. \tag{8.16}$$

The secular terms are removed in each order by imposing the periodicity condition $X(T + 2\pi) = X(T)$, i.e., $x_n(T + 2\pi) = x_n(T)$, for $n = 0, 1, 2, \ldots$. The initial conditions are $x_0(0) = 1$, $\dot{x}_0 = 0$, and $x_n(0) = 0$, $\dot{x}_n(0) = 0$ for $n = 1, 2, \ldots$.

The following recipe carries out Lindstedt's method for the given NLODE. The maximum power, $N = 2$, of ϵ to be kept in the expansion is entered. The following **alias** command produces the output symbol ϵ when e is entered.

```
> restart: with(plots): N:=2: alias(epsilon=e):
```
The NLODES (8.14) and (8.15) are entered in *ode1a* and *ode1b*, respectively.
```
> ode1a:=diff(y(t),t,t)+y(t)+e*y(t)^5=0;
```

$$ode1a := (\frac{d^2}{dt^2} y(t)) + y(t) + \epsilon y(t)^5 = 0$$

```
> ode1b:=Omega^2*diff(X(T),T,T)+X(T)+e*X(T)^5=0;
```

$$ode1b := \Omega^2 (\frac{d^2}{dT^2} X(T)) + X(T) + \epsilon X(T)^5 = 0$$

The perturbation expansions for $X(T)$ and Ω, given in (8.16), are inputted.
```
> X(T):=sum('x||n(T)*e^n','n'=0..N);
```

$$X(T) := x0(T) + x1(T)\epsilon + x2(T)\epsilon^2$$

```
> Omega:=1+sum('omega||n*e^n','n'=1..N);
```

$$\Omega := 1 + \omega1\,\epsilon + \omega2\,\epsilon^2$$

The above series are automatically substituted into *ode1b*, which is expanded.
```
> ode2:=expand(ode1b):
```
Powers of ϵ are collected in *ode2*, and terms higher than ϵ^N set equal to 0.
```
> ode3:=simplify(collect(ode2,e),{e^(N+1)=0}):
```
A functional operator **eq** is formed to set the coefficient of the nth power of ϵ on the lhs of *ode3* equal to zero.
```
> eq:=n->coeff(lhs(ode3),e,n)=0:
```

Using the functional operator in the following sequence command produces the linear ODEs *eq0*, *eq1*, and *eq2*, corresponding to ϵ^0, ϵ^1, and ϵ^2, respectively.

```
>  eqs:=seq(eq||n=eq(n),n=0..N); assign(eqs);
```

$$eqs := eq0 = ((\frac{d^2}{dT^2}\, x0(T)) + x0(T) = 0),$$

$$eq1 = (x1(T) + (\frac{d^2}{dT^2}\, x1(T)) + x0(T)^5 + 2\,\omega 1\,(\frac{d^2}{dT^2}\, x0(T)) = 0),$$

$$eq2 = ((\frac{d^2}{dT^2}\, x2(T)) + 5\, x0(T)^4\, x1(T) + 2\,\omega 1\,(\frac{d^2}{dT^2}\, x1(T)) + x2(T)$$

$$+ \,\omega 1^2\,(\frac{d^2}{dT^2}\, x0(T)) + 2\,\omega 2\,(\frac{d^2}{dT^2}\, x0(T)) = 0)$$

An operator `sol` is created to analytically solve the *n*th order ODE, using the Laplace transform method. The order *n* and initial value *A* of x_n must be given.

```
>  sol:=(n,A)->dsolve({eq||n,x||n(0)=A,D(x||n)(0)=0},x||n(T),
                   method=laplace):
```

Setting $A=1$, the zeroth-order ODE is solved and the solution assigned.

```
>  sol0:=sol(0,1); assign(sol0):
```

$$sol0 := x0(T) = \cos(T)$$

The 1st ODE is solved with $A=0$, trig terms are combined, and *sol1* assigned.

```
>  sol1:=combine(sol(1,0),trig); assign(sol1);
```

$$sol1 := x1(T) = \frac{5}{128}\cos(3\,T) - \frac{1}{24}\cos(T) + \frac{1}{384}\cos(5\,T)$$

$$- \frac{5}{16}\,T\sin(T) + T\sin(T)\,\omega 1$$

The secular terms, $-\frac{5}{16}\,T\sin(T) + T\sin(T)\,\omega 1$, occur and must be removed from $x1(T)$ to maintain periodicity, i.e., to have $x1(T + 2\,\pi) = x1(T)$. This implies that we must have $\omega 1 = 5/16$. This frequency is extracted and the secular term removed from $x1(T)$ in the following command lines.

```
>  omega||1:=solve(coeff(x||1(T),sin(T))=0,omega||1);
   x||1(T):=subs(sin(T)=0,x||1(T));
```

$$\omega 1 := \frac{5}{16} \qquad x1(T) := \frac{1}{384}\cos(5\,T) + \frac{5}{128}\cos(3\,T) - \frac{1}{24}\cos(T)$$

Then, the second-order ODE is solved, and the solution assigned.

```
>  sol2:=combine(sol(2,0),trig); assign(sol2):
```

$$sol2 := x2(T) = \frac{1}{98304}\cos(9\,T) + \frac{95}{294912}\cos(7\,T) - \frac{5}{192}\cos(3\,T)$$

$$+ \frac{3791}{147456}\cos(T) + T\sin(T)\,\omega 2 + \frac{215}{3072}\,T\sin(T)$$

Secular terms occur in *sol2*, which are removed by choosing $\omega 2 = -215/3072$.

```
>  omega||2:=solve(coeff(x||2(T),sin(T))=0,omega||2);
```

```
x||2(T):=subs(sin(T)=0,x||2(T));
```

$$\omega2 := \frac{-215}{3072}$$

$$x2(T) := \frac{1}{98304}\cos(9\,T) + \frac{95}{294912}\cos(7\,T) - \frac{5}{192}\cos(3\,T) + \frac{3791}{147456}\cos(T)$$

The perturbation solution is now completely determined to second order in ϵ. Setting $\epsilon = 1/2$ and $T = \Omega\,t$, the Lindstedt perturbation solution X is displayed. This would be a tedious result to derive by hand.

```
>   e:=1/2: X:=subs(T=Omega*t,x||0(T)+e*x||1(T)+e^2*x||2(T));
```

$$X := \frac{581327}{589824}\cos(\frac{13993\,t}{12288}) + \frac{1}{768}\cos(\frac{69965\,t}{12288}) + \frac{5}{384}\cos(\frac{13993\,t}{4096})$$

$$+ \frac{1}{393216}\cos(\frac{41979\,t}{4096}) + \frac{95}{1179648}\cos(\frac{97951\,t}{12288})$$

A graph of X, as well as the $\epsilon = 0$ solution, is created over the time interval $t=0$ to 20.

```
>   gr1:=plot({X,cos(t)},t=0..20):
```

The original NLODE, *ode1a*, is numerically solved and plotted over the same time interval, size 12 black circles being used for the numerical curve.

```
>   Y:=dsolve({ode1a,y(0)=1,D(y)(0)=0},y(t),numeric,
            output=listprocedure):
```

```
>   gr2:=odeplot(Y,[t,y(t)],0..20,style=point,symbol=circle,
            symbolsize=12,numpoints=120,color=black):
```

The two graphs are superimposed, the resulting picture being shown in Figure 8.9. The Lindstedt curve fits the numerical data quite well, and differs appreciably from the $\epsilon = 0$ solution.

```
>   display({gr1,gr2});
```

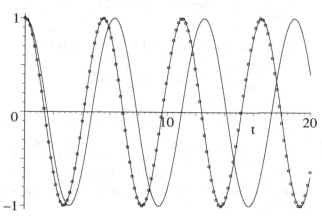

Figure 8.9: Lindstedt curve lies on numerical circles. Other curve: $\epsilon = 0$ solution.

8.3.3   Krylov–Bogoliubov Have A Say

He who says he hates every kind of flattery, and says it in earnest, certainly does not yet know every kind of flattery.
G. C. Lichtenberg, German physicist and philosopher, (1742–99)

For $a = 0$, equation (8.13) reduces to

$$\ddot{y} + y + \epsilon \dot{y}^3 = 0, \tag{8.17}$$

the initial condition being $y(0) = 1$, $\dot{y}(0) = 0$. For ϵ small, we saw that the Poisson perturbation expansion broke down for this NLODE, generating secular terms which grew with time. Since the solution of this equation is not periodic, the Lindstedt procedure cannot be used to remove the secular terms.

A procedure developed by Krylov and Bogoliubov [KB43] can be employed to solve NLODEs of the general structure

$$\ddot{x} + \omega_0^2 x + \epsilon f(x, \dot{x}) = 0, \tag{8.18}$$

with ϵ a small parameter. Equation (8.17) is of this structure with $x = y$, $\omega_0 = 1$ and $f = \dot{x}^3$. For $\epsilon = 0$, a general solution of (8.18) would be of the form $x = a \sin(\omega_0 t + \phi)$, with a and ϕ constants determined by the initial conditions. If x is the displacement, the velocity would be $\dot{x} = a \omega_0 \cos(\omega_0 t + \phi)$. In the Krylov–Bogoliubov (KB) method[4], exactly the same forms are assumed for x and $\dot{x}$, but the amplitude a and phase factor ϕ are allowed to become time-dependent. Provided that $a(t)$ and $\phi(t)$ are "slowly varying", their structures are determined by solving [EM00] [Zwi89],

$$\dot{a} = -\frac{\epsilon}{2\pi\omega_0} \int_0^{2\pi} f(a \sin\psi, a\omega_0 \cos\psi) \, \cos\psi \, d\psi,$$
$$\dot{\phi} = \frac{\epsilon}{2\pi a\omega_0} \int_0^{2\pi} f(a \sin\psi, a\omega_0 \cos\psi) \, \sin\psi \, d\psi, \tag{8.19}$$

where $\psi \equiv \omega_0 t + \phi$. The criterion for being slowly varying is that

$$\left| \frac{\dot{a}}{a} \right| \frac{2\pi}{\omega_0} \ll 1, \quad \text{and} \quad \left| \frac{\dot{\phi}}{\omega_0} \right| \ll 1. \tag{8.20}$$

Recalling that $\omega_0 = 1$, for equation (8.17) we have $f = (a \cos\psi)^3$. Since the relevant integrand $\cos^3\psi \sin\psi$ is an odd function of ψ, $\dot{\phi} = 0$, so ϕ is a constant. We need only determine the form of $a(t)$ and make sure that the solution $x = a(t) \sin(t + \phi)$ satisfies the initial condition. We shall now do this for the same initial condition as in Recipe **08-3-1**, taking $x(0) = 1$ and $\dot{x}(0) = 0$. The KB solution will be compared to the numerical solution for the same value of ϵ as in that recipe, viz., $\epsilon = 0.4$.

Use of the following `alias` command allows us to enter `e` in the input and generate ϵ in the output.

---

[4] Also referred to as the *method of averaging* [Zwi89].

```
> restart: with(plots): alias(epsilon=e):
```
The form of f is entered in terms of ψ,
```
> f:=(a(t)*cos(psi))^3;
```

$$f := a(t)^3 \cos(\psi)^3$$

and the integration in the $\dot{a}$ equation in (8.19) carried out.
```
> eq:=diff(a(t),t)=-(e/(2*Pi))*int(f*cos(psi),psi=0..2*Pi);
```

$$eq := \frac{d}{dt}\, a(t) = -\frac{3}{8}\, \epsilon\, a(t)^3$$

The resulting first-order ODE eq is analytically solved for $a(t)$,
```
> sol:={dsolve(eq,a(t))};
```

$$sol := \left\{ a(t) = \frac{2}{\sqrt{3\,\epsilon t + 4\,\_C1}},\ a(t) = -\frac{2}{\sqrt{3\,\epsilon t + 4\,\_C1}} \right\}$$

generating two answers with _C1 an arbitrary integration constant. The rhs of the positive square root solution (the first one here) is selected to form the solution x. The two constants _C1 and ϕ remain to be determined.
```
> x:=rhs(sol[1])*sin(t+phi);
```

$$x := \frac{2\sin(t + \phi)}{\sqrt{3\,\epsilon t + 4\,\_C1}}$$

Taking $\epsilon = 0.4$, we set $x(0) = 1$ in initial condition $ic1$,
```
> e:=0.4: ic1:=eval(x,t=0)=1;
```

$$ic1 := \frac{1}{2}\, \frac{\sqrt{4}\sin(\phi)}{\sqrt{\_C1}} = 1$$

and $\dot{x}(0) = 0$ in $ic2$.
```
> ic2:=eval(diff(x,t),t=0)=0;
```

$$ic2 := -\frac{0.07500000000\,\sqrt{4}\sin(\phi)}{\_C1^{(3/2)}} + \frac{1}{2}\, \frac{\sqrt{4}\cos(\phi)}{\sqrt{\_C1}} = 0$$

Then $ic1$ and $ic2$ are numerically solved for ϕ and _C1. The solution $sol2$ is assigned, and the final KB solution, x, displayed.
```
> sol2:=fsolve({ic1,ic2},{phi,_C1}); assign(sol2): x:=x;
```

$$sol2 := \{\_C1 = 0.9769696007,\ \phi = -11.14792061\}$$

$$x := \frac{2\sin(t - 11.14792061)}{\sqrt{1.2\,t + 3.907878403}}$$

The KB solution is plotted over the time interval $t = 0$ to 20.
```
> gr1:=plot(x,t=0..20,labels=["t","x"]):
```
The NLODE is entered,
```
> ode:=diff(y(t),t,t)+y(t)+e*diff(y(t),t)^3=0;
```

$$ode := \left(\frac{d^2}{dt^2}\, y(t)\right) + y(t) + 0.4\left(\frac{d}{dt}\, y(t)\right)^3 = 0$$

solved numerically subject to the same initial condition,

```
>  Y:=dsolve({ode,y(0)=1,D(y)(0)=0},y(t),numeric,
      output=listprocedure):
```
and plotted using 50 size 12 black circles to represent the numerical curve.
```
>  gr2:=odeplot(Y,[t,y(t)],0..20,style=point,symbol=circle,
      symbolsize=12,numpoints=50,color=black):
```
The KB and numerical solutions are superimposed with the `display` command,
```
>  display([gr1,gr2]);
```
the resulting picture being shown in Figure 8.10.

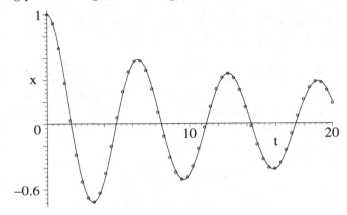

Figure 8.10: Circles: numerical solution. Curve: KB solution.

The KB solution is in quite good agreement with the exact (numerical) solution.

8.3.4   A Ritzy Approach

We must never confuse elegance with snobbery.
Yves Saint Laurent, French couturier, *Ritz, no. 85* (London, 1984)

The perturbation and Krylov–Bogoliubov methods can be applied to a NLODE
when the nonlinear terms are small compared to the linear parts of the equation.
The Ritz method is more general, not being restricted to small nonlinearities.
Consider a general ODE,

$$f(x, \dot{x}, \ddot{x}, ..., t) = 0, \tag{8.21}$$

where f is some nonlinear function of its arguments. In the spirit of the
Rayleigh–Ritz method for estimating eigenvalues, we try to guess at a trial
wave function Φ which comes "close" to satisfying (8.21). Since, in general, Φ
will not be an exact solution, substitution of Φ into (8.21) will not generate
0 on the right-hand side, but some time-dependent contribution $e(t)$, referred
to as the *residual*, i.e., $f(\Phi, \dot{\Phi}, \ddot{\Phi}, ...t) = e(t)$. Paralleling the method of least
squares for fitting experimental data, the *Ritz method* calculates the *total error*

$$E = \int_a^b e^2(t)\, dt, \qquad (8.22)$$

over a range a to b of interest. One then tries to choose a Φ which minimizes the total error. As in the Rayleigh–Ritz method, one or more adjustable parameters may be included and the error is minimized with respect to these parameters.

As an example, let's apply the Ritz method to the hard spring equation,

$$\ddot{x} + x + x^3 = 0, \qquad (8.23)$$

with $x(0) \equiv A = 2$, $\dot{x}(0) = 0$. Although we have already solved the hard spring equation analytically in terms of Jacobian elliptic functions, the Ritz solution will be compared to the numerical solution. As a trial function, let's take

$$\Phi = C_1 \cos(\omega t) + (A - C_1) \cos(3\omega t), \qquad (8.24)$$

with C_1 and ω adjustable parameters. At $t = 0$, $\Phi(0) = A$, $\dot{\Phi}(0) = 0$, so the initial conditions are satisfied.

The value of A is entered along with the upper limit $T = n\,2\pi/|\omega|$ in the time range $t = 0$ to T "of interest" that will be used for plotting purposes. For sake of definiteness, let's take $n = 4$, so that $T = 8\pi/|\omega|$, with ω to be determined.

```
> restart: with(plots): A:=2: n:=4: T:=n*2*Pi/abs(omega):
```
The trial function Φ is entered.
```
> Phi:=C1*cos(omega*t)+(A-C1)*cos(3*omega*t);
```

$$\Phi := C1 \cos(\omega t) + (2 - C1)\cos(3\omega t)$$

A functional operator e for generating the residual of the NLODE (8.23) is formed. Entering $e(\text{Phi})$ then will produce the residual.
```
> e:=x->diff(x,t,t)+x+x^3;
```

$$e := x \rightarrow (\frac{d^2}{dt^2}\, x) + x + x^3$$

The total error is calculated, the assumption that $\omega > 0$ being provided to facilitate the integration. A lengthy expression in terms of $C1$ and ω results.
```
> Error:=int(e(Phi)^2,t=0..T) assuming omega>0;
```

$$
\begin{aligned}
Error := \pi(&1256\, C1^4 + 4032\, \omega^2\, C1 - 1376\, C1 + 2416\, C1^2 + 2592\, \omega^4 \\
& - 390\, C1^5 - 2304\, \omega^2 + 55\, C1^6 + 544 - 2592\, C1\, \omega^4 + 656\, C1^2\, \omega^4 \\
& + 1728\, C1^3\, \omega^2 - 3712\, C1^2\, \omega^2 - 2256\, C1^3 - 312\, C1^4\, \omega^2)/(2\, \omega)
\end{aligned}
$$

The total error must be minimized with respect to ω and $C1$. To this end, the derivatives of the *Error* with respect to ω and $C1$ are set equal to zero in *cond1* and *cond2*. The very lengthy algebraic equations are not shown here.
```
> cond1:=diff(Error,omega)=0; cond2:=diff(Error,C1)=0;
```
The two conditions are numerically solved for $C1$, ω and the solution assigned.
```
> sol:=fsolve({cond1,cond2},{C1,omega}); assign(sol):
```

$$sol := \{\omega = -1.978539105,\ C1 = 1.937051656\}$$

The fully evaluated total error, trial function Φ, and time T, are now displayed.

> `Error:=evalf(abs(Error)); Phi:=Phi; T:=evalf(T);`

$$Error := 0.2146672855$$

$$\Phi := 1.937051656\cos(1.978539105\,t) + 0.062948344\cos(5.935617315\,t)$$

$$T := 12.70267601$$

The trial function Φ is plotted over the time interval $t=0$ to T. Since Φ is our "best" approximation to the exact solution x, the ordinate is labeled as x.

> `gr1:=plot(Phi,t=0..T,labels=["t","x"]):`

The NLODE is solved numerically, subject to the same initial condition,

> `Y:=dsolve({e(y(t))=0,y(0)=A,D(y)(0)=0},y(t),numeric,`
>   `output=listprocedure):`

and plotted over the same time range. A point style is chosen, the numerical curve being represented by 150 size 12 black circles.

> `gr2:=odeplot(Y,[t,y(t)],0..T,style=point,symbol=circle,`
>   `symbolsize=12,numpoints=150,color=black):`

The graphs of the Ritz and numerical solutions are superimposed,

> `display([gr1,gr2]);`

the resulting picture being shown in Figure 8.11. The Ritz solution is in excel-

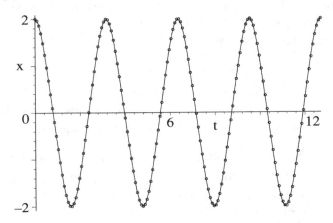

Figure 8.11: Circles: numerical solution. Curve: Ritz solution.

lent agreement with the numerical points over the time range of the plot. It should be mentioned that there are several variations on the Ritz approximation method, which are discussed in Zwillinger's book.

8.4 Nonlinear PDEs

Most of the nonlinear PDEs that my graduate students and I have had to solve in our research in nonlinear optics and fluid dynamics have involved either seeking special solutions, or more often using numerical methods. Here, a few examples of the former are presented, the latter being covered in Chapter 9.

8.4.1 John Scott Russell's Chance Interview

The boisterous sea of liberty is never without a wave.
Thomas Jefferson, American president, (1743–1826)

Shortly after Jefferson's death, the great Scottish naval architect, John Scott Russell, made the following important scientific observation:

> *I was observing the motion of a boat which was rapidly drawn along a narrow channel by a pair of horses, when the boat suddenly stopped – not so the mass of water in the channel which it had put in motion; it accumulated round the prow of the vessel in a state of violent agitation, then suddenly leaving it behind, rolled forward with great velocity, assuming the form of a large solitary elevation, a rounded smooth and well-defined heap of water, which continued its course along the channel apparently without change of form or diminution of speed. I followed it on horseback, and overtook it still rolling on at a rate of some eight or nine miles an hour, preserving its original figure some thirty feet long and a foot to a foot and a half in height. Its height gradually diminished, and after a chase of one or two miles I lost it in the windings of the channel. Such, in the month of August 1834, was my first chance interview with that singular and beautiful phenomenon . . .*

The relevant nonlinear PDE describing the water waves for the above situation was derived several decades later by Korteweg and deVries, the Korteweg-deVries (KdV) equation taking the form [KdV95]

$$\frac{\partial \psi}{\partial t} + \psi \frac{\partial \psi}{\partial x} + \frac{\partial^3 \psi}{\partial x^3} = 0, \tag{8.25}$$

with x the distance, t the time, and ψ the transverse displacement of the water from equilibrium. Damping of the water waves is completely neglected here. What Scott Russell observed was a special localized solution of the KdV equation, referred to as a *solitary wave* solution. The form of the solitary wave may be obtained by assuming that $\psi(x,t) = U(z = x - ct)$, subject to the asymptotic boundary conditions that $U(z)$ and all its derivatives vanish as $|z| \to \infty$. The parameter c will be the speed of the solitary wave. In this recipe, U will be derived and the solitary wave animated for two different values of c.

The plots and PDEtools packages are loaded, the former containing the animation command, while the PDEtools package has the dchange command which will be used to transform the variables in the KdV equation.

```
>   restart: with(plots): with(PDEtools):
```
The KdV equation (8.25) is entered.

```
>   pde:=diff(psi(x,t),t)+psi(x,t)*diff(psi(x,t),x)
        +diff(psi(x,t),x,x,x)=0;
```

$$pde := (\frac{\partial}{\partial t}\,\psi(x,\,t)) + \psi(x,\,t)\,(\frac{\partial}{\partial x}\,\psi(x,\,t)) + (\frac{\partial^3}{\partial x^3}\,\psi(x,\,t)) = 0$$

The variable transformation $x = z + c\tau$, $t = \tau$, and $\psi(x,t) = U(z)$ is entered, x, t and $\psi(x,t)$ being the "old" variables, while z, τ, and $U(z)$ are the "new" variables. This transformation will have the effect of reducing the nonlinear pde to a nonlinear ODE.

```
>   tr:={x=z+c*tau,t=tau,psi(x,t)=U(z)};
```

$$tr := \{x = z + c\tau,\, t = \tau,\, \psi(x,\,t) = U(z)\}$$

The variable transformation tr is applied to pde using the dchange command.

```
>   ode1:=dchange(tr,pde,[z,tau,U(z)]);
```

$$ode1 := -(\frac{d}{dz}\,U(z))\,c + U(z)\,(\frac{d}{dz}\,U(z)) + (\frac{d^3}{dz^3}\,U(z)) = 0$$

The lhs of $ode1$ is now integrated. Since $U(z)$ and its second derivative must vanish at $z = \infty$, the integration constant is equal to zero. The integrated result is set equal to zero, yielding the second order NLODE $ode2$.

```
>   ode2:=int(lhs(ode1),z)=0;
```

$$ode2 := -c\,U(z) + \frac{1}{2}\,U(z)^2 + (\frac{d^2}{dz^2}\,U(z)) = 0$$

Then $ode2$ is analytically solved for $U(z)$, yielding two implicit solutions.

```
>   sol:=dsolve(ode2,U(z));
```

$$sol := \int^{U(z)} \frac{3}{\sqrt{-3\,\_a^3 + 9\,c\,\_a^2 + 9\,\_C1}}\, d\_a - z - \_C2 = 0,$$

$$\int^{U(z)} -\frac{3}{\sqrt{-3\,\_a^3 + 9\,c\,\_a^2 + 9\,\_C1}}\, d\_a - z - \_C2 = 0$$

The positive square root solution (the first one here) is now differentiated with respect to z, and dU/dz is isolated to the lhs of the slope equation.

```
>   slope:=isolate(diff(sol[1],z),diff(U(z),z));
```

$$slope := \frac{d}{dz}\,U(z) = \frac{1}{3}\,\sqrt{-3\,U(z)^3 + 9\,c\,U(z)^2 + 9\,\_C1}$$

Since $U(z)$ and $dU(z)/dz$ vanish at $z = \infty$, the integration constant $\_C1$ can be set equal to zero. The radical simplification command is also applied in $ode3$.

```
>   ode3:=radsimp(eval(slope,_C1=0));
```

$$ode3 := \frac{d}{dz} U(z) = \frac{1}{3}\sqrt{3}\, U(z)\sqrt{-U(z) + 3c}$$

The first order ODE *ode3* is analytically solved.

> sol2:=dsolve(ode3,U(z));

$$sol2 := z + \frac{2\,\mathrm{arctanh}(\dfrac{1}{3}\dfrac{\sqrt{-3\,U(z) + 9c}}{\sqrt{c}})}{\sqrt{c}} + \_C1 = 0$$

Then *sol2* is solved for $U(z)$. The integration constant $\_C1$ determines the location of the peak of the solitary wave and can be set equal to zero without loss of generality.

> U:=eval(solve(sol2,U(z)),_C1=0);

$$U := 3c - 3\tanh(\frac{z\sqrt{c}}{2})^2 c$$

Converting U to a sine/cosine form and simplifying with the trig option gives us the final form of the solitary wave solution in terms of the variable z.

> U:=simplify(convert(U,sincos),trig);

$$U := \frac{3c}{\cosh(\dfrac{z\sqrt{c}}{2})^2}$$

U has a maximum height at $z = 0$ which is proportional to c, a width which decreases with increasing c, and goes to zero as $|z| \to \infty$. We can convert back to the original variables by substituting $z = x - ct$ into U. ψ, given in the output of the following command line, is the solitary wave observed so long ago by John Scott Russell. This is clearly a localized pulse which travels unchanged in shape in the positive x direction with velocity c. Since the height is proportional to c, this implies that taller KdV solitary waves travel faster than shorter ones. We can confirm this by animating the solitary wave profile for two different velocities. First, let's apply the **unapply** command to ψ, turning it into a functional operator **f** depending on the value of c.

> psi:=subs(z=x-c*t,U); f:=unapply(psi,c):

$$\psi := \frac{3c}{\cosh(\dfrac{(x - ct)\sqrt{c}}{2})^2}$$

Taking $c = 1$ and $c = 3$, two solitary waves are animated by entering **f(1)** and **f(3)** in the animate command. On executing the following command line on the computer, and clicking on the plot, the animation will begin.

> animate({f(1),f(3)},x=-10..40,t=0..10,numpoints=200,
      frames=50,thickness=2,axes=framed,labels=["x","psi"]);

The initial profile of the animation is shown in Figure 8.12. As time progresses, the shorter solitary wave lags behind the taller one.

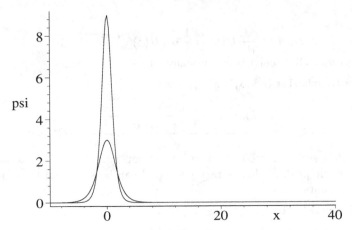

Figure 8.12: Initial frame of the solitary wave animation.

This difference in speeds between solitary waves of different amplitudes will be used in the next chapter to numerically test their stability against collisions with one another. If they survive unchanged, except perhaps for a phase shift, they are called *solitons*.

8.4.2 There is a Similarity

Ought not education to bring out and fortify the differences rather than the similarities?
Virginia Woolf, British novelist, referring to the gender issue, (1882–1941).

In the previous recipe, the number of independent variables was reduced from two to one, thus converting a nonlinear PDE to an ODE for which a physically important exact solution could be obtained. It was a simple application of the *similarity method*. Now I will show you a slightly more complicated similarity solution, solving a nonlinear diffusion equation by taking a different algebraic combination of the independent variables.

Consider the one-dimensional nonlinear diffusion equation,

$$\frac{\partial C}{\partial t} = \frac{\partial}{\partial x}\left(D(C)\frac{\partial C}{\partial x}\right), \quad D(C) = C^n, \tag{8.26}$$

with the diffusion coefficient D no longer constant, but a function of the concentration C. Several special cases of (8.26) have been studied in the literature, viz., $n = 3$ to model the spreading of thin liquid films under the action of gravity [Buc77], $n \geq 1$ to describe the percolation of gas through porous media [Mus37], and $n = 6$ to model certain radiative heat transfer [LP80].

After loading the PDEtools and plots library packages,

```
>   restart: with(PDEtools): with(plots):
```

the nonlinear diffusion equation (8.26) is entered. The inert form of the derivative operator is used to prevent the rhs from being explicitly differentiated.

> pde:=Diff(C(x,t),t)-Diff(C(x,t)^n*diff(C(x,t),x),x)=0;

$$pde := (\frac{\partial}{\partial t} C(x, t)) - (\frac{\partial}{\partial x} (C(x, t)^n (\frac{\partial}{\partial x} C(x, t)))) = 0$$

New variables z, τ, and $U(z)$ are introduced, related to the old variables x, t, and $C(x,t)$ by the transformation $x = z\tau^m$, $t = \tau$, and $C(x,t) = U(z)/\tau^m$, with the parameter m to be determined.

> tr:={x=z*tau^m,t=tau,C(x,t)=U(z)/tau^m};

$$tr := \{x = z\tau^m, \, t = \tau, \, C(x, t) = \frac{U(z)}{\tau^m}\}$$

The transformation is applied to pde using the dchange command, the result then being multiplied by $-\tau^{m+1}$ and expanded.

> ode:=expand(-tau^(m+1)*dchange(tr,pde,[z,tau,U(z)]));

$$ode := (\frac{d}{dz} U(z)) \, z \, m + U(z) \, m + \frac{\tau(\frac{U(z)}{\tau^m})^n \, n \, (\frac{d}{dz} U(z))^2}{(\tau^m)^2 \, U(z)}$$

$$+ \frac{\tau(\frac{U(z)}{\tau^m})^n \, (\frac{d^2}{dz^2} U(z))}{(\tau^m)^2} = 0$$

To remove all τ factors from ode and thus really produce an ODE, we must have $(\tau^m)^2 = \tau(1/\tau^m)^n$ which is now entered.

> eq:=(tau^m)^2=tau*(1/tau^m)^n;

$$eq := (\tau^m)^2 = \tau(\frac{1}{\tau^m})^n$$

Then eq is solved for m, yielding $m = 1/(n+2)$. This m value will be automatically substituted into ode.

> m:=solve(eq,m);

$$m := \frac{1}{2+n}$$

Dividing ode by m, simplifying symbolically with the power option, and then applying the general simplification command, yields the formidable looking NLODE given in $ode2$

> ode2:=simplify(simplify(ode/m,power,symbolic));

$$ode2 := (\frac{d}{dz} U(z)) \, z + U(z) + 2 \, U(z)^{(n-1)} \, n \, (\frac{d}{dz} U(z))^2 + U(z)^n \, (\frac{d^2}{dz^2} U(z)) \, n$$

$$+ U(z)^{(n-1)} \, n^2 \, (\frac{d}{dz} U(z))^2 + 2 \, U(z)^n \, (\frac{d^2}{dz^2} U(z)) = 0$$

A general analytic solution to $ode2$ is sought using the dsolve command, the result then being simplified symbolically with the power option.

> sol:=simplify(dsolve(ode2,U(z)),power,symbolic);

$sol := U(z) = \_b(\_a)$ &where

$$[\{\_b(\_a)^n \,(\frac{d}{d\_a}\_b(\_a)) + \frac{\_a\_b(\_a) + 2\_C1 + \_C1\,n}{2+n} = 0\},$$

$$\{\_a = z,\ \_b(\_a) = U(z)\},\ \{z = \_a,\ U(z) = \_b(\_a)\}]$$

A general solution for $U(z)$ is not produced in sol, but $U(z) = \_b(\_a)$ where $\_b(\_a)$ satisfies a first order NLODE. A solution can be generated if the integration constant $\_C1$ is set equal to zero. The following operand command is used to extract the resulting ODE from sol.

```
>  _C1:=0: ode3:=op([2,2,1],sol)[1];
```

$$ode3 := \_b(\_a)^n \,(\frac{d}{d\_a}\_b(\_a)) + \frac{\_a\_b(\_a)}{2+n} = 0$$

Then, $ode3$ is analytically solved for $\_b(\_a)$.

```
>  sol2:=dsolve(ode3,_b(_a));
```

$$sol2 := \_b(\_a) = \frac{(\dfrac{-\_a^2\,n + 4\_C2 + 2\_C2\,n}{2+n})^{(\frac{1}{n})}}{2^{(\frac{1}{n})}}$$

The integration constant $\_C2$ controls the height and width of the solution. Here, I will take $\_C2 = 1$. Substituting $\_a = z$ on the rhs of $sol2$ yields $U(z)$.

```
>  U:=subs({_a=z,_C2=1},rhs(sol2));
```

$$U := \frac{(\dfrac{-z^2\,n + 4 + 2\,n}{2+n})^{(\frac{1}{n})}}{2^{(\frac{1}{n})}}$$

The above solution only exists for z values such that $(-z^2\,n + 4 + 2\,n)/(2+n) \geq 0$. Outside this range, we can take $U = 0$, which clearly satisfies the NLODE. Substituting $z = x/t^m$ into U/t^m gives us a solution, labeled c, in terms of the original variables.

```
>   c:=subs(z=x/t^m,U/t^m);
```

$$c := \frac{\left(\dfrac{-\dfrac{x^2\,n}{(t^{(\frac{1}{2+n})})^2} + 4 + 2\,n}{2+n}\right)^{(\frac{1}{n})}}{2^{(\frac{1}{n})}\,t^{(\frac{1}{2+n})}}$$

The x range for which c is valid is found by setting $c = 0$ and solving for x. The result is simplified with the radical simplification command and labeled X.

```
>  X:=radsimp(solve(c=0,x));
```

$$X := \frac{\sqrt{2}\,\sqrt{n\,(2+n)}\,t^{(\frac{1}{2+n})}}{n}$$

As a specific example, let's take $n = 3$, so the solution can be used to describe the spreading of a thin liquid film under the action of gravity. Other n values can be chosen, if so desired. The total time range is taken to be $T = 1500$, and X is evaluated at this time and assigned the name $X0$.

```
>  n:=3: T:=1500: X0:=eval(X,t=T):
```

The complete concentration profile C is then described by the following piecewise function.

```
>  C:=piecewise(abs(x)<X,c,abs(x)>X,0);
```

$$C := \begin{cases} \dfrac{2^{(2/3)}\left(-\dfrac{3\,x^2}{5\,t^{(2/5)}} + 2\right)^{(1/3)}}{2\,t^{(1/5)}} & |x| < \dfrac{\sqrt{2}\,\sqrt{15}\,t^{(1/5)}}{3} \\[4ex] 0 & \dfrac{\sqrt{2}\,\sqrt{15}\,t^{(1/5)}}{3} < |x| \end{cases}$$

A 3-dimensional plot of C is now produced,

```
>  plot3d(C,x=-X-5..X+5,t=1..T,numpoints=1500,axes=box,labels
   =["x","t","C"],orientation=[40,70],tickmarks=[3,3,3]);
```

the resulting picture being shown in Figure 8.13.

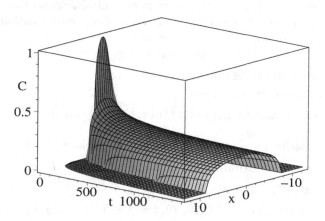

Figure 8.13: Similarity solution of the nonlinear diffusion equation for $n = 3$.

The solution captures some of the experimentally observed features of spreading thin films. Unlike the situation for the linear (D constant) diffusion equation, there is now a sharp interface separating the regions of nonzero and zero concentration. Further, the interface propagates with finite speed, in contrast with the infinite speed of linear diffusion. The propagation of this interface can be observed by executing the following **animate** command, clicking on the resulting computer plot, and on the start arrow.

```
>  animate(C,x=-X0..X0,t=1..T,frames=50,numpoints=500,
   labels=["x","C"]);
```

For more on similarity methods, you are referred to Bluman and Cole [BC74].

8.4.3 Creating Something Out Of Nothing

Say nothing good of yourself, you will be distrusted;
say nothing bad of yourself, you will be taken at your word.
Joseph Roux, French priest, writer, (1834–86)

The sine-Gordon equation (SGE),

$$\frac{\partial^2 u}{\partial x^2} - \frac{\partial^2 u}{\partial t^2} = \sin u, \qquad (8.27)$$

is a model equation for describing the motion of a Bloch domain wall between
two ferromagnetic domains. A solitary wave solution to the SGE, describing the
Bloch wall motion, could be obtained in a similar manner to that for the KdV
equation. An alternate way is to make use of an *auto-Bäcklund transformation*.
Given a solution of a nonlinear PDE, such a transformation allows us to find a
different solution of the same PDE. More, generally a *Bäcklund transformation*
may enable one to use the solution of one nonlinear PDE to determine the
solution of another nonlinear PDE. Bäcklund transformations are difficult to
find, so in this recipe I will merely confirm the auto-Bäcklund transformation
for the SGE and use it to create a non-trivial solitary wave solution of (8.27),
starting with the trivial null ($u=0$) solution.

After loading the PDEtools and plots library packages,

>   `restart: with(PDEtools): with(plots):`

the sine-Gordon equation is entered.

>   `sge:=diff(u(x,t),x,x)-diff(u(x,t),t,t)=sin(u(x,t));`

$$sge := (\frac{\partial^2}{\partial x^2}\, u(x,\, t)) - (\frac{\partial^2}{\partial t^2}\, u(x,\, t)) = \sin(u(x,\, t))$$

Introducing the variable transformation $x = X + T$, $t = X - T$, $u(x,t) = U(X,T)$,

>   `tr:={x=X+T,t=X-T,u(x,t)=U(X,T)};`

$$tr := \{u(x,\, t) = U(X,\, T),\, t = X - T,\, x = X + T\}$$

and using the dchange command, the SGE takes the form shown in *sge2*.

>   `sge2:=dchange(tr,sge,[X,T,U(X,T)]);`

$$sge2 := \frac{\partial^2}{\partial X\,\partial T}\, U(X,\, T) = \sin(U(X,\, T))$$

From Zwillinger [Zwi89], the auto-Bäcklund transformation for the SGE is given
by *ab1* and *ab2*, where a is an arbitrary parameter.

>   `ab1:=diff(V(X,T),X)=diff(U(X,T),X)+2*a*sin((V(X,T)+U(X,T))/2);`

$$ab1 := \frac{\partial}{\partial X}\, V(X,T) = (\frac{\partial}{\partial X}\, U(X,T)) + 2\,a\sin(\frac{1}{2}\, V(X,T) + \frac{1}{2}\, U(X,T))$$

>   `ab2:=diff(V(X,T),T)=-diff(U(X,T),T)+(2/a)*sin((V(X,T)-U(X,T))/2);`

$$ab2 := \frac{\partial}{\partial T}\, V(X,T) = -(\frac{\partial}{\partial T}\, U(X,T)) + \frac{2\sin(\frac{1}{2}\, V(X,T) - \frac{1}{2}\, U(X,T))}{a}$$

If U is a solution of the SGE, and V satisfies the transformation, then V is also a solution of the SGE. Let us confirm the transformation. First, differentiating *ab1* and *ab2* with respect to T and X, respectively, and subtracting, yields 0 since $\partial^2 V/\partial T\partial X = \partial^2 V/\partial X\partial T$.

> `pde1:=diff(ab1,T)-diff(ab2,X);`

Substituting *ab1* and *ab2* into *pde1*/2, and applying the `combine` command with the trig option, we find that U satisfies the transformed SGE, *sge2*.

> `sge3:=combine(subs({ab1,ab2},pde1/2),trig);`

$$sge3 := 0 = (\frac{\partial^2}{\partial X\,\partial T}\,U(X,\,T)) - \sin(U(X,\,T))$$

Similarly, let's differentiate *ab1* and *ab2* with respect to T and X, respectively, and add the results.

> `pde2:=diff(ab1,T)+diff(ab2,X);`

Then, on dividing both the lhs and rhs of *pde2* by 2, substituting *ab1* and *ab2* into the latter and applying the `combine` command with the trig option, we find that V also satisfies the transformed SGE.

> `sge4:=lhs(pde2)/2=combine(subs({ab1,ab2},rhs(pde2))/2,trig);`

$$sge4 := \frac{\partial^2}{\partial X\,\partial T}\,V(X,\,T) = \sin(V(X,\,T))$$

Having confirmed the auto-Bäcklund transformation, let's set $U(X,T) = 0$ and calculate $ab1 + a^2\,ab2$ in *pde3*.

> `U(X,T):=0: pde3:=ab1+a^2*ab2;`

$$pde3 := (\frac{\partial}{\partial X}\,V(X,\,T)) + a^2\,(\frac{\partial}{\partial T}\,V(X,\,T)) = 4\,a\sin(\frac{1}{2}\,V(X,\,T))$$

Using the `pdsolve` command, a general solution V of *pde3* is obtained (not displayed here), expressed in terms of the arctangent function.

> `V:=rhs(pdsolve(pde3,V(X,T)));`

The solution is put into a simpler form by using the operand command on V. Here $\_F1(T - a^2\,X)$ is an arbitrary function of the argument.

> `V:=2*arctan(op([2,2],V)/op([2,1],V));`

$$V := -2\arctan\left(\frac{1}{2}\,\frac{e^{(4\,X\,a+4\,\_F1(T-a^2\,X)\,a)} - 1}{e^{(2\,X\,a+2\,\_F1(T-a^2\,X)\,a)}}\right)$$

Setting $T = (x - t)/2$ and $X = (x + t)/2$ transforms us back to the original independent variables. As an example, let's take $\_F1(T - a^2\,X) = T - a^2\,X$.

> `T:=(x-t)/2:  X:=(x+t)/2:  _F1(T-a^2*X):=T-a^2*X:`

Then, on simplifying, V takes the form shown below.

> `V:=simplify(V); s:=unapply(V,a):`

$$V := -2\arctan(\frac{1}{2}\,(e^{(-2\,a\,(-2\,x+a^2\,x+a^2\,t))} - 1)\,e^{(a\,(-2\,x+a^2\,x+a^2\,t))})$$

Having used the `unapply` command to change V into a functional operator s in terms of the parameter a, which controls the solitary wave velocity, the solution is animated for $a = -0.9$. The opening frame is shown on the left of Figure 8.14, the profile changing in a localized region from $V = -\pi$ to $+\pi$.

```
>   animate(s(-0.9),x=-10..10,t=0..20,frames=100,
    numpoints=500,thickness=2,labels=["x","V"]);
```

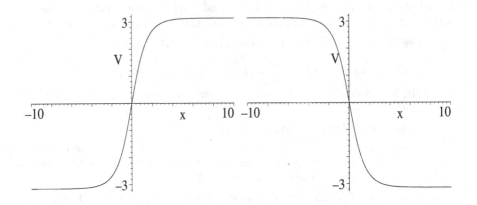

Figure 8.14: Left: sine-Gordon kink soliton. Right: Anti-kink soliton.

This profile is another example of a solitary wave solution (a *"kink" solitary wave*), propagating in the positive x direction with constant shape and velocity in the animation. It can be demonstrated that this solitary wave is collisionally stable so is referred to as a sine-Gordon *kink soliton*. If we take $a = 0.9$ in the recipe, an *anti-kink soliton* solution results as shown on the right of Figure 8.14.

8.4.4 Portrait of a Nerve Impulse

Is it a fact – or have I dreamt it – that, by means of electricity, the world of matter has become a great nerve, vibrating thousands of miles in a breathless point of time?
Nathaniel Hawthorne, American author, (1804–64)

A graphical technique for determining whether or not a nonlinear PDE has a solitary wave solution is now illustrated. One of the most important classes of systems to display nonlinear diffusion is found in the nerve fibers of animals. During the first half of the nineteenth century, physiologists assumed nervous activity to be propagated with the speed of light. Helmholtz [Hel50] decided to make a direct measurement of the nerve signal propagation speed in a frog's sciatic nerve, and obtained a speed of 32 meters per second. Nonlinear diffusion was not understood at that time and Helmholtz conjectured that the relatively small velocity was due to motion of the material particles. The correct understanding of what was going on eluded neurophysiologists until well into the twentieth century. Attributing the result to A.F. Huxley, Nagumo [NYA65] reported a solitary wave solution to the following nerve fiber equation,

$$\frac{\partial^2 \phi}{\partial x^2} = \frac{\partial \phi}{\partial t} + \phi\,(\phi - A)\,(\phi - 1), \tag{8.28}$$

with A a constant, and the solitary wave having a velocity $V = (1 - 2\,A)/\sqrt{2}$. Equation (8.28) is just the linear diffusion equation to which a cubic nonlinear term has been added for the electric potential ϕ. In the absence of the nonlinear term, the diffusion velocity is infinite.

By assuming that $\phi(x,t) = X(z)$ with $z = x - V\,t$ and creating a phase-plane portrait, $Y = dX/dz$ vs. $X(z)$, we shall graphically see that an anti-kink solitary wave solution to (8.28) does exist. This graphical approach has also been successfully used by one of my former graduate students to establish the existence of solitary wave solutions to a very formidable nonlinear optical PDE where an analytical solution was not possible.[5]

Loading the DEtools library package, let's take, say $A = 1/4$ and calculate the velocity V given by Nagumo.

```
>   restart: with(DEtools):
>   A:=1/4: V:=(1-2*A)/sqrt(2);
```

$$V := \frac{\sqrt{2}}{4}$$

Assuming $\phi(x,t) = X(z = x - V\,t)$, eq. (8.28) reduces to the pair of ODEs,

$$X'(z) = P(X,Y) = Y, \quad Y'(z) = Q(X,Y) = -V\,Y + X\,(X - A)\,(X - 1).$$

Functional operators P and Q are formed to calculate the rhs of these ODEs.

```
>   P:=(X,Y)->Y;  Q:=(X,Y)->-V*Y+X*(X-A)*(X-1);
```

$$P := (X, Y) \to Y \quad Q := (X, Y) \to -V\,Y + X\,(X - A)\,(X - 1)$$

The derivatives $a = \dfrac{\partial P}{\partial X}$, $b = \dfrac{\partial P}{\partial Y}$, $c = \dfrac{\partial Q}{\partial X}$, and $d = \dfrac{\partial Q}{\partial Y}$ are carried out.

```
>   a:=diff(P(X,Y),X); b:=diff(P(X,Y),Y);  c:=diff(Q(X,Y),X);
    d:=diff(Q(X,Y),Y);
```

Setting $P(X,Y) = 0$ and $Q(X,Y) = 0$ and solving for X and Y, yields

```
>   sol:=solve({P(X,Y)=0,Q(X,Y)=0},{X,Y});
```

$$sol := \{Y = 0,\ X = 0\},\ \{Y = 0,\ X = \frac{1}{4}\},\ \{Y = 0,\ X = 1\}$$

three stationary points at $(X = 0,\ Y = 0)$, $(1/4, 0)$ and $(1, 0)$. Operators for calculating $p = -(a + d)$ and $q = a\,d - b\,c$ for the ith stationary point are created.

```
>   p:=i->evalf(eval(-(a+d),sol[i])):
    q:=i->evalf(eval(a*d-b*c,sol[i])):
```

For each stationary point, the following do loop evaluates p, q, and $r = p^2 - 4\,q$, and, based on the values of these quantities, identifies the nature of the stationary point.

---

[5]J.P. Ogilvie, *Nonlocal Solitons in Photorefractive Materials*, M. Sc. thesis, Simon Fraser University, 1996.

```
> for i from 1 to 3 do
> sol[i]; p||i:=p(i); q||i:=q(i); r||i:=simplify(p||i^2-4*q||i);
> if q||i<0 then s||i:=saddle;
  elif q||i>0 and p||i>0 and r||i<0 then  s||i:=stablefocal;
  elif q||i>0 and p||i>0 and r||i>=0 then s||i:=stablenodal;
  elif q||i>0 and p||i<0 and r||i<0  then s||i:=unstablefocal;
  elif q||i>0 and p||i<0 and r||i>=0 then s||i:=unstablenodal;
  elif q||i>0 and p||i=0 then sing||i:=vortex_or_focal;
  else s||i:=higherorder; end if: s||i;
> end do;
```

$$\{Y = 0,\ X = 0\}\quad p1 := 0.3535533905\quad q1 := -0.2500000000$$

$$r1 := 1.125000000\quad saddle$$

$$\{Y = 0,\ X = \frac{1}{4}\}\quad p2 := 0.3535533905\quad q2 := 0.1875000000$$

$$r2 := -0.6250000001\quad stable focal$$

$$\{Y = 0,\ X = 1\}\quad p3 := 0.3535533905\quad q3 := -0.7500000000$$

$$r3 := 3.125000000\quad saddle$$

So, $(0,0)$ and $(1,0)$ are saddle points, while $(1/4,0)$ is a stable focal point. To create the phase-plane portrait, the system of ODEs is now entered.

```
> sys:=diff(X(z),z)=P(X(z),Y(z)),diff(Y(z),z)=Q(X(z),Y(z));
```

$$sys := \frac{d}{dz}X(z) = Y(z),\quad \frac{d}{dz}Y(z) = -\frac{1}{4}\sqrt{2}\,Y(z) + X(z)\,(X(z) - \frac{1}{4})\,(X(z) - 1)$$

Instead of using the **phaseportrait** command to create the tangent field with trajectories corresponding to specified initial conditions, we can form a functional operator **gr** using **DEplot** to accomplish the same objective. The **DEplot** command has the advantage that the X and Y ranges for the plot can be independently fixed. An initial condition is chosen close to the stationary point at $(1,0)$. The corresponding trajectory will be approximately that corresponding to the solitary wave. For other A values, this number will have to be altered if you wish to draw the trajectory corresponding to the solitary wave. The scene variables have to be specified.

```
> gr:=(x,y)->DEplot([sys],[X(z),Y(z)],z=0..25,
     [[X(0)=0.9999,Y(0)=-0.0001]],X=-0.1..1.1,Y=-0.25..0.25,
     stepsize=.01,scene=[x,y],arrows=MEDIUM,color=black,
     dirgrid=[30,30],linecolour=red):
```

Making use of the above graphing operator, the phase-plane portrait (X vs. Y) is produced along with X vs. z for the given initial condition.

```
> gr(X(z),Y(z)); gr(z,X(z));
```

The resulting pictures are shown in Figure 8.15. The behavior of the tangent field in the phase-plane portrait shown on the left is clearly consistent with the stationary point analysis given earlier. The trajectory corresponding to the

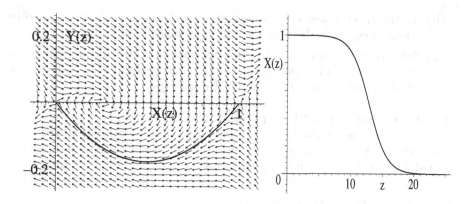

Figure 8.15: Left: Phase-plane portrait with saddle points at $(0,0)$, $(1,0)$, and stable focal point at $(1/4, 0)$. Right: X vs. z for separatrix trajectory.

chosen initial condition is approximately that of the separatrix from the saddle point at $(1,0)$ to the saddle point at the origin. The separatrix separates two distinct types of "flow". Above the separatrix, the tangent arrows wind onto the stable focal point at $(1/4, 0)$, while below the separatrix, the arrows flow off to infinity. The shape of $X(z)$ for the separatrix trajectory is shown on the right of the figure, clearly approximating that of an anti-kink solitary wave.

For larger values of A, the nature of the stationary points changes, but a separatrix (solitary wave) solution still exists. The initial condition needed to plot the separatrix must be altered accordingly.

8.5 Supplementary Recipes

08-S01: A Bunch of Bernoulli equations
For each of the following ODEs, confirm that it is a Bernoulli equation, analytically solve the ODE for $y(0) = 1$, and plot $y(x)$ over a suitable range.

(a) $y' + x^2 y = x^4 y^3$;

(b) $y - y' = 3 y^3 e^{-2x}$;

(c) $y^3 + 3 y^2 y' = 4$.

08-S02: The Riccati Equation
Another first-order nonlinear ODE which can be analytically solved is Riccati's equation which has the general structure
$$y' + a y^2 + f_1(x) y + f_2(x) = 0.$$

(a) By introducing a new dependent variable $z = e^{a \int_0^x y(X) \, dX}$, show that Riccati's equation can be reduced to the linear ODE
$$z'' + f_1(x) z' + a f_2(x) z = 0,$$
and the solution to Riccati's equation given by $y = z'/(a z)$.

(b) Taking $f_1 = 1/x$ and $f_2 = 1/a$, use the transformation in part (a) to solve Riccati's equation. Identify the functions which result.

(c) Solve Riccati's equation with $f_1 = 1/x$ and $f_2 = 1/a$ directly using the dsolve command. Making use of the infolevel command, show that Maple identifies the ODE as of the Riccati type.

08-S03: Period of the Plane Pendulum

The equation of motion of the undamped plane pendulum is $\ddot{\theta} + \omega^2 \sin\theta = 0$, where the angle θ is measured from the vertical. Analytically determine the period T of the plane pendulum for arbitrary angular amplitude A, with $|A| < \pi$. Taking $\omega = 1$, plot T versus A over the range $A = 0$ to π. Discuss the result.

08-S04: The Child–Langmuir Law

In a vacuum diode, electrons (each of charge q and mass m) are emitted from a hot planar cathode at $x = 0$, held at zero potential, and accelerated across a gap to a parallel planar anode at $x = d$, held at a positive potential V_0. The cloud of moving electrons within the gap (called the space charge) quickly builds up to the point where it reduces the electric field at the cathode to zero. Thereafter, a steady current I flows between the plates. If the cathode and anode plate areas $A \gg d$, edge effects can be neglected, and the potential V between the plates is governed by the 1-dimensional Poisson equation, $d^2V(x)/dx^2 = -\rho(x)/\epsilon_0$, where $\rho(x)$ is the electron charge density and ϵ_0 is the permittivity of free space.

(a) Assuming that the electrons start from rest at the cathode and that steady-state prevails, determine $V(x)$ between the plates. Plot $V(x)$ along with the form that the potential would have without space charge.

(b) Show that the current I satisfies the Child–Langmuir law $I = K V_0^{3/2}$, where the constant K remains to be determined in terms of V_0, A, q, m, d, and ϵ_0. With space charge present, a nonlinear relation exists between the current and voltage, i.e., Ohm's law is not satisfied.

08-S05: Soft Spring

The displacement $x(t)$ of a soft spring from equilibrium satisfies the NLODE

$$\ddot{x} + \omega_0^2 \left(1 - b^2 x^2\right) x = 0,$$

with $\omega_0 = 1$ and $b = 2$. Create a phase-plane portrait showing the tangent field in the $y \equiv \dot{x}$ vs. x plane and trajectories for the initial conditions $(x(0) = 0.4, y(0) = 0.01)$, $(x(0) = -0.941, y(0) = 0.9)$, and $(x(0) = 0.941, y(0) = -0.9)$. Take the time range to be $t = 0$ to 20, the time stepsize to be 0.01, the grid to be 30 by 30, and the x and y ranges to each be from -1 to $+1$. Plot x vs t for the three trajectories, and discuss the results.

08-S06: Gnits vs. Gnots

The gnits (N_1 per unit area) and gnots (N_2 per unit area) are competing for the same food supply, and are described by the following system of NLODES,

$$\dot{N}_1 = (a_1 - b_1 N_1 - c_1 N_2) N_1, \quad \dot{N}_2 = (a_2 - b_2 N_1 - c_2 N_2) N_2,$$

with $a_1 = 4$, $b_1 = 0.0002$, $c_1 = 0.0004$, $a_2 = 1.4$, $b_2 = 0.00015$, and $c_2 = 0.00005$.

(a) Find and identify the stationary points of this system.

(b) Produce a phase-plane portrait which includes all four stationary points, shows the tangent field, and the approximate separatrix trajectories which divide the phase plane into different possible outcomes for the gnit and gnot populations. What are these possible outcomes?

(c) Plot N_1 vs. t and N_2 vs. t for the above trajectories.

08-S07: The Vibrating Eardrum

In the era before the introduction of computer algebra into the teaching of my nonlinear physics class, I used to assign the following problem and would find that a substantial fraction of the students would make algebraic mistakes and get the wrong answer for the general case. This no longer happens in this computer algebra age.

Consider the equation of motion of the vibrating eardrum,

$$\ddot{x} + \omega_0^2 x + \epsilon x^2 = 0,$$

subject to the initial conditions $x(0) = A0$, $\dot{x}(0) = 0$. Using the Lindstedt perturbation method, determine the perturbation solution to second order in the parameter ϵ. Taking $\omega_0 = 2$, $\epsilon = 1$, and $A0 = 1/3$ plot the perturbation solution along with the numerical solution and discuss the results.

08-S08: Van der Pol Transient Growth

Taking ϵ to be small, and assuming $x(0) = A$ and $\dot{x}(0) = 0$, derive the KB solution to the Van der Pol equation. Plot the KB and numerical solutions together for $A = 1$ and $\epsilon = 1/10$ over the time interval 0 to 60, using circles to represent the numerical curve and a solid line for the KB solution.

08-S09: Another Ritzy Solution

Derive a KB solution to the nonlinear diode equation $\ddot{x} + x + x^5 = 0$, with $x(0) = 1$. Graphically compare the KB solution with the numerical answer. You should experiment with the form of the trial wave function until you obtain a small total error and hence a good fit of the curves.

08-S10: Portrait of a Dark Soliton

The *nonlinear Schrödinger equation*,

$$i \frac{\partial E}{\partial z} \pm \frac{1}{2} \frac{\partial^2 E}{\partial \tau^2} + |E|^2 E = 0,$$

can be used to describe picosecond duration optical envelope solitons in transparent optical fibers governed by the Kerr refractive index $n = n_0 + n_2 |\phi|^2$, with $n_0 > 0$, $n_2 > 0$. Here E is proportional to the electric field amplitude ϕ, z to the distance coordinate Z in the direction of propagation, and τ to $t - Z/v_g$, where t is the time and v_g is the group velocity. The plus-minus sign corresponds to anomalous or normal dispersion, respectively. In terms of the light intensity, which is proportional to $|E|^2$, so-called "bright" solitons (the intensity is a localized peak against a zero intensity background) occur for the plus sign, while "dark" solitons (the intensity is a localized dip to zero in a background of constant non-zero intensity) occur for the minus sign. The solitary wave solutions are found by assuming a "stationary" solution of the form $E(z, \tau) = X(\tau)\, e^{i\beta z}$ with X real and β real and positive. Taking $\beta = 1$, determine the nature of the stationary points for the dark case and create a phase-plane diagram of $X(\tau)$ versus $Y = dX/d\tau$ showing the tangent field and the two separatrixes which divide the phase plane into regions of qualitatively different flows. Noting that the intensity is proportional to $X(\tau)^2$, show that the two separatrixes correspond to dark solitary waves. It can be shown that these solitary waves are collisionally stable, and therefore are solitons.

08-S11: Bright Soliton Solution

Referring to **08-S10**, determine the analytic form $X(\tau)$ of the bright solitary wave solution and plot the intensity X^2 versus τ for $\beta = 1$. The bright solitary wave solution can be shown to be collisionally stable, so is a bright soliton.

Chapter 9

Numerical Methods

To this point, when a numerical solution to an ODE was sought, we simply resorted to the numeric option of Maple's `dsolve` command. In the first section of this chapter, we will briefly explore the underlying principles on which the numerical algorithms for solving linear and nonlinear ODEs are based. In the second section, some numerical recipes for solving linear and nonlinear PDEs are presented. If you wish to learn more about numerical methods, consult, e.g., *Numerical Analysis* [BF89] or *Numerical Recipes*[PFTV90].

9.1 Ordinary Differential Equations

As noted earlier, the second-order Van der Pol equation
$$\ddot{x} - \epsilon \left(1 - x^2\right) \dot{x} + x = 0, \quad \epsilon > 0, \tag{9.1}$$
can be written as the following pair of first-order ODEs,
$$\dot{x} = y, \quad \dot{y} = \epsilon \left(1 - x^2\right) y - x \equiv F(x, y). \tag{9.2}$$
More generally, an nth-order ODE can be decomposed into a system of n first-order equations. Numerical algorithms for solving such a system are based on replacing each first derivative and each function (e.g.,F) on the rhs with some finite difference approximation based on a finite Taylor series expansion.

For small h, $y(x \pm h)$ can be Taylor expanded in powers of h about x, viz.,
$$y(x \pm h) = y(x) \pm h y'(x) + \frac{1}{2!} h^2 y''(x) \pm \frac{1}{3!} h^3 y'''(x) + \cdots \tag{9.3}$$
where $y' \equiv dy/dx$, etc. For time-dependent problems, x is replaced with t. Taking the $+$ sign in (9.3), and dropping terms of order h^2 and higher, yields the *forward difference approximation* (FWDA) to $y'(x)$,
$$y'(x) = (1/h)[y(x + h) - y(x)] + O(h), \tag{9.4}$$
while the minus sign yields the *backward difference approximation* (BWDA),
$$y'(x) = (1/h)[y(x) - y(x - h)] + O(h). \tag{9.5}$$
Explicit (implicit) numerical schemes are based on the FWDA (BWDA).

9.1.1   Joe's Problem Revisited

One man's remorse is another man's reminiscence.
Ogden Nash, American poet, *A Clean Conscience Never Relaxes*, 1938

Recall that in the tale of Joe and the Van der Pol Scroll, the `phaseportrait` command was used to easily plot the solution of the VdP equation. In the present recipe, the VdP equation is again solved, but now using *Euler's method*, the simplest of the explicit schemes. The FWDA is used for the time derivatives in (9.2), thus connecting, say, the kth time step to step $k + 1$, and the right-hand sides are approximated by their values at the kth step. Explicitly, on setting $t_{k+1} = t_k + h$, the Euler algorithm for numerically solving (9.2) is

$$x_{k+1} = x_k + h\,y_k, \qquad y_{k+1} = y_k + h\,F(x_k, y_k). \tag{9.6}$$

For given values of h, x_0, and y_0, equations (9.6) are iterated forward in time.

So, let's begin our recipe. The plots package is loaded because we shall be using the `pointplot` and `odeplot` commands to plot the numerical solution.

```
>   restart: with(plots):
```

18 digits accuracy is specified, so a comparison can be made at a specific time with Maple's "dial-up" version of the Euler method which uses 18 digits. To ensure a reasonable approximation of the Euler algorithm to the exact ODEs, a small value of h, $h = 0.01$ s, is chosen. $n = 6000$ iterations are considered, so the total time is $0.01 \times 6000 = 60$ s.

```
>   Digits:=18: h:=0.01: n:=6000:
```

At $t_0 = 0$, let's take $x_0 = 0.2$, $y_0 = 0.1$, and choose $\epsilon = 20$.

```
>   t[0]:=0: x[0]:=0.2: y[0]:=0.1: epsilon:=20:
```

An operator is formed to calculate the function F.

```
>   F:=(x,y)->epsilon*(1-x^2)*y-x:
```

An important aspect in comparing numerical algorithms is to know how long it takes the algorithm to execute. Here, the time will be short, but let's record it anyways. The `time()` command records the total cpu time since the beginning of the Maple session. Here it marks the beginning of our Euler iteration.

```
>   begin:=time():
```

The iterative process is carried out with the following do loop. In the body of the loop, we have the time increment, $t_{k+1} = t_k + h$, followed by equations (9.6). The triplet of numbers, t_k, x_k, y_k on the kth time step are grouped as a Maple list in pt_k for 3-dimensional plotting purposes, and the loop is ended.

```
>   for k from 0 to n do
>   t[k+1]:=t[k]+h;
>   x[k+1]:=x[k]+h*y[k];
>   y[k+1]:=y[k]+h*F(x[k],y[k]);
>   pt[k]:=[t[k],x[k],y[k]]:
>   end do:
```

The time at the end of the do loop is recorded and the beginning time subtracted
to give us the elapsed cpu time for the algorithm.

```
> cpu_time:=time()-begin;
```

$$cpu\_time := 0.980$$

It took[1] about 1 s to generate all the data points needed for plotting. A 3-
dimensional plot of t vs. x vs. y is produced, using the sequence of n data
points, with the pointplot3d command. The orientation [-90,0] produces x vs.
t as shown on the left of Figure 9.1. Each cross represents a data point. In the
slowly varying regions, the data points are so thick that they blend into a solid
line. If you wish to see the phase plane portrait, y vs. x, simply rotate the
computer plot by clicking on it and dragging with your mouse.

```
> pointplot3d([seq(pt[j],j=0..n)],axes=normal,symbol=cross,
    symbolsize=8,color=red,labels=["t","x","y"],
    orientation=[-90,0]);
```

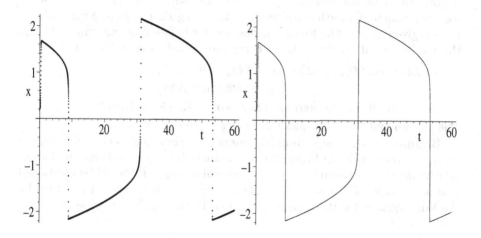

Figure 9.1: Euler solution of VdP eq. Left:"First principles"; Right:"Dial-up".

Maple has built-in options in its dsolve command to implement some common
specific numerical algorithms, the (forward) Euler method being one of them.
Let's now illustrate this "dial-up" approach, entering the system of first-order
ODEs for the VdP oscillator.

```
> sys:=diff(X(t),t)=Y(t),diff(Y(t),t)=F(X(t),Y(t));
```

$$sys := \frac{d}{dt} X(t) = Y(t), \ \frac{d}{dt} Y(t) = 20 \left(1 - X(t)^2 \right) Y(t) - X(t)$$

The dependent variables are specified along with the initial condition.

```
> vars:={X(t),Y(t)}: ic:=X(0)=0.2,Y(0)=0.1:
```

---

[1]All cpu times quoted in this chapter are for a 1 GHz personal computer.

The `dsolve` command is applied to *sys*, subject to the initial condition, a numerical solution being sought with stepsize *h*. To obtain the Euler approximation, rather than Maple's default scheme, the option `method=classical[foreuler]` is specified. The output is given as a listprocedure, so that the numerical data can be extracted for comparison with the first-principles calculation.

```
>  sol:=dsolve({sys,ic},vars,numeric,stepsize=h,
         method=classical[foreuler],output=listprocedure):
```

The numerical solution, $X(t)$ vs. t, is then plotted over the time interval $t=0$ to $hn=60$, using the `odeplot` command. The same number of points is taken as in the first-principles calculation. The result is shown on the right of Figure 9.1, a line style having been selected.

```
>  odeplot(sol,[t,X(t)],0..h*n,axes=normal,style=line,
         numpoints=n,labels=["t","x"]);
```

The two pictures appear to be identical, but we can confirm the agreement quantitatively by considering a particular time, say, $t=1$. The solution *sol* is used to evaluate X at arbitrary time t in XX. Using XX, the dial-up value of x at $t=1$ is given in *x1*. The first-principles answer is found by entering `pt[100]`, the x value at $0.01 \times 100 = 1.00$ s being the second entry in the output.

```
>  XX:=eval(X(t),sol): x1:=XX(1); pt[100];
```

$$x1 := 1.64892899170276830$$

$$[1.00, 1.64892899170276830, -0.0478758403582165596]$$

The dial-up and first-principles values of x agree to 18 digits!

Because ϵ was increased from 5 in our earlier recipe (**08-2-1**) to 20 here, the relaxation oscillations in Figure 9.1 are characterized by even longer periods of relative inaction between the abrupt temporal changes. It should be mentioned that an example of a relaxation oscillator in nature (although not governed by the VdP equation) is the famous geyser Old Faithful in Yellowstone Park.

9.1.2 Survival of the Fittest

The price which society pays for the law of competition ... is great. But, whether the law be benign or not, ... it is here; we cannot evade it; no substitutes for it have been found; and while the law may be sometimes hard for the individual, it is best for the race, because it ensures the survival of the fittest in every department.
Andrew Carnegie, American industrialist, philanthropist, (1835–1919)

In terms of a Taylor series expansion, the Euler method is of $O(h)$ accurate. This necessitates taking h very small, which leads to a longer computation time than for an algorithm which is of higher order accuracy in h, thus permitting a bigger h to be used. In this recipe, the Volterra–Lotka (V-L) equations will be solved with the *modified Euler* method, which is of $O(h^2)$ accuracy, as well as with the less accurate Euler method.

The V-L competition equations are used in mathematical biology to describe the interaction of biological populations, such as, e.g., rats and ferrets. Letting $r(t)$ and $f(t)$ be the rat and ferret population numbers per unit area at time t, the V-L equations are

$$\dot{r} = R(r, f) = a_1\, r - a_2\, rf, \qquad \dot{f} = F(r, f) = -b_1\, f + b_2\, rf, \qquad (9.7)$$

where the coefficients a_1, a_2, b_1, and b_2 are all positive. In (9.7), it is assumed that the ferrets survive by eating only rats, which munch on garbage. If the rats and ferrets did not interact ($a_2 = b_2 = 0$), the rat (ferret) equation would become $\dot{r} = a_1\, r$ ($\dot{f} = -b_1\, f$), which has an exponentially growing (decaying) solution for the rat (ferret) number. Inclusion of the nonlinear interaction terms in each ODE will dramatically change the nature of the solution.

Setting $t_{k+1} = t_k + h$, the modified Euler algorithm for (9.7) is

$$r_{k+1} = r_k + h\,(R_1[k] + R_2[k])/2, \qquad f_{k+1} = f_k + h\,(F_1[k] + F_2[k])/2, \qquad (9.8)$$

with $R_1[k] \equiv R(r_k, f_k)$, $F_1[k] \equiv F(r_k, f_k)$, $R_2[k] \equiv R(r_k + hR_1[k], f_k + hF_1[k])$, and $F_2[k] \equiv F(r_k + hR_1[k], f_k + hF_1[k])$.

That this scheme is of $O(h^2)$ accuracy can be confirmed by expanding (9.8) in powers of h and showing that the resulting equations agree to order h^2 with the Taylor expansions of $r_{k+1} = r(t_{k+1} = t_k + h)$, $f_{k+1} = f(t_{k+1})$.

As an illustrative example, let's take the nominal values $a_1 = 2$, $a_2 = 0.02$, $b_1 = 1$, $b_2 = 0.01$, with $r_0 = 300$ and $f_0 = 100$ at $t_0 = 0$.

```
> restart: with(plots):
> a1:=2: a2:=0.02: b1:=1: b2:=0.01:
> t[0]:=0: r[0]:=300: f[0]:=100:
```

10 digits accuracy is taken and for the modified Euler method the stepsize $h = 0.06$ and $n = 500$ iterations are considered. For the Euler method, a smaller stepsize $H = h/3 = 0.02$ is chosen and a correspondingly larger number $N = 3\,n = 1500$ iterations taken to make the total time for both solutions the same.

```
> Digits:=10: h:=0.06: n:=500: H:=h/3: N:=3*n:
```

Functional operators are formed to calculate R and F, i.e., the rhs of eqs. (9.7).

```
> R:=(r,f)->a1*r-a2*r*f: F:=(r,f)->-b1*f+b2*r*f:
```

Using spacecurve, an operator gr is formed to plot the numerical solution for either method as a red curve in 3 dimensions. The numerical points p must be specified as well as the upper limit n in the sequence command. The orientation is taken to be [0,90], so that a phase-plane portrait, f vs. r, results. Of course, the 3-dimensional plots can be rotated on the computer screen.

```
> gr:=(p,n)->spacecurve([seq(p[k],k=0..n)],color=red,axes=
            normal,labels=["t","r","f"],orientation=[0,90]):
```

The Euler solution will be obtained first. The start time for the algorithm is recorded in begin1,

```
> begin1:=time():
```

and the Euler method applied to the V-L equations, the stepsize being H and
N the number of iterations .

```
>   for k from 0 to N do
>   t[k+1]:=t[k]+H;
>   r[k+1]:=r[k]+H*R(r[k],f[k]);
>   f[k+1]:=f[k]+H*F(r[k],f[k]);
>   pt1[k]:=[t[k],r[k],f[k]];
>   end do:
```

The cpu time for the above algorithm is determined.

```
>   cpu_time1:=time()-begin1;
```

$$cpu\_time1 := 0.208$$

It took about 0.2 seconds cpu time to execute the Euler method. Using gr and
the numerical points pt1, the Euler solution of the V-L equations is plotted,
the resulting picture being shown on the left of Figure 9.2.

```
>   gr(pt1,N);
```

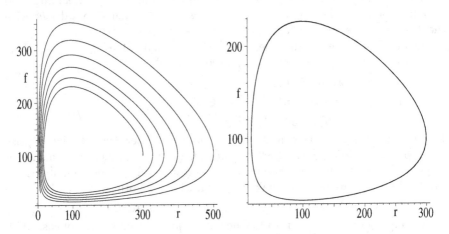

Figure 9.2: Left: Euler solution. Right: Modified Euler solution.

From this picture, one might conclude that the trajectory is an unstable spiral,
unwinding from the starting point $r_0 = 300$, $f_0 = 100$. Let's see what happens
when the modified Euler method is now applied to the V-L equations. Again
the starting and ending times for the algorithm are recorded.

```
>   begin2:=time():
>   for k from 0 to n do
>   t[k+1]:=t[k]+h;
>   R1[k]:=R(r[k],f[k]); F1[k]:=F(r[k],f[k]);
>   R2[k]:=R(r[k]+h*R1[k],f[k]+h*F1[k]);
```

```
>    F2[k]:=F(r[k]+h*R1[k],f[k]+h*F1[k]);

>    r[k+1]:=r[k]+h*(R1[k]+R2[k])/2;

>    f[k+1]:=f[k]+h*(F1[k]+F2[k])/2;

>    pt2[k]:=[t[k],r[k],f[k]];

>    end do:

>    cpu_time2:=time()-begin2;
```

$$cpu\_time2 := 0.164$$

Because a larger stepsize was used with fewer iterations, the cpu time for the modified Euler method is slightly less here than for the Euler algorithm. Again using gr, and considering the numerical points pt2, the resulting trajectory is as shown on the right of Figure 9.2.

```
>    gr(pt2,n);
```

Because the cumulative numerical error for the modified Euler method is less, a more accurate curve is obtained. It appears that the correct solution is a cyclic variation in the rat and ferret population numbers. This is in dramatic contrast to the situation where there is no interaction. Although the parameter values were artificial and there are limitations[2] to the V-L equations, such cyclic behavior has been observed in a wide variety of competing biological species.

Since the modified Euler and Euler algorithms are actually equivalent to finite Taylor expansions in h, it would not be surprising that the algorithms break down for large values of h. What may surprise you is that the breakdown can occur for values of h considerably less than 1, the breakdown showing up as a *numerical instability* with the solution diverging to infinity. Without changing any of the other parameters you should see what happens in this recipe as you increase the value of h. You may have to cut down on n to avoid numerical overflow. Numerical instabilities for sufficiently large h are an inherent feature of explicit schemes.

9.1.3 A Chemical Reaction

What a man calls his "conscience" is merely the mental action that follows a sentimental reaction after too much wine or love.
Helen Rowland, American journalist, *A Guide to Men*, 1922

The Euler and modified Euler methods are the simplest examples of single-step explicit methods for numerically solving ordinary differential equations. They represent the lowest approximations in terms of accuracy of the Rungé–Kutta (RK) methods ([LS71][PFTV90]). Historically, the fourth-order (accurate to order h^4) RK method has been the "work horse" of these methods as it combines reasonable accuracy with reasonable speed.

---

[2]For example, if there were no interaction, the rat population would not grow indefinitely if confined to a given area because of limited resources, diseases, use of poisons, etc. Further, ferrets eat other creatures such as rabbits.

Given the general ODE, $dx/dt = f(t, x)$, the 4th-order RK algorithm is

$$x_{k+1} = x_k + [K_1 + 2K_2 + 2K_3 + K_4]/6, \qquad (9.9)$$

with $K_1 = hf(t_k, x_k)$, $K_2 = hf(t_k+h/2, x_k+K_1/2)$, $K_3 = hf(t_k+h/2, x_k+K_2/2)$, and $K_4 = hf(t_k+h, x_k+K_3)$. The algorithm is easily generalized to higher-order ODEs or systems of first-order ODEs.

As a simple example, consider the irreversible chemical reaction

$$2\,K_2Cr_2O_7 + 2\,H_2O + 3\,S \rightarrow 4\,KOH + 2\,Cr_2O_3 + 3\,SO_2$$

with initially N_1 molecules of potassium dichromate ($K_2Cr_2O_7$), N_2 molecules of water (H_2O), and N_3 atoms of sulphur (S). The number x of potassium hydroxide (KOH) molecules at time t seconds is given by the rate equation

$$dx/dt = k(2N_1 - x)^2(2N_2 - x)^2(4N_3/3 - x)^3$$

with $k = 1.64 \times 10^{-20}\ s^{-1}$ and $x(0) = 0$.

Taking $N_1 = 1000$, $N_2 = 2000$, $N_3 = 5000$, stepsize $h = 0.001$ s, and $n = 200$ iterations, we will solve the rate equation using the fourth-order RK algorithm and graphically compare the solution with that obtained using Maple's `dsolve` command with the option `method=classical[rk4]`. We will also determine the number of KOH molecules at 0.1 s.

After loading the plots package and setting the accuracy at 10 digits,

```
>   restart: with(plots): Digits:=10:
```

an operator F is created for calculating the rhs of the rate equation.

```
>   F:=x->k*((2*N1-x)^2)*((2*N2-x)^2)*((4*N3/3-x)^3);
```

$$F := x \rightarrow k\,(2\,N1 - x)^2\,(2\,N2 - x)^2\,(\frac{4\,N3}{3} - x)^3$$

The various parameter values are entered in the next two command lines.

```
>   k:=1.64*10^(-20): h:=0.001: n:=200:
```

```
>   x[0]:=0: t[0]:=0: N1:=1000: N2:=2000: N3:=5000:
```

After recording the start time, the 4th-order RK algorithm is applied to the rate equation, making use of F. A plotting point, `pt`, is formed on each step.

```
>   begin:=time():
>   for j from 0 to n do
>   K1[j]:=h*F(x[j]);
>   K2[j]:=h*F(x[j]+K1[j]/2);
>   K3[j]:=h*F(x[j]+K2[j]/2);
>   K4[j]:=h*F(x[j]+K3[j]);
>   x[j+1]:=x[j]+(K1[j]+2*K2[j]+2*K3[j]+K4[j])/6;
>   t[j+1]:=t[j]+h;
>   pt[j]:=[t[j],x[j]];
>   end do:
```

The cpu time for executing the algorithm is determined, and a graph of the numerical points created in `gr1`. The points are represented by size 12 blue circles. To avoid a messy overlap, only every third point is plotted.

```
>  cpu_time:=time()-begin;
```
$$cpu\_time := 0.153$$

```
>  gr1:=pointplot([seq(pt[3*j],j=0..n/3)],symbol=circle,
        symbolsize=12,color=blue):
```
To determine the number of KOH molecules at 0.1 s, the plotting point at $n/2 = 100$ is determined and the **round** command applied to the second entry.

```
>  Pt:=pt[n/2]; N:=round(Pt[2]);
```
$$Pt := [0.100, 1586.746962] \quad N := 1587$$

At 0.1 s, there are 1587 KOH molecules. Now the rate equation is solved using the `dsolve` command. The relevant ODE is entered, again making use of F.

```
>  ode:=diff(X(t),t)=F(X(t));
```
Taking the same stepsize h and initial condition $X(0) = 0$, *ode* is numerically solved for $X(t)$, with the option `method=classical[rk4]`.

```
>  sol:=dsolve({ode,X(0)=0},X(t),numeric,stepsize=h,
        method=classical[rk4]):
```
This solution is then plotted in `gr2` with `odeplot`, using a line style.

```
>  gr2:=odeplot(sol,[t,X(t)],0..h*n,numpoints=n,style=line):
```
The two graphs, `gr1` and `gr2`, are then displayed in Figure 9.3, the circles lying exactly on top of the solid line as expected.

```
>  display({gr1,gr2},labels=["t","x"]);
```

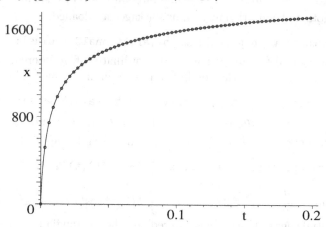

Figure 9.3: Circles: "First principles" solution. Line: "Dial-up" solution.

9.1.4 Parametric Excitation

It is the unknown that excites the ardor of scholars, who,
in the known alone, would shrivel up with boredom.
Wallace Stevens, American poet, (1879–1955)

In numerically solving some nonlinear ODEs, regions of solution space can be encountered where the solution is changing relatively slowly and a larger stepsize could be used while maintaining a reasonable accuracy, while in other regions the solution is rapidly changing and a smaller stepsize should be used. In such situations, an algorithm which adapts its stepsize according to the solution "terrain" is desirable. One of the most popular adaptive step schemes is the Rungé–Kutta–Fehlberg [Feh70] fourth-fifth (RKF 45) method, which uses both the fourth- and fifth-order RK algorithms and adapts the stepsize according to the difference between the two solutions. Fortunately, one of the options in the numerical **dsolve** command is precisely this scheme. To invoke it, simply include the option **method=rkf45**.

As an illustrative example of using the **rkf45** option, consider the following interesting problem. The pivot point for the simple pendulum (a mass m attached to a light rod of length r making an angle θ with the vertical) is undergoing vertical oscillations given by $A\sin(\omega t)$. Using the Lagrangian approach, show that the relevant equation of motion is

$$\ddot{\theta} + [W^2 - A\omega^2 \sin(\omega t)/r]\sin\theta = 0,$$

with $W = \sqrt{g/r}$, where g is the gravitational acceleration. This nonlinear ODE with a time-dependent coefficient is an example of a *parametric excitation*. Taking $A = 1$ m, $r = 9.8$ m, $g = 9.8$ m/s^2, $\theta(0) = 2\pi/3$ rads, and $\dot{\theta}(0) = 0$ rads/s, numerically solve the equation of motion using the RKF 45 method for $\omega = 0.1$ and 1.2 rads/s. Plot $\theta(t)$ over the time interval $t = 100$ to 300 s.

The plots and VariationalCalculus packages are loaded,

```
> restart: with(plots): with(VariationalCalculus):
```

and the horizontal (x) and vertical (y) coordinates of m entered. The coordinates are measured from the equilibrium position of the suspended mass.

```
> x:=r*sin(theta(t)); y:=r*(1-cos(theta(t)))+A*sin(omega*t);
```

$$x := r\sin(\theta(t)) \qquad y := r\,(1 - \cos(\theta(t))) + A\sin(\omega t)$$

The kinetic energy $T = \frac{1}{2}m(\dot{x}^2 + \dot{y}^2)$ is calculated and simplified.

```
> T:=simplify((m/2)*(diff(x,t)^2+diff(y,t)^2));
```

$$T := \frac{1}{2}m(r^2\,(\frac{d}{dt}\,\theta(t))^2 + 2\,r\sin(\theta(t))\,(\frac{d}{dt}\,\theta(t))\,A\cos(\omega t)\,\omega + A^2\cos(\omega t)^2\,\omega^2)$$

The potential energy $V = m\,g\,y$ is entered and the Lagrangian $L = T - V$ formed.

```
> V:=m*g*y: L:=T-V;
```

$$L := \frac{1}{2}m(r^2\,(\frac{d}{dt}\,\theta(t))^2 + 2\,r\sin(\theta(t))\,(\frac{d}{dt}\,\theta(t))\,A\cos(\omega\,t)\,\omega$$
$$+ A^2\cos(\omega\,t)^2\,\omega^2) - m\,g\,(r\,(1 - \cos(\theta(t))) + A\sin(\omega\,t))$$

The Euler–Lagrange command is applied to L, simplified, and equated to 0.

> eq:=simplify(EulerLagrange(L,t,theta(t))[1])=0;

$$eq := -m\,r\,(g\sin(\theta(t)) + r\,(\frac{d^2}{dt^2}\,\theta(t)) - \sin(\theta(t))\,A\sin(\omega\,t)\,\omega^2) = 0$$

The output in *eq* is the equation of motion. Because L contained an explicit time-dependence, no first integral was produced here. Multiplying *eq* by -1 and dividing by $m\,r^2$ produces the result in *eq2*.

> eq2:=expand(-eq/(m*r^2));

$$eq2 := \frac{g\sin(\theta(t))}{r} + (\frac{d^2}{dt^2}\,\theta(t)) - \frac{\sin(\theta(t))\,A\sin(\omega\,t)\,\omega^2}{r} = 0$$

Next, we substitute $g = r\,W^2$ into *eq2*.

> eq3:=subs(g=r*W^2,eq2);

and obtain the desired NLODE by collecting the $\sin(\theta(t))$ terms in *eq3*.

> eq4:=collect(eq3,sin(theta(t)));

$$eq4 := (W^2 - \frac{A\sin(\omega\,t)\,\omega^2}{r})\sin(\theta(t)) + (\frac{d^2}{dt^2}\,\theta(t)) = 0$$

The given parameter values are entered and the frequency $W = \sqrt{g/r}$ calculated.

> A:=1: r:=9.8: g:=9.8: W:=sqrt(g/r);

$$W := 1.000000000$$

The initial condition is entered.

> ic:=theta(0)=2*Pi/3,D(theta)(0)=0;

$$ic := \theta(0) = \frac{2\,\pi}{3},\ D(\theta)(0) = 0$$

Applying the unapply command to *eq4* turns the ODE into a functional operator, where the frequency ω must be supplied.

> ode:=unapply(eq4,omega):

An operator sol is formed for numerically solving the ODE for $\theta(t)$, using the RKF 45 method, for a given ω and subject to the initial condition.

> sol:=omega->dsolve({ode(omega),ic},theta(t),numeric,
        method=rkf45):

An operator p is created to plot the numerical solution ($\theta(t)$ vs. t) for a specified value of ω. The time range is taken from $t = 100$ to 300, the lower limit being such as to eliminate the transient. To avoid an incorrect set of figures, 3000 plotting points are chosen.

> p:=omega->odeplot(sol(omega),[[t,theta(t)]],100..300,
        numpoints=3000,thickness=2):

Entering p(0.1) and p(1.2) yields the pictures on the left and right of Fig. 9.4.

>   p(0.1); p(1.2);

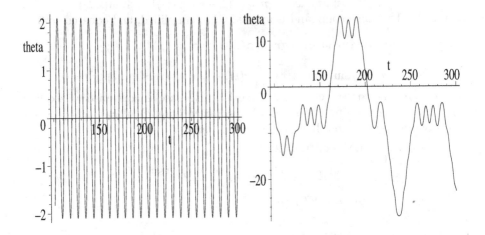

Figure 9.4: RKF 45 solution for $\omega = 0.1$ (left) and $\omega = 1.2$ (right).

For $\omega = 0.1$, a periodic response of the pendulum is obtained, while for $\omega = 1.2$ the solution appears to be highly irregular or chaotic. Other values of ω yield different results.

9.1.5   A Stiff System

A stiff apology is a second insult ....
The injured party does not want to be compensated because he has been wronged; he wants to be healed because he has been hurt.
G. K. Chesterton, British author, (1874–1936)

A *stiff* ODE, or ODE system, is one for which there are two or more very different time or spatial scales for the independent variable. The shortest characteristic time τ (or distance) acts as an approximate boundary between numerical stability and instability for any explicit numerical scheme with a fixed stepsize h. When h is less than τ, the scheme is stable, but as h is increased above τ the scheme becomes increasingly unstable, with floating point overflow eventually occuring. Numerical algorithms having an adaptive stepsize, such as RKF45, circumvent this problem by adjusting the stepsize appropriately. In this recipe, we shall flesh these ideas out by numerically solving a stiff system of linear ODEs with both the fourth-order RK and the RKF 45 algorithms. This system will also be solved analytically so as to illustrate the existence of two widely different characteristic times and to provide us with a check on the accuracy of our numerical algorithms.

Consider the following pair of coupled first-order linear ODEs with time-dependent forcing terms present in each equation,

$$\dot{x} = 9\,x + 24\,y + 5\cos t - \frac{1}{3}\sin t, \quad \dot{y} = -24\,x - 51\,y - 9\cos t + \frac{1}{3}\sin t.$$

Suppose that the initial condition is $x(0) = 4/3$, $y(0) = 2/3$. Let's begin the recipe by entering the ODE system and the initial condition.

```
> restart: with(plots):
> sys:=diff(x(t),t)=9*x(t)+24*y(t)+5*cos(t)-(1/3)*sin(t),
        diff(y(t),t)=-24*x(t)-51*y(t)-9*cos(t)+(1/3)*sin(t);
```

$$sys := \frac{d}{dt}\,x(t) = 9\,x(t) + 24\,y(t) + 5\cos(t) - \frac{1}{3}\sin(t),$$

$$\frac{d}{dt}\,y(t) = -24\,x(t) - 51\,y(t) - 9\cos(t) + \frac{1}{3}\sin(t)$$

```
> ic:=x(0)=4/3,y(0)=2/3:
```

The analytic forms of $x(t)$ and $y(t)$ follow on applying the dsolve command to *sys*, subject to the initial condition.

```
> sol:=dsolve({sys,ic},{x(t),y(t)});
```

$$sol := \{x(t) = -e^{(-39\,t)} + 2\,e^{(-3\,t)} + \frac{1}{3}\cos(t), y(t) = 2\,e^{(-39\,t)} - e^{(-3\,t)} - \frac{1}{3}\cos(t)\}$$

There are two widely differing characteristic times in the transient part of the solution, viz., $\tau_1 = 1/39 \simeq 0.03$ and $\tau_2 = 1/3 \simeq 0.3$. The shorter time τ_1 will set an approximate boundary between the stability or the lack thereof of the RK4 scheme. For later comparison, let's plot the exact solution, first using the following operator F to select the forms of x and y separately from *sol*.

```
> F:=v->rhs(select(has,sol,v(t))[1]):
```

Then, using F with $v = x$ and y, the exact analytic solution is plotted, using the spacecurve command, in the 3-dimensional t vs. x vs. y space.

```
> gr:=spacecurve([t,F(x),F(y)],t=0..20,numpoints=2000):
```

An operator sol2 is formed to numerically solve the ODE system, for a specified stepsize h, using the fourth-order RK algorithm.

```
> sol2:=h->dsolve({sys,ic},{x(t),y(t)},numeric,stepsize=h,
        method=classical[rk4]):
```

An operator G is created to plot the numerical solution s for a specified h and upper time limit T. The latter is included because for $h \gg \tau_1$, but less than τ_2, numerical overflow will occur if we try to plot the RK4 solution for large T.

```
> G:=(s,h,T)->odeplot(s(h),[t,x(t),y(t)],t=0..T,numpoints=2000,
        style=line,axes=box,labels=["t","x","y"]):
```

Using G, the RK4 numerical solution is plotted for $h = 0.05$ and $T = 20$ and superimposed on the exact solution (plotted in gr) with the display command.

```
> display({gr,G(sol2,0.05,20)});
```

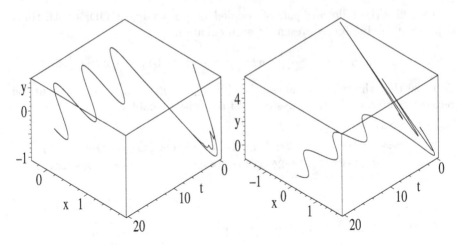

Figure 9.5: Left: Exact & RK4 curves ($h=.05$). Right: RK4 & RKF45 ($h=.08$).

The resulting picture is shown on the left of Figure 9.5, the exact solution being the completely smooth curve, the RK4 numerical solution deviating away from the exact result during the transient interval but locking onto the steady-state oscillatory solution for larger times. The slight deviation of the RK4 solution away from the exact curve is a precursor of the onset of numerical instability which occurs if h is increased much further. Before showing what happens, let's form another operator sol3 to numerically solve the ODE system for a given h using the RKF 45 algorithm.

```
>  sol3:=h->dsolve({sys,ic},{x(t),y(t)},numeric,method=rkf45):
```

Now the time stepsize h is increased to 0.08, which is still substantially less than the other characteristic time $\tau_2 \simeq 0.3$ in the problem. The RK4 and RKF 45 solutions for $h = 0.08$ are now superimposed with the display command, the resulting curves being shown on the right of Figure 9.5.

```
>  display({G(sol2,0.08,0.21),G(sol3,0.08,20)});
```

The jagged diverging curve is the RK4 solution, the smooth one the RKF 45 result. For the RKF 45 solution, derived with sol3, the upper time limit is still 20, but for the RK4 solution, derived with sol2, T has been shortened to 0.21 to avoid floating point overflow. Noting that the x and y scales of the two viewing boxes in Figure 9.5 are different, the RKF 45 result agrees with the analytical result.

It should be mentioned that other numerical algorithm options are available for solving stiff ODE systems. The reader should consult Maple's Help under the topic heading "dsolve,numeric" for information about these options.

9.1.6 A Strange Attractor

Wickedness is a myth invented by good people to account for the curious attractiveness of others.
Oscar Wilde, Anglo-Irish playwright, author, (1854–1900)

Implicit numerical algorithms, based on replacing the exact first derivative with the BWDA, are stable for any fixed stepsize h, although they may not be too accurate if h is made too large. For nonlinear ODEs, the implicit schemes involve solving simultaneous nonlinear algebraic equations at each step. In practice, the algorithms are often made *semi-implicit* by linearizing the algebraic equations. In this recipe, a semi-implicit scheme is developed for the Rössler system [Ros76], which is famous for its "top-hat" *strange attractor*.

The Rössler equations are

$$\dot{x} = -(y+z) \equiv X, \quad \dot{y} = x + ay \equiv Y, \quad \dot{z} = b + z(x-c) \equiv Z, \qquad (9.10)$$

with x, y, and z real and a, b, c real, positive, constants. They were introduced by Rössler to illustrate how a simple 3-dimensional ODE system can exhibit periodic and chaotic solutions, depending on the parameter values.

Using the BWDA in the Euler method, with k replaced by $k{+}1$, (9.10) yields

$$x_{k+1} = x_k + h X_{k+1}, \qquad y_{k+1} = y_k + h Y_{k+1}, \qquad z_{k+1} = z_k + h Z_{k+1}. \quad (9.11)$$

This resembles the Euler algorithm that was used before, except now the rhs of (9.10) are evaluated at the "new" time step instead of the "old" one. This is an example of an implicit scheme. If all the ODEs are linear, it is easy to solve the equations, expressing x_{k+1}, y_{k+1}, and z_{k+1} completely in terms of the old known values. For the Rössler system, Z_{k+1} contains the nonlinear term, $z_{k+1} x_{k+1}$, involving the new unknown values. It is customary to convert implicit schemes to semi-implicit ones by Taylor expanding the nonlinear terms around the old values. For example, for the nonlinear term in Z, we write

$$f \equiv z_{k+1} x_{k+1} = f(x_k, z_k) + (x_{k+1} - x_k)\left(\frac{\partial f}{\partial x}\right)_{x_k, z_k} + (z_{k+1} - z_k)\left(\frac{\partial f}{\partial z}\right)_{x_k, z_k} + \cdots$$

$$= x_k z_k + (x_{k+1} - x_k) z_k + (z_{k+1} - z_k) x_k + \cdots.$$

Substituting this expansion into (9.11) and collecting terms yields

$$x_{k+1} + h y_{k+1} + h z_{k+1} = x_k,$$

$$-h x_{k+1} + (1 - ha) y_{k+1} = y_k, \qquad (9.12)$$

$$-h z_k x_{k+1} + (1 + hc - h x_k) z_{k+1} = hb + (1 - h x_k) z_k.$$

This system of three linear algebraic equations is then solved on each time step.

The semi-implicit scheme derived above is only of $O(h)$ accuracy. To derive a second-order accurate implicit scheme, we can average[3] the old and the new

---

[3] This is the basis of the Crank–Nicolson scheme for solving nonlinear PDEs.

on the rhs, i.e., replace equations (9.11) with

$$x_{k+1} = x_k + \frac{h}{2}\left(X_k + X_{k+1}\right), \quad y_{k+1} = y_k + \frac{h}{2}\left(Y_k + Y_{k+1}\right), \quad z_{k+1} = z_k + \frac{h}{2}\left(Z_k + Z_{k+1}\right).$$
$$(9.13)$$

Using the same expansion for the nonlinear term, a second-order semi-implicit algorithm for solving the Rössler equations then is

$$x_{k+1} + \frac{1}{2}h\,y_{k+1} + \frac{1}{2}h\,z_{k+1} = x_k - \frac{1}{2}h\,(y_k + z_k),$$
$$-\frac{1}{2}h\,x_{k+1} + \left(1 - \frac{1}{2}h\,a\right)y_{k+1} = \frac{1}{2}h\,x_k + \left(1 + \frac{1}{2}h\,a\right)y_k, \qquad (9.14)$$
$$-\frac{1}{2}h\,z_k\,x_{k+1} + \left(1 + \frac{1}{2}h\,c - \frac{1}{2}h\,x_k\right)z_{k+1} = h\,b + \left(1 - \frac{1}{2}h\,c\right)z_k.$$

In this recipe, we will iterate equations (9.14), taking 10 digits,

>   restart: with(plots): Digits:=10:

with $a = b = 0.2$, $c = 5.0$, $h = 0.05$ and $n = 3000$ iterations.

>   a:=0.2: b:=0.2: c:=5.0: h:=0.05: n:=3000:

The initial condition at $t_0 = 0$ is $x_0 = -1$, $y_0 = 0$, $z_0 = 0$.

>   t[0]:=0: x[0]:=-1: y[0]:=0: z[0]:=0:

After recording the start time, a do loop is created to iterate eqs. (9.14)

>   begin:=time():

>   for k from 0 to n do

The rhs of each equation is entered in A1[k], A2[k], and A3[k].

>   A1[k]:=x[k]-0.5*h*(y[k]+z[k]);

>   A2[k]:=0.5*h*x[k]+(1+0.5*h*a)*y[k];

>   A3[k]:=h*b+(1-0.5*h*c)*z[k];

The three equations of (9.14) are entered in E1, E2, and E3.

>   E1:=x[k+1]+0.5*h*y[k+1]+0.5*h*z[k+1]=A1[k];

>   E2:=-0.5*h*x[k+1]+(1-0.5*h*a)*y[k+1]=A2[k];

>   E3:=-0.5*h*z[k]*x[k+1]+(1+0.5*h*c-0.5*h*x[k])*z[k+1]=A3[k];

The three algebraic equations, E1, E2, E3 are numerically solved for x_{k+1}, y_{k+1}, and z_{k+1} and the solution then assigned.

>   sol[k+1]:=(fsolve({E1,E2,E3},{x[k+1],y[k+1],z[k+1]}));

>   assign(sol[k+1]);

The values of x_{k+1}, y_{k+1}, z_{k+1}, and $t_{k+1} = t_k + h$, are recorded,

>   x[k+1],y[k+1],z[k+1]; t[k+1]:=t[k]+h;

and the plotting point (x_k, y_k, z_k) formed for the kth time step.

>   pt[k]:=[x[k],y[k],z[k]];

>   end do:

On ending the do loop, the cpu time for the loop is calculated

>   cpu:=time()-begin;

$$cpu := 10.242$$

and found to be about 10 seconds. Using the `spacecurve` command with a line style and zhue coloring, the solution is plotted and shown in Figure 9.6.

```
>   spacecurve([seq(pt[j],j=0..n)],style=line,shading=zhue,
    orientation=[45,60],axes=boxed,labels=["x","y","z"]);
```

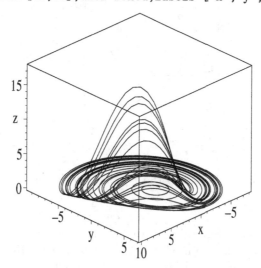

Figure 9.6: Rössler's strange attractor.

After an initial transient period, the trajectory is attracted to a localized region, where it continually traces out a new path, i.e., the motion is chaotic. This localized chaotic motion, resembling a man's "top hat", is an example of what mathematicians call a strange attractor. Starting with some other initial values of x_0, y_0, and z_0, you will find that the trajectory is eventually attracted to the top hat region. For other values of the parameters, periodic motions can also be observed. As an exercise, try varying the parameter c, holding all other parameter values fixed.

9.2 Partial Differential Equations

Finite difference approximations can also be used to numerically solve PDEs. For example, our first recipe involves finding the steady-state temperature $T(x, y)$ inside a thin rectangular plate with specified temperatures on the four edges. $T(x, y)$ will satisfy the 2-dimensional form of Laplace's equation,

$$\frac{\partial^2 T}{\partial x^2} + \frac{\partial^2 T}{\partial y^2} = 0. \tag{9.15}$$

Our approach is to divide the x-y plane into a rectangular grid or "mesh", as shown in Figure 9.7, with each small rectangle having sides of length h and k.

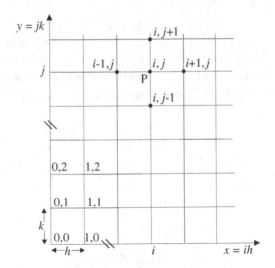

Figure 9.7: Subdividing the x-y plane with a numerical mesh.

The coordinates of a typical mesh (intersection) point P are $x = i\,h$, $y = j\,k$, with i, $j = 0$, 1, 2, ... The mesh points may be labeled by a pair of integers, the point P being indicated by $(i,\,j)$, and the temperature at that point by $T_{i,j} \equiv T(x = i\,h, y = j\,k)$. From Chapter 2, each second derivative in Laplace's equation can be replaced with a central difference approximation (CDA), viz.,

$$\frac{(T_{i+1,j} - 2\,T_{i,j} + T_{i-1,j})}{h^2} + \frac{(T_{i,j+1} - 2\,T_{i,j} + T_{i,j-1})}{k^2} = 0, \qquad (9.16)$$

or, on setting $r = (h/k)^2$ and rearranging,

$$2\,(1 + r)\,T_{i,j} - T_{i+1,j} - T_{i-1,j} - r\,T_{i,j+1} - r\,T_{i,j-1} = 0. \qquad (9.17)$$

For a given point P, the mesh points involved in (9.17) are as shown in Figure 9.7. If the temperature is specified at the mesh points making up the boundary of the plate, equation (9.17) can be used to calculate the temperature at each of the internal mesh points, as will now be demonstrated.

9.2.1   Steady-State Temperature Distribution

If you can't stand the heat, get out of the kitchen.
Harry S. Truman, former American president, (1884–1972)

A thin rectangular plate, with $0 \le x \le L1 = 1/2$ and $0 \le y \le L2 = 1$, has the following temperature distributions along its four edges: $T(x, 0) = 500\,x\,(L1 - x)$, $T(x, L2) = 700\,x\,(L1 - x)$, $T(0, y) = 0$, $T(L1, y) = 1000\,y\,(L2 - y)$. Dividing the plate into a numerical mesh with 15 steps in the x direction and 30 steps in the y direction, determine and plot the temperature profile inside the plate.

The plots library package is loaded and the start time recorded.

```
>  restart: with(plots): begin:=time():
```

The values $L1 = 1/2$, $L2 = 1$, and the numbers of steps, $m = 15$ and $n = 30$, are entered. The corresponding stepsizes $h = L1/m$ and $k = L2/n$ are calculated, along with the ratio $r = (h/k)^2$.

```
>  L1:=1/2: L2:=1: m:=15: n:=30: h:=L1/m; k:=L2/n; r:=(h/k)^2;
```

$$h := \frac{1}{30} \qquad k := \frac{1}{30} \qquad r := 1$$

The x and y coordinates of the mesh points are generated.

```
>  Xcoords:=seq(x[i]=i*h,i=0..m): Ycoords:=seq(y[j]=j*k,j=0..n):
```

The temperature distributions along the four edges are evaluated at the boundary mesh points by setting $x = i\,h$ and $y = j\,k$.

```
>  bc1:=seq(T[i,0]=500*i*h*(L1-i*h),i=0..m):
>  bc2:=seq(T[i,n]=700*i*h*(L1-i*h),i=0..m):
>  bc3:=seq(T[0,j]=0,j=0..n):
>  bc4:=seq(T[m,j]=1000*j*k*(L2-j*k),j=0..n):
```

The x and y mesh point coordinates and the 4 boundary conditions are assigned.

```
>  assign(Xcoords,Ycoords,bc1,bc2,bc3,bc4):
```

A functional operator f is formed to calculate the lhs of equation (9.17).

```
>  f:=(i,j)->2*(1+r)*T[i,j]-T[i+1,j]-T[i-1,j]-r*T[i,j+1]
            -r*T[i,j-1];
```

$$f := (i, j) \rightarrow 2\,(1+r)\,T_{i,j} - T_{i+1,j} - T_{i-1,j} - r\,T_{i,j+1} - r\,T_{i,j-1}$$

Making use of f and a nested sequence command, the equations which have to be solved at the $(m - 1) \times (n - 1)$ internal mesh points are generated.

```
>  eqs:={seq(seq(f(i,j)=0,i=1..m-1),j=1..n-1)}:
```

The $(m - 1) \times (n - 1)$ unknown temperature variables $T_{i,j}$ are entered.

```
>  vars:={seq(seq(T[i,j],i=1..m-1),j=1..n-1)}:
```

The mesh equations are then numerically solved for the variables, the answers being given to 6 digits. The solution is then assigned.

```
>  sol:=evalf(fsolve(eqs,vars),6): assign(sol):
```

We now create the plotting points $(x_i,\, y_j,\, T_{i,j})$ for $i = 0$ to m and $j = 0$ to n.

```
>  pts:=seq(seq([x[i],y[j],T[i,j]],i=0..m),j=0..n):
```

The numerical points are plotted with the `pointplot3d` command, being represented by size 6 circles which are colored with the `zhue` shading option. The resulting temperature profile inside the 3-dimensional viewing box is as shown in Figure 9.8. The box can be rotated in the usual manner.

```
>  pointplot3d([pts],symbol=circle,symbolsize=6,shading=zhue,
     axes=boxed,orientation=[55,60],labels=["x","y","T"]);
```

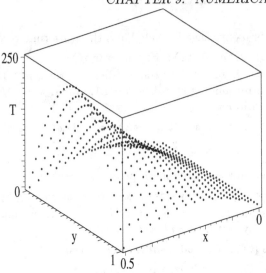

Figure 9.8: Numerically obtained temperature profile inside plate.

The cpu time for the entire recipe is about 13 seconds.

```
>   cpu:=time()-begin;
```

$$cpu := 13.361$$

9.2.2   1-Dimensional Heat Flow

In general, the art of government consists in taking as much money
as possible from one party of the citizens to give to the other.
Francois Voltaire, French philosopher, (1696–1778)

One-dimensional heat flow is governed by the temperature diffusion equation

$$\frac{\partial T}{\partial t} = \sigma \frac{\partial^2 T}{\partial x^2}, \tag{9.18}$$

with x the spatial coordinate, t the time, and σ the heat diffusion coefficient.

Setting $y = \sigma t$, a numerical algorithm can be formed by replacing the second order x derivative with a CDA (spatial step, h) and the first order y derivative by a FWDA (y step, k), viz.,

$$\frac{T_{i,j+1} - T_{i,j}}{k} = \frac{T_{i+1,j} - 2\,T_{i,j} + T_{i-1,j}}{h^2}, \tag{9.19}$$

or, on setting $r = k/h^2$ and rearranging,

$$T_{i,j+1} = r\,T_{i-1,j} + (1 - 2\,r)\,T_{i,j} + r\,T_{i+1,j}. \tag{9.20}$$

In terms of a "mesh diagram" for the x-y plane, the mesh points involved in (9.20) are as shown in Figure 9.9. The unknown temperature $T_{i,j+1}$ is to be explicitly determined on the $j + 1$st time step from the three known values $T_{i-1,j}$, $T_{i,j}$, and $T_{i+1,j}$ on the previous jth step.

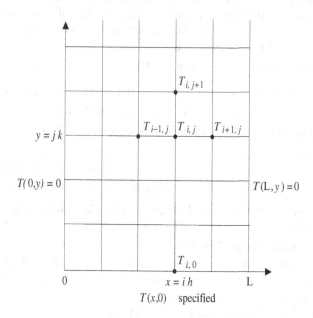

Figure 9.9: Numerical mesh for the heat flow recipe.

One starts with the bottom row ($j=0$) which corresponds to the specified initial temperature distribution inside, say, a thin rod of length L, and calculates T at each internal mesh point of the first ($j=1$) time row. With $T_{i,1}$ all known, one proceeds to calculate T on the second time row, and so on. To implement the algorithm, we must also specify the boundary conditions at the ends of the rod. For example, we might have $T(0, y) = T(L, y) = 0$ as in the figure.

Suppose that $T(x,0) = 200\,x^3\,(L - x)$ with $L = 1$ m. To implement the algorithm, the LinearAlgebra package is loaded so a matrix multiplication can be used for calculating T at the internal mesh points. Let's subdivide the range $x = 0$ to $x = L$ by taking $N = 25$ internal grid points. The relevant matrix A for implementing the rhs of (9.20) will be an $N \times N$ *tridiagonal* matrix, with $1 - 2r$ along the main diagonal, r along the first subdiagonals, and zeros everywhere else. For $N > 10$, the matrix will not be explicitly displayed unless `rtablesize` is set to N or larger in the `interface` command. Here it is set to infinity.

```
> restart: with(LinearAlgebra): interface(rtablesize=infinity):
```

Entering N and L, the spatial stepsize is $h = L/(N+1)$ which is now evaluated. The explicit scheme becomes numerically unstable if $r = k/h^2 > 1/2$, so let's take $r = 0.05$. Then the y stepsize $k = r\,h^2$ is calculated. The number of plots to be produced in the iteration is given by `numplots`.

```
> N:=25: L:=1.0: h:=L/(N+1); r:=0.05: k:=r*h^2; numplots:=100:
```
$$h := 0.03846153846 \qquad k := 0.00007396449705$$

Using the BandMatrix command, the relevant matrix A is entered. The diagonal
entries are given by $1 - 2r = 0.90$, the subdiagonal entries by $r = 0.05$. The
number 1 indicates that there is one subdiagonal and N gives the size of the
matrix, i.e., $N \times N$. For brevity, the output shown here in the text has been
truncated, only 5 of the 25 rows being shown.

```
> A:=BandMatrix([r,1-2*r,r],1,N);
```

$$A := \begin{bmatrix} 0.90, 0.05, 0, 0, 0, 0, 0, 0, 0, 0, 0, 0, 0, 0, 0, 0, 0, 0, 0, 0,0, 0, 0, 0, 0, 0, 0 \\ 0.05, 0.90, 0.05, 0, 0, 0, 0, 0, 0, 0, 0, 0, 0, 0, 0, 0, 0, 0,0, 0, 0, 0, 0, 0, 0, 0 \\ 0, 0.05, 0.90, 0.05, 0, 0, 0, 0, 0, 0, 0, 0, 0, 0, 0, 0, 0, 0, 0,0, 0, 0, 0, 0, 0, 0 \\ [..] \\ 0, 0, 0, 0, 0, 0, 0, 0, 0, 0, 0, 0, 0, 0, 0, 0, 0, 0, 0, 0, 0, 0, 0.05, 0.90, 0.05 \\ 0, 0, 0, 0, 0, 0, 0, 0, 0, 0, 0, 0, 0, 0, 0, 0, 0, 0, 0, 0, 0, 0, 0, 0.05, 0.90 \end{bmatrix}$$

A functional operator f is formed for generating the values of the initial temperature profile at the internal spatial grid points.

```
> f:=x->evalf(200*x^3*(L-x)):
```

Using f, the values of the temperature at the internal x mesh points on the
zeroth time row are formed into a column vector v. Again, only a few of the 25
entries are shown here in the text.

```
> v:=Vector([[seq(f(i*h),i=1..N)]]);
```

$$v := \begin{bmatrix} 0.01094149364 \\ 0.08403067118 \\ 0.2717867022 \\ \cdots \\ \cdots \\ 21.01335738 \\ \cdots \\ \cdots \\ 12.10041666 \\ 6.838433534 \end{bmatrix}$$

To establish the vertical scale of our plot, the following max command is used
to select the largest temperature value from the sequence of entries in v.

```
> vmax:=max(seq(v[i],i=1..N));
```
$$vmax := 21.01335738$$

The start time for the following do loop, which carries out the requisite matrix
multiplication, is recorded.

```
> begin:=time():
```

The do loop runs from $n=1$ to numplots. The matrix multiplication is carried
out in T, a plot being produced on every 20th time step. For plotting purposes

the end points are included in the list of plotting points. The points will be joined by straight lines.

```
>   for n from 1 to numplots do
>   T:=A^(20*n) . v;
>   p[n]:=plot([[0,0],seq([i/(N+1),T[i]],i=1..N),[1,0]],
        tickmarks=[2,2],labels=["x","y"]);
>   end do:
```

The cpu time for the do loop is about 20 s.

```
>   cpu:=time()-begin;
```

$$cpu := 20.037$$

Making use of the `display` command with the `insequence=true` option, the heat diffusion process in the rod is animated. The opening frame of the animation is shown in Figure 9.10. Execute the recipe to see what happens.

```
>   plots[display]([seq(p[i],i=1..numplots)],insequence=true,
        view=[0..L,0..vmax+1]);
```

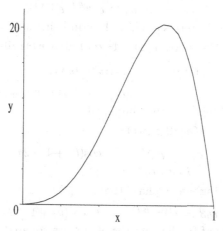

Figure 9.10: Opening frame of the heat diffusion animation.

9.2.3   Von Neumann Stability Analysis

No civilization would ever have been possible
without a framework of stability....
Hannah Arendt, American political philosopher, (1906–75)

That the explicit scheme used in the heat diffusion equation recipe **09-2-2** is *conditionally stable*, i.e., stable for $r = k/h^2 \leq 1/2$, could be proved by executing the recipe with varying values of r and looking at the numerical output. Alternately, one can use the *Von Neumann stability analysis* as will be demonstrated in the following recipe.

Recall that the relevant algorithm is of the structure

$$T_{m,n+1} = r\,T_{m-1,n} + (1 - 2\,r)\,T_{m,n} + r\,T_{m+1,n} \qquad (9.21)$$

with m and n indexing the spatial and time steps, respectively. Assume that $T_{m,n} = T^0_{m,n} + U_{m,n}$, where T^0 is the exact solution of the difference scheme and U represents a small numerical error due to roundoff, etc. Substituting this form into (9.21) yields an identical finite difference scheme for U. The question is "Does U grow or decay with time?", being unstable (stable) if it grows (decays). To answer this question, we assume that $U_{m,n}$ can be represented by the following exponential form,

$$U_{m,n} = e^{i\,m\,\theta}\,e^{i\,n\,\lambda}, \qquad (9.22)$$

with θ real and $\lambda = \alpha + i\,\beta$. It is assumed that both α and β are real.

>   restart: assume(alpha::real,beta::real):

An operator U is formed to calculate $U_{m,n}$ for arbitrary subscripts m and n.

>   U:=(m,n)->exp(I*m*theta)*exp(I*n*lambda);

$$U := (m,\,n) \to e^{(m\,\theta\,I)}\,e^{(n\,\lambda\,I)}$$

Eq. (9.21) is entered in terms of U, each term being automatically evaluated.

>   eq:=U(m,n+1)=r*U(m-1,n)+(1-2*r)*U(m,n)+r*U(m+1,n);

$$eq := e^{(m\,\theta\,I)}\,e^{((n+1)\,\lambda\,I)} = r\,e^{((m-1)\,\theta\,I)}\,e^{(n\,\lambda\,I)}$$
$$+ (1 - 2\,r)\,e^{(m\,\theta\,I)}\,e^{(n\,\lambda\,I)} + r\,e^{((m+1)\,\theta\,I)}\,e^{(n\,\lambda\,I)}$$

Then, eq is divided by $U_{m,n}$ and simplified.

>   eq2:=simplify(eq/U(m,n));

$$eq2 := e^{(\lambda\,I)} = 2\,r\cos(\theta) + 1 - 2\,r$$

We substitute $\lambda = \alpha + i\,\beta$ into $eq2$.

>   eq3:=subs(lambda=alpha+I*beta,eq2);

$$eq3 := e^{((\alpha + \beta\,I)\,I)} = 2\,r\cos(\theta) + 1 - 2\,r$$

The absolute values of the lhs and rhs of $eq3$ are equated, and simplified.

>   eq4:=simplify(abs(lhs(eq3)))=abs(rhs(eq3));

$$eq4 := e^{(-\beta)} = |2\,r\cos(\theta) + 1 - 2\,r|$$

For stability, $\beta \geq 0$ is required, because if $\beta < 0$, then $U_{m,n} \sim e^{|\beta|\,n}$ diverges as n (i.e., time) increases. For $\beta \geq 0$, $e^{-\beta} \leq 1$, so we must have the rhs of $eq4$ less than or equal to 1 for stability. The rhs of $eq4$ is turned into an operator g in terms of r by using the unapply command.

>   g:=unapply(rhs(eq4),r);

$$g := r \to |2\,r\cos(\theta) + 1 - 2\,r|$$

Using g, the rhs of $eq4$ is plotted over the range $\theta = -2.5\,\pi$ to $2.5\,\pi$ for $r = 0.49$ and 0.51, being represented on the computer screen by red and blue curves, respectively. A horizontal green line is also plotted at a vertical height 1. The corresponding black and white picture is shown in Figure 9.11.

```
> plot([g(.49),g(.51),1],theta=-2.5*Pi..2.5*Pi,
  color=[red,blue,green],numpoints=1000);
```

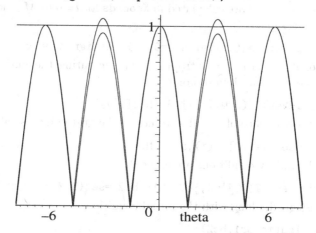

Figure 9.11: The $r=0.51$ curve > 1 for some θ, the $r=0.49$ curve doesn't.

Since the $r=0.51$ curve exceeds 1 for some ranges of θ, the numerical scheme is unstable for this r value. The $r=0.49$ curve doesn't exceed 1, so the algorithm is stable. By varying r between 0.49 and 0.51, you may use the recipe to check that the critical value of r for numerical stability or instability is $1/2$.

9.2.4 Sometimes It Pays to be Backwards

Diaper backward spells repaid. Think about it.
M. McLuhan, Canadian communications theorist, *Vancouver Sun, June 1969*

An *unconditionally stable* algorithm can be obtained for $\partial T/\partial y = \partial^2 T/\partial x^2$ by approximating the y derivative with the BWDA. In this case, on still replacing the x derivative with the standard CDA, the numerical algorithm becomes

$$(1+2r)\,T_{i,j} - r\,T_{i+1,j} - r\,T_{i-1,j} = T_{i,j-1}. \tag{9.23}$$

Let's use equation (9.23) to resolve the example given in recipe **09-2-2**, which involved determining the evolution of the temperature distribution in a thin rod of length $L=1$ m, with boundary conditions $T(0,y)=T(L,y)=0$, and an initial profile $T(x,0)=f(x)=200\,x^3\,(L-x)$.

We take $M=30$ spatial steps, so $h=L/M=1/30$. Choosing the time (y) step to be $k=0.002$ produces a ratio $r=k/h^2=1.8$. Not only is r larger than $1/2$, it's greater than 1! Nevertheless, as you will see, the scheme remains stable, although as r is increased the algorithm becomes increasingly less accurate. The total number of time steps is taken to be $N=250$.

```
> restart: with(plots):
> L:=1: M:=30: h:=L/M; k:=0.002; r:=k/h^2; N:=250:
```

$$h := \frac{1}{30} \qquad k := 0.002 \qquad r := 1.800$$

The values of $x_i = i\,h$ are calculated in Xcoords for $i=0$ to M, and the sequence of numbers assigned.

```
>  Xcoords:=seq(x[i]=i*h,i=0..M): assign(Xcoords):
```

A functional operator f is introduced for generating the initial temperature value at the ith spatial mesh point.

```
>  f:=i->evalf(200*(x[i]^3*(L-x[i]))):
```

Using f, the sequence of initial T values at the grid points is calculated.

```
>  ic:=seq(T[i,0]=f(i),i=0..M):
```

The two boundary conditions are entered,

```
>  bc1:=seq(T[0,j]=0,j=0..N): bc2:=seq(T[M,j]=0,j=0..N):
```

and are assigned, along with the initial condition.

```
>  assign(ic,bc1,bc2):
```

An operator F is formed for implementing the algorithm (9.23).

```
>  F:=(i,j)->(1+2*r)*T[i,j]-r*T[i+1,j]-r*T[i-1,j]-T[i,j-1];
```

$$F := (i, j) \rightarrow (1 + 2\,r)\,T_{i,j} - r\,T_{i+1,j} - r\,T_{i-1,j} - T_{i,j-1}$$

The start time for executing the algorithm in the do loop is recorded.

```
>  begin:=time():
```

We now generate and numerically solve the algorithm equations, $F(i,j)=0$, at each of the internal grid points on each of the j time steps, for $j=1$ to N.

```
>  for j from 1 to N do
>  eqs[j]:={seq(F(i,j)=0,i=1..M-1)};
>  sol[j]:=fsolve(eqs[j],{seq(T[i,j],i=1..M-1)});
>  assign(sol[j]):
>  end do:
```

The cpu time for the do loop is now displayed.

```
>  cpu:=time()-begin;
```

$$cpu := 14.110$$

A graphing operator gr is created for producing the sequence ($i=0$ to M) of plotting points $(x_i, T_{i,j})$ on the jth time step and plotting them.

```
>  gr:=j->plot([seq([x[i],T[i,j]],i=0..M)]):
```

The temperature diffusion process is animated with the display command, the insequence=true option being included. Click on the computer plot and the start arrow in the tool bar to see the animation.

```
>  display([seq(gr(j),j=0..N)],insequence=true,labels=["x","T"]);
```

9.2.5 Daniel Still Separates, I Now Iterate

The ultimate result of shielding men from the effects of folly
is to fill the world with fools.
Herbert Spencer, British philosopher, sociologist, (1820–1903)

To numerically solve the 1-dimensional wave equation, we can replace each
second derivative with a CDA, viz.,

$$\frac{\partial^2 U}{\partial x^2} = \frac{1}{c^2}\frac{\partial^2 U}{\partial t^2} \implies \frac{U_{i+1,j} - 2U_{i,j} + U_{i-1,j}}{h^2} = \frac{1}{c^2}\frac{U_{i,j+1} - 2U_{i,j} + U_{i,j-1}}{k^2},$$

$$\implies U_{i,j+1} = 2(1-r)U_{i,j} + r(U_{i+1,j} + U_{i-1,j}) - U_{i,j-1}, \quad r = (ck/h)^2. \quad (9.24)$$

In this case, the unknown $U_{i,j+1}$ depends explicitly on U values on the previous
two steps, as shown schematically in Figure 9.12, where x is horizontal and
t vertical. To start the iteration scheme (9.24), we must therefore know the

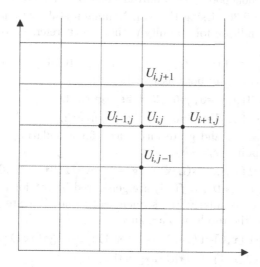

Figure 9.12: Mesh points involved in wave equation algorithm (9.24).

values of U on both the $j=0$ and $j=1$ time steps. To see how we deal with
this issue, let's consider the following possible scenario.

Young Daniel is playing with a light horizontal string of length L fixed at
$x=0$ and L, i.e., $U(0,t)=U(L,t)=0$. Before he manages to separate it into its
individual strands, his older cousin Justine cleverly manages to give it an initial
profile $U(x,0) = f(x) = Ax^3(L-x)$ and transverse velocity $\dot{U}(x,0) = g(x) = Bx(L-x)^2$. Our task is to iterate the numerical algorithm (9.24) to determine
the subsequent motion of the string and then animate the string vibrations,
taking $L=1$ m, $A=10$ m^{-3}, $B=5$ m^{-2}, and $c=3$ m/s.

On the zeroth time row, we have $U_{i,0}=f(x_i)$. Using the forward difference
approximation, the U values on the first time row, i.e., $U_{i,1}$, are found thus:

$$\frac{\partial U(x_i,0)}{\partial t} = \frac{U_{i,1} - U_{i,0}}{k} = g(x_i), \quad \text{so} \quad U_{i,1} = f(x_i) + k\,g(x_i) + O(k^2). \quad (9.25)$$

An even better approximation for $U_{i,1}$, which we shall use in our recipe, is

$$U_{i,1} = f(x_i) + k\,g(x_i) + \frac{r}{2}\,[f(x_{i+1}) - 2\,f(x_i) + f(x_{i-1})] + O(k^3). \quad (9.26)$$

To begin the recipe, the values of L, c, A, and B are entered. The total time is taken to be $T=2$ s, and the number of time and spatial steps are $N=200$ and $M=20$, respectively.

```
> restart: with(plots):
> L:=1: T:=2: M:=20: N:=200: c:=3: A:=10: B:=5:
```

The spatial step $h=L/M$, time step $k=T/N$, and $r=(c\,k/h)^2$ are calculated.

```
> h:=evalf(L/M); k:=evalf(T/N); r:=evalf((c*k/h)^2);
```

$$h := 0.05000000000 \qquad k := 0.01000000000 \qquad r := 0.3600000000$$

Here, we have $r = 0.36$. Using the Von Neumann stability analysis, it can be shown that the condition for stability is that $r \leq 1$, referred to as the *Courant stability condition*.

The boundary conditions $U_{0,j} = 0$ and $U_{M,j} = 0$ for $j=0$ to N are entered in bc1 and bc2, which are both assigned.

```
> bc1:=seq(U[0,j]=0,j=0..N): assign(bc1):
> bc2:=seq(U[M,j]=0,j=0..N): assign(bc2):
```

Functional operators f and g are introduced for calculating $f(x)$ and $g(x)$ at the internal grid points $x_i = i\,h$.

```
> f:=x->evalf(A*x^3*(L-x)): g:=x->evalf(B*x*(L-x)^2):
```

The sequence of values $U_{i,0} = f(i\,h)$ are generated in ic1 for $i=1...M-1$ and the result assigned. Similarly, the sequence of values $U_{i,1}$ are generated using (9.26) in ic2, and the result also assigned.

```
> ic1:=seq(U[i,0]=f(i*h),i=1..M-1): assign(ic1):
> ic2:=seq(U[i,1]=f(i*h)+k*g(i*h)
       +(r/2)*(f((i+1)*h)-2*f(i*h)+f((i-1)*h)),i=1..M-1):
> assign(ic2):
```

The time at the beginning of the do loop is recorded.

```
> begin:=time():
```

The algorithm (9.24) is iterated from $j=1$ to $N-1$ and the solution assigned.

```
> for j from 1 to N-1 do
> sol:=seq(U[i,j+1]=2*(1-r)*U[i,j]+r*(U[i+1,j]+U[i-1,j])
       -U[i,j-1],i=1..M-1);
> assign(sol):
> end do:
```

The cpu time for the do loop is found to be about 0.4 s.

```
> cpu:=time()-begin;
```

$$cpu := 0.404$$

An operator **gr** is formed for plotting the spatial profile on the *j*th time step.

```
>  gr:=j->plot([seq([i*h,U[i,j]],i=0..M)],labels=["x","U"]):
```

The motion is animated by including insequence=true in the display command. The opening frame of the animation is shown in Figure 9.13.

```
>  display(seq(gr(j),j=0..N),insequence=true);
```

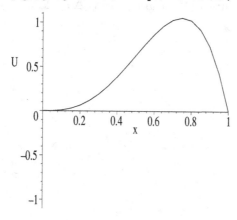

Figure 9.13: Opening frame of the vibrating string animation.

9.2.6 Interacting Laser Beams

The exercise of power is determined by thousands of interactions between the world of the powerful and that of the powerless, all the more so because these worlds are never divided by a sharp line: everyone has a small part of himself in both.
Vclav Havel, Czech playwright, president, (b. 1936)

The interaction of two intense laser pulses as they pass through each other in opposite directions in a certain resonant absorbing fluid can be described [RE76] by the following normalized PDEs for the laser intensities *u* and *v*:

$$\frac{\partial u}{\partial x} + \frac{\partial u}{\partial y} = -g_1\,u\,v - b\,u, \qquad \frac{\partial v}{\partial x} - \frac{\partial v}{\partial y} = -g_2\,u\,v + b\,v. \qquad (9.27)$$

Here *x* is the distance inside the fluid medium, *y* the time, $g_1 > 0$ and $g_2 > 0$ are the "gain" coefficients, and $b \geq 0$ the absorption coefficient. The *u* (*v*) beam travels in the positive (negative) *x* direction.

Our goal is to numerically solve this set of nonlinear PDEs for the intensities, using the *method of characteristics*, which is based on finding "characteristic" directions along which the PDEs can be reduced to ODEs. Since $u = u(x, y)$ and $v = v(x, y)$, then $du = (\partial u/\partial x)\,dx + (\partial u/\partial y)\,dy$ and $dv = (\partial v/\partial x)\,dx + (\partial v/\partial y)\,dy$.

Making use of these results, equations (9.27) can be rewritten as

$$\frac{du}{dx} + (1 - \frac{dy}{dx})\frac{\partial u}{\partial y} = f_1, \qquad \frac{dv}{dx} - (1 + \frac{dy}{dx})\frac{\partial v}{\partial y} = f_2, \qquad (9.28)$$

with $f_1 \equiv -g_1\,u\,v - b\,u$ and $f_2 \equiv -g_2\,u\,v + b\,v$. If we work along a line whose slope is $dy/dx = 1$, then $du/dx = f_1$, while along a line of slope $dy/dx = -1$ one has $dv/dx = f_2$. The characteristic directions $dy/dx = \pm1$ form a diamond-shaped grid as in Figure 9.14, the grid spacing $\Delta y = \Delta x = h$. To solve for $u(x,y)$

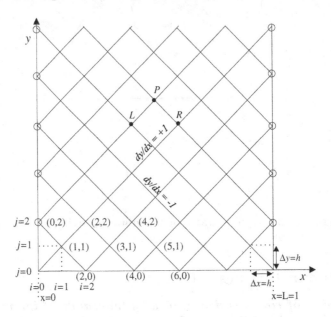

Figure 9.14: Diamond-shaped grid for solving (9.27).

and $v(x,y)$, we impose the following initial and boundary conditions:

- $u(x,0) = v(x,0) = 0$ for $0 < x < L = 1$, i.e., no pulses initially inside fluid,

- $u(0,y) = v(1,y) = \sin(2\pi y)$ for $0 \le y \le y_0 = 1/2$ and zero for $y > y_0$, i.e., positive half-sine wave pulses fed in at $x = 0$ and $x = L = 1$.

Starting on the bottom time row, we move along the characteristic directions $dy/dx = 1$ (e.g., from the point L to P) and $dy/dx = -1$ (from R to P) in calculating the changes in u and v, respectively. Using, say, the Euler approximation, the values of u and v at P are given by

$$u_P = u_L + h\,(f_1)_L, \qquad v_P = v_R - h\,(f_2)_R. \qquad (9.29)$$

Taking, say, $g_1 = 0.4$, $g_2 = 20$, $b = 0.5$, we implement (9.29) in the following recipe and animate the numerical result. The fluid sample length $L = 1$ is divided into $M = 120$ steps, so that the x (or y) stepsize is $h = L/M = 1/120$.

The leading edge of each pulse will just make contact in $N = M/2 = 60$ time steps. If we let i and j index the x and y steps, as in Figure 9.14, the grid coordinates for $j=0$, 2, ... are $(0,j)$, $(2,j)$, $(4,j)$, ..., while for $j=1$, 3, ... they are $(1,j)$, $(3,j)$, $(5,j)$, ... For later convenience, we set $m=2i$ and $n=2j$.

```
> restart: with(plots): begin:=time():
> L:=1: M:=120: h:=L/M; N:=M/2;  m:=2*i: n:=2*j:
```

$$h := \frac{1}{120} \qquad N := 60$$

The values of g_1, g_2, and b are entered. A functional operator S is introduced to calculate the input laser shape on the boundaries.

```
> g1:=0.4; g2:=20; b:=0.5; S:=j->sin(n*evalf(Pi)/N);
```

$$g1 := 0.4 \quad g2 := 20 \quad b := 0.5 \quad S := j \rightarrow \sin(\frac{n\,\mathrm{evalf}(\pi)}{N})$$

Using S, the boundary conditions are applied at $i=0$ and M in bc1 and bc2.

```
> bc1:=seq(u[0,n]=S(j),j=0..N/2),seq(u[0,n]=0,j=N/2+1..3*N):
> bc2:=seq(v[M,n]=S(j),j=0..N/2),seq(v[M,n]=0,j=N/2+1..3*N):
```
The initial condition is entered and assigned along with bc1 and bc2.

```
> ic:=seq(u[m,0]=0,i=0..M/2),seq(v[m,0]=0,i=0..M/2):
> assign(bc1,bc2,ic):
```
Functional operators are formed to evaluate f_1 and f_2 for a given i,j.

```
> f1:=(i,j)->-g1*u[i,j]*v[i,j]-b*u[i,j]:
  f2:=(i,j)->-g2*u[i,j]*v[i,j]+b*v[i,j]:
```
A double do loop is used to iterate (9.29). The conditional if...then statement is included so that i starts at $i0=0$ for $j=0$, 2, 4, ... and $i0=1$ for $j=1$, 3, ...

```
> for j from 0 to 3*N do
> if j mod 2=0 then i0:=0 else i0:=1 end if;
> for i from i0 to M by 2 do
> u[i+1,j+1]:=u[i,j]+h*f1(i,j);
> v[i-1,j+1]:=v[i,j]-h*f2(i,j);
> end do: end do:
```
An operator gr is formed to plot u and v on the jth time step.

```
> gr:=j->plot([[seq([m*h,u[m,n]],i=0..M/2)],[seq([m*h,v[m,n]],
    i=0..M/2)]],color=[red,blue],labels=["x","u,v"],thickness=2):
```
The numerically obtained profiles are animated with the display command with the insequence=true option. Click on the computer plot and the start arrow to initiate the animation.

```
> display([seq(gr(j),j=0..2*N)],insequence=true);
```
The cpu time for the entire recipe is given below.

```
> cpu:=time()-begin;
```

$$cpu := 7.752$$

9.2.7 KdV Solitons

In this recipe, we shall numerically investigate the collision of a taller faster solitary wave overtaking a shorter slower one. The standard algorithm for numerically integrating the KdV equation (8.25) is obtained as follows. Returning to the general Taylor expansion (9.3) of $y(x \pm h)$, adding the plus and minus results yields the central difference approximation to the first derivative, viz.,

$$y' = \frac{y(x+h) - y(x-h)}{2h} + O(h^2). \tag{9.30}$$

This approximation is used for the first derivatives in (8.25), viz.,

$$\frac{\partial U}{\partial t} \Rightarrow \frac{(U_{i,j+1} - U_{i,j-1})}{2k}, \qquad \frac{\partial U}{\partial x} \Rightarrow \frac{(U_{i+1,j} - U_{i-1,j})}{2h},$$

where k and h are the time and spatial stepsizes, respectively.

A central difference approximation to the third derivative can be obtained by modifying the first part of **Recipe 02-1-2** as follows. A functional operator t is formed to Taylor expand $y(x + a h)$ to 5th order in h for arbitrary a. The order of term is removed with the `convert( ,polynom)` command.

```
>   restart:
>   t:=a->y(x+a*h)=convert(taylor(y(x+a*h),h,5),polynom):
```

Then, using the operator, the sum $y(x+2h) - 2y(x+h) + 2y(x-h) - y(x-2h)$ is calculated in *eq1*, yielding $2h^3 y'''$ on the rhs. The error here is $O(h^5)$.

```
>   eq1:=t(2)-2*t(1)+2*t(-1)-t(-2);
```

$$eq1 := y(x + 2h) - 2y(x + h) + 2y(x - h) - y(x - 2h) = 2(D^{(3)})(y)(x) h^3$$

A finite difference approximation with error $O(h^2)$ follows for the third derivative on using the `isolate` command.

```
>   eq2:=isolate(eq1,D[1,1,1](y)(x));
```

$$eq2 := (D^{(3)})(y)(x) = -\frac{1}{2} \frac{-y(x+2h) + 2y(x+h) - 2y(x-h) + y(x-2h)}{h^3}$$

So, we will make the central difference approximation,

$$\frac{\partial^3 U}{\partial x^3} \Rightarrow \frac{(U_{i+2,j} - 2U_{i+1,j} + 2U_{i-1,j} - U_{i-2,j})}{2h^3}.$$

Finally, let's use the central difference approximation $(U_{i+1,j} + U_{i,j} + U_{i-1,j})/3$, which also has an error $O(h^2)$, for the U term in the KdV equation. Putting all these approximations together, the algorithm for the KdV equation is

$$U_{i,j+1} = U_{i,j-1} - \frac{k}{h} \frac{(U_{i+1,j} + U_{i,j} + U_{i-1,j})}{3} (U_{i+1,j} - U_{i-1,j})$$

$$- \frac{k}{h^3} (U_{i+2,j} - 2U_{i+1,j} + 2U_{i-1,j} - U_{i-2,j}). \tag{9.31}$$

Using the Von Neumann stability analysis, it can be shown that this algorithm is numerically stable for $k/h^3 < 2/(3\sqrt{3}) = 0.3849$.

Note that the algorithm connects the $j+1$ time step to steps j and $j-1$, so to iterate (9.31) the U values must be known on the $j=0$ and $j=1$ time steps.

The former are given by $U_{i,0} = f(x_i)$, where $f(x)$ is the input profile. For the U values on the first step, we shall use the result (9.25), i.e., $U_{i,1} = f(x_i) + k\, g(x_i)$, where $g(x)$ is the initial transverse velocity.

We begin the main part of our recipe by considering $M = 150$ spatial steps and $N = 300$ time steps. The starting time of the recipe is also recorded.

```
>   restart: with(plots): M:=150: N:=300: begin:=time():
```

A functional operator F is formed to calculate a time-dependent solitary wave profile centered at time $t = 0$ at $x = X$ and having speed c.

```
>   F:=(X,c)->3*c *(sech((sqrt(c)/2)*((x-X)-c*t)))^2:
```

The spatial stepsize is taken to be $h = 1$, the time stepsize $k = 0.25$, so $k/h^3 = 0.25$ and the scheme will be stable. For $f(x)$, we will consider two separated solitary waves, a taller one having speed $c1 = 0.8$ and a shorter one with speed $c2 = 0.1$. The taller, faster, solitary wave will be initially centered at $X1 = M/3 = 50$, the shorter, slower, wave at $X2 = M/2 = 75$. As time progresses, the taller solitary wave will overtake the shorter one and a collision take place.

```
>   h:=1: k:=.25: c1:=.8; c2:=.1; X1:=M/3; X2:=M/2;
```
$$c1 := 0.8 \qquad c2 := 0.1 \qquad X1 := 50 \qquad X2 := 75$$

Then F(X1,c1)+F(X2,c2) is entered and the time derivative taken in g.

```
>   f:= F(X1,c1)+F(X2,c2): g:=diff(f,t):
```

On setting $t = 0$, the initial profile is given by $f = 3\,c1\ \mathrm{sech}(\frac{\sqrt{c1}}{2}(x - X1))^2 + 3\,c2\ \mathrm{sech}(\frac{\sqrt{c2}}{2}(x - X2))^2$ which is now plotted in Figure 9.15.

```
>   t:=0: plot(f,x=0..M);
```

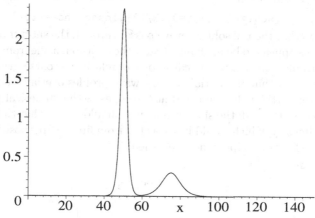

Figure 9.15: Initial profile.

To evaluate f at the spatial mesh points on the zeroth time row and g at the mesh points on the first time row, they are turned into functional operators in terms of x by using the unapply command.

```
>   f2:=unapply(f,x): g2:=unapply(g,x):
```

Using f2 and g2, $U_{i,0}$ and $U_{i,1}$ are calculated in ic1 and ic2 for $i = 0$ to M.

```
>  ic1:=seq(U(i,0)=evalf(f2(i)),i=0..M):
>  ic2:=seq(U(i,1)=evalf(f2(i))+k*g2(i),i=0..M):
```

To avoid unknown U values creeping into the do loop, we will set all U values equal to zero for $i = 0$ to M and $j = 2$ to N. These "initialized" zeros will be overwritten as the do loop progresses.

```
>  init:=seq(seq(U(i,j)=0,i=0..M),j=2..N):
```

Then *ic1*, *ic2*, and *init* are assigned.

```
>  assign(ic1,ic2,init):
```

The algorithm (9.31) is then iterated from $i = 2$ to $M - 2$ and $j = 1$ to N. We must start i at 2 and end at $M - 2$ to avoid unknown U values entering the iterative calculation.

```
>  for j from 1 to N do;
>  for i from 2 to M-2 do
>  U(i,j+1):=U(i,j-1)-(k/(3*h))*(U(i+1,j)+U(i,j)+U(i-1,j))
                *(U(i+1,j)-U(i-1,j))-(k/h^3)*(U(i+2,j)-2*U(i+1,j)
                +2*U(i-1,j)-U(i-2,j));
>  end do: end do:
```

An operator **gr** is created to plot $U_{i,j}$ for $i = 2$ to $M - 2$ on the jth time step.

```
>  gr:=j->plot([seq([i,U(i,j)],i=2..M-2)]):
```

Every second graph is displayed and the sequence of pictures animated with the insequence=true option. Click on the computer plot and on the start arrow to initiate the animation.

```
>  display([seq(gr(2*j),j=0..N/2)],insequence=true);
```

It is observed that the tall solitary wave passes through the shorter one and both solitary waves appear to be unchanged after the collision, aside from a very small numerical ripple which arises because of the relatively coarse grid which was used and the fact that the initial solitary wave profiles overlapped slightly and their tails were artificially truncated at the edges of the numerical grid. As the tall pulse passes through the shorter one, you should notice that the amplitudes do not add linearly, which would be expected from linear superposition. Finally, the cpu time for the recipe is now recorded.

```
>  cpu:=time()-begin;
```
$$cpu := 29.913$$

9.3 Supplementary Recipes

09-S01: White Dwarf Equation

Chandresekhar's theory of white dwarf stars produces the NLODE [Cha39]

$$x\,y''(x) + 2\,y'(x) + x\,(y^2 - C)^{3/2} = 0, \quad \text{with} \ \ y(0) = 1, \ y'(0) = 0.$$

Making use of the first principles and dial-up Euler methods with $h = .01$ and 18-digits accuracy, numerically compute $y(x)$ over the range $0 \le x \le 4$ with

$C = 0.1$ and plot both numerical results together. Show that the numerical value of $y(x = 1)$ is exactly the same for both methods. Hint: Start at $x = 0.01$ to avoid any problem at the origin.

09-S02: Spruce Budworm Infestation

The sudden outbreak of the spruce budworm which can rapidly defoliate a forest and kill the trees can be described [LJH78] by the dimensionless ODE

$$\dot{x}(\tau) = rx(1 - \frac{x}{K}) - \frac{x^2}{1 + x^2}.$$

Here $x(\tau)$ is proportional to the budworm population number at time τ and the growth coefficient r and carrying capacity parameter K are positive constants. The first term describes the growth of the budworm population with a saturation effect included due to the finite forest available, while the last term models the decrease in population due to bird predation. Taking $K = 300$, use the first-principles Euler method with $h = 0.01$ and $x(0) = 0.5$ to determine the time evolution of the budworm population over the time interval $t = 0$ to 50 for $r = 0.1$, $r = 0.5$, and $r = 1.0$. Plot and discuss the results.

09-S03: A Math Example

Consider the first-order nonlinear ODE

$$dy/dx = \sqrt{x}\,(y - x)\,(y - 2)$$

with $y(0) = 1$. Show that an analytic closed form of solution is not obtainable with the dsolve command. Taking $h = 0.01$ and 10-digit accuracy, solve for $y(x)$ out to $x = 5$ using the modified Euler method and plot every 10th numerical point.

09-S04: Hermione Hippo

Hermione Hippo swims across the muddy Mombopo river by steadily aiming at a tree directly across the river on the water's edge. The Mombopo is 1 km wide and has a speed of 1.5 km/hour while Hermione's speed is 2 km/hour. In Cartesian coordinates, Hermione is initially at $(x = 1, y = 0)$ while the tree is at $(0, 0)$. Derive Hermione's equations of motion in Cartesian coordinates. Using the first-principles 4th-order RK method with $h = 0.001$, numerically solve these equations to determine how long it takes Hortense to reach the tree. Determine the analytic solution $y(x)$ for Hortense's path across the river and plot the analytic and numerical solutions together in the same figure.

09-S05: The Oregonator

A kinetic model ([FKN72],[FN74],[EM00]) describing the oscillations in a certain chemical system is the Oregonator system of nonlinear ODEs, viz.,

$$\epsilon \dot{x} = x + y - q x^2 - x y, \qquad \dot{y} = -y + 2 h z - x y, \qquad p \dot{z} = x - z,$$

where x, y, and z are the dimensionless concentrations of three chemical species and ϵ, p, q, and h are positive parameters. Taking $\epsilon = 0.003$, $p = 2$, $q = 0.006$, $h = 0.75$, $x(0) = 30$, $y(0) = 10$, and $z(0) = 30$, numerically solve the system of equations using the RKF45 method with the option maxfun=0. Then use the odeplot command with numpoints=3000 to plot (a) x vs. y vs. z for $t = 0..50$, (b) x vs. t, y vs. t, and z vs. t in the same figure, for $t = 0..30$. Numerically show that a closed loop results in the x-y-z space no matter what initial conditions are chosen. The Oregonator is an example of a 3-dimensional limit cycle.

09-S06: Lorenz's Butterfly

The Lorenz system [Lor63] of nonlinear ODES are given by

$$\dot{x} = \sigma\,(y - x), \qquad \dot{y} = r\,x - y - x\,z, \qquad \dot{z} = x\,y - b\,z,$$

with x, y, z real and σ, r, and b real, positive constants. Derive a second-order-accurate semi-implicit numerical scheme for the Lorenz system. Taking $\sigma = 10$, $b = 8/3$, $r = 28$, $x(0) = 2$, $y(0) = 5$, $z(0) = 5$, a stepsize $h = 0.01$, and $n = 4000$ iterations, iterate the algorithm. Plot x vs. y vs. z, using a zhue shading. The resulting figure should resemble the wings of a butterfly. After a transient interval, the trajectory remains confined to the region of the wings, but never retraces the same path. This is another example of a strange chaotic attractor.

09-S07: A Stiff Harmonic Oscillator

Consider the following heavily overdamped forced harmonic oscillator equation,

$$\ddot{x} + 50\,\dot{x} + x = 2\,\sin t,$$

with the initial condition $x(0) = 0$, $\dot{x}(0) = 10$. Solve the ODE analytically and determine the two widely different characteristic times in the solution. Create operators to numerically solve the ODE using the 4th-order Rungé–Kutta dial-up method and plot the numerical solution for arbitrary stepsize h. By plotting the analytic and numerical solution together, determine the approximate h value at which numerical instability sets in. This critical h value should be roughly comparable to the shortest characteristic time.

09-S08: Courant Stability Condition

Prove the Courant stability condition for the linear wave equation.

09-S09: Poisson's Equation

Consider Poisson's equation

$$\frac{\partial^2 V}{\partial x^2} + \frac{\partial^2 V}{\partial y^2} = x\,e^y, \quad 0 \le x \le 2,\ 0 \le y \le 1,$$

with $V(0, y) = 0$, $V(2, y) = 2\,e^y$, $V(x, 0) = x$, and $V(x, 1) = e\,x$, as the boundary conditions. Using a central difference approximation for each second derivative at a mesh point (i, j) and evaluating the rhs of Poisson's equation at this point, derive a finite difference scheme for solving the equation. Dividing the x-interval into 30 steps and the y interval into 15 steps, determine V at each mesh point and plot the numerical solution.

09-S10: Crank–Nicolson Method

The Crank–Nicolson (CN) method results on averaging the forward and backward difference methods. For the normalized linear temperature equation, the forward and backward schemes connecting the j and $j+1$ time steps are,

$$\frac{(T_{i,j+1} - T_{i,j})}{k} = \frac{(T_{i+1,j} - 2T_{i,j} + T_{i-1,j})}{h^2},$$

$$\frac{(T_{i,j+1} - T_{i,j})}{k} = \frac{(T_{i+1,j+1} - 2T_{i,j+1} + T_{i-1,j+1})}{h^2}.$$

Adding the two equations, dividing by 2, and rearranging, yields,

$$-r\,T_{i-1,j+1} + 2\,(1+r)\,T_{i,j+1} - r\,T_{i+1,j+1} = r\,T_{i-1,j} + 2\,(1-r)\,T_{i,j} + r\,T_{i+1,j}$$

with $r = k/h^2$. This finite difference CN formula relates three unknown values of T on the $j+1$ time row to three known values on the j time row. Modify recipe **09-2-2** to implement the CN formula, taking all parameters the same, except now with $r = 0.6$. As you may verify the explicit scheme in **09-2-2** is numerically unstable for this r value, but the CN scheme is stable.

09-S11: Klein–Gordon Equation

The transverse vibrations of a horizontal string embedded in a vertical elastic membrane which exerts a Hooke's law restoring force on the string can be described by the *Klein–Gordon equation*,

$$\frac{\partial^2 U}{\partial x^2} - \frac{\partial^2 U}{\partial t^2} = a\,U,$$

with a a positive constant. If such a string of length $L = 1$ is fixed at both ends and has the initial shape $U(x,0) = x^4\,(L-x)^2$ and transverse velocity $\dot{U}(x,0) = x\,(L-x)$, numerically determine the transverse displacement of the string for $t \geq 0$ using a matrix approach and animate the motion. Take $a = 100$, spatial stepsize $h = 0.025$ and time stepsize $k = 0.01$.

Bibliography

[AS72]   M. Abramowitz and I. A. Stegun. *Handbook of Mathematical Functions with Formulas, Graphs, and Mathematical Tables*. National Bureau of Standards, Washington, DC, 1972.

[AW00]   G. B. Arfken and H. J. Weber. *Mathematical Methods for Physicists*, 5th ed. Academic Press, New York, 2000.

[Bay99]   W. E. Baylis. *Electrodynamics: A Modern Geometric Approach*. Birkhäuser, Boston, MA, 1999.

[BC74]   G. W. Bluman and J. D. Cole. *Similarity Methods for Differential Equations*. Springer-Verlag, New York, 1974.

[BF89]   R. L. Burden and J. D. Faires. *Numerical Analysis, 4th ed*. PWS-Kent, Boston, 1989.

[Boa83]   M. L. Boas. *Mathematical Methods in the Physical Sciences*, 2nd ed. John Wiley, New York, 1983.

[Bra68]   T. C. Bradbury. *Theoretical Mechanics*. Wiley, New York, 1968.

[Buc77]   J. Buckmaster. Viscous sheets advancing over dry beds. *Journal of Fluid Mechanics*, 81:735, 1977.

[Cha39]   S. Chandresekhar. *An Introduction to the Study of Stellar Structure*. Dover Reprint, Chicago, 1939.

[Cha03]   B. W. Char. *Maple 9 Learning Guide*. Waterloo Maple, Waterloo, Canada, 2003.

[Dav62]   H. T. Davis. *Introduction to Nonlinear Differential and Integral Equations*. Dover, New York, 1962.

[DL03]   C. Doran and A. Lasenby. *Geometric Algebra for Physicists*. Cambridge University Press, Cambridge, UK, 2003.

[EM00]   R. H. Enns and G. C. McGuire. *Nonlinear Physics with Maple for Scientists and Engineers*, 2nd ed. Birkhäuser, Boston, MA, 2000.

[FC99]   G. R. Fowles and G. L Cassiday. *Analytical Mechanics*, 6th ed. Saunders College, Orlando, FL, 1999.

[Feh70]   E. Fehlberg. Klassische runge-kutta formeln vierter und niedrigerer ordnung mit schrittweiten-kontrolle und ihre andwendung auf wärmeleitungsprobleme. *Computing*, 6:61, 1970.

[FKN72]   R. J. Field, E. Körös, and R. M. Noyes. Oscillations in chemical systems, Part 2. Thorough analysis of temporal oscillations in the bromate–cerium–malonic acid system. *Journal of the American Chemical Society*, 94:8649, 1972.

[FN74]   R. J. Field and R. M. Noyes. Oscillations in chemical systems, IV. Limit cycle behavior in a model of a real chemical reaction. *Journal of Chemical Physics*, 60:1877, 1974.

[GPS02]   H. Goldstein, C. Poole, and J. Safco. *Classical Mechanics, 3rd ed.* Addison Wesley, New York, 2002.

[Gri95]   D. J. Griffiths. *Introduction to Quantum Mechanics*. Prentice-Hall, Englewood Cliffs, N.J., 1995.

[Gri99]   D. J. Griffiths. *Introduction to Electrodynamics,* 3rd ed. Prentice-Hall, Upper Saddle River, N.J., 1999.

[Hel50]   H. Helmholtz. Messungen über den zeitlichen verlauf der zuchung animalischer muskeln und die fortplanzungsgeschwindigkeit der reizung in der nerven. *Arch. Anat. Physiol.*, 276, 1850.

[Hes99]   D. Hestenes. *New Foundations for Classical Mechanics, 2nd ed.* Kluwer Academic, Dordrecht,The Netherlands, 1999.

[Hil57]   F. B. Hildebrand. *Advanced Calculus for Engineers*. Prentice-Hall, Englewood Cliffs, N. J., 1957.

[IM69]   J. Irving and N. Mullineux. *Mathematics in Physics and Engineering*. Academic Press, New York, 1969.

[KB43]   N. Krylov and N. Bogoliubov. *Introduction to Nonlinear Mechanics*. Princeton University Press, Princeton, 1943.

[KdV95]   D. J. Korteweg and G. de Vries. On the change of form of long waves advancing in a rectangular canal, and a new type of long stationary wave. *Philosophical Magazine*, 39:422, 1895.

[LJH78]   D. Ludwig, D. D. Jones, and C. S. Holling. Qualitative analysis of insect outbreak systems: the spruce budworm and forest. *J. Anim. Ecol.*, 47:315, 1978.

[Lor63]   E. N. Lorenz. Deterministic nonperiodic flow. *J. Atmospheric Sci.*, 20:130, 1963.

[LP80]   E. W. Larsen and G. C. Pomraning. Asymptotic analysis of non-linear marshak waves. *SIAM Journal of Applied Mathematics*, 39:201, 1980.

[LS71]   L. Lapidus and J. H. Seinfeld. *Numerical Solution of Ordinary Differential Equations*. Academic Press, New York, 1971.

[MF53]   P. M. Morse and H. Feshbach. *Methods of Theoretical Physics*. McGraw Hill, New York, 1953.

[MGH+03a]   M. B. Monagan, K. O. Geddes, K. M. Heal, G. Labahn, S. M. Vorkoetter, J. McCarron, and P. DeMarco. *Maple 9 Advanced Programming Guide*. Waterloo Maple, Waterloo, Canada, 2003.

[MGH+03b]   M. B. Monagan, K. O. Geddes, K. M. Heal, G. Labahn, S. M. Vorkoetter, J. McCarrona, and P. DeMarco. *Maple 9 Introductory Programming Guide*. Waterloo Maple, Waterloo, Canada, 2003.

[MM57]   H. Margenau and G. M. Murphy. *The Mathematics of Physics and Chemistry, 2nd ed.* D. Van Nostrand, New York, 1957.

[Mor48]   P. M. Morse. *Vibration and Sound*. McGraw-Hill, New York, 1948.

[MT95]   J. B. Marion and S. T. Thornton. *Classical Dynamics of Particles and Systems, 4th ed.* Saunders College, Orlando, FL, 1995.

[Mus37]   M. Muskat. *The Flow of Homogeneous Fluids Through Porous Media*. McGraw Hill, New York, 1937.

[MW71]   J. Mathews and R. L. Walker. *Mathematical Methods of Physics, 2nd ed.* Addison-Wesley, New York, 1971.

[NYA65]   J. Nagumo, S. Yoshizawa, and S. Arimoto. Bistable transmission lines. *Trans. IEEE on Circuit Theory*, CT-12:400, 1965.

[PFTV90]   W. H. Press, B. P. Flannery, S. A. Teukolsky, and W. T. Vetterling. *Numerical Recipes*. Cambridge University Press, Cambridge, 1990.

[PLA92]   M. Peastrel, R. Lynch, and A. Armenti. Terminal velocity of a shuttlecock in vertical fall. In Angelo Armenti (Jr.), editor, *The Physics of Sports*. American Institute of Physics, New York, 1992.

[RE76]   S. S. Rangnekar and R. H. Enns. Numerical solution of the transient gain equations for stimulated backward scattering in absorbing fluids. *Canadian Journal of Physics*, 54:1564, 1976.

[Ros76]   O. E. Rossler. An equation for continous chaos. *Physics Letters*, 57A:397, 1976.

[Sco87]   D. E. Scott. *An Introduction to Circuit Analysis*. McGraw-Hill, New York, 1987.

[Spi71]   M. Spiegel. *Advanced Mathematics for Engineers and Scientists*. McGraw Hill, New York, 1971.

[SR66]   I. S. Sokolnikoff and R. M. Redheffer. *Mathematics of Physics and Modern Engineering*. McGraw-Hill, New York, 1966.

[Ste87]   J. Stewart. *Calculus*. Brooks/Cole, Pacific Grove, CA, 1987.

[Ste94]   B. Stearn. *Simply Heartsmart Cooking*. Random House, Toronto, Ont., Canada, 1994.

[Wal84]   P. R. Wallace. *Mathematical Analysis of Physical Problems*. Dover, New York, 1984.

[Wie73]   S. Wieder. *The Foundations of Quantum Theory*. Academic Press, New York, 1973.

[Zwi89]   D. Zwillinger. *Handbook of Differential Equations*. Academic Press, San Diego, 1989.

Index

Printed in the United States
By Bookmasters

Printed in the United States
By Bookmasters